D1271393

Methods in Enzymology

Volume 376
CHROMATIN AND CHROMATIN REMODELING ENZYMES
Part B

METHODS IN ENZYMOLOGY

EDITORS-IN-CHIEF

John N. Abelson   Melvin I. Simon

DIVISION OF BIOLOGY
CALIFORNIA INSTITUTE OF TECHNOLOGY
PASADENA, CALIFORNIA

FOUNDING EDITORS

Sidney P. Colowick and Nathan O. Kaplan

Methods in Enzymology

Volume 376

Chromatin and Chromatin Remodeling Enzymes

Part B

EDITED BY

C. David Allis

THE ROCKEFELLER UNIVERSITY
NEW YORK, NEW YORK

Carl Wu

NATIONAL CANCER INSTITUTE
BETHESDA, MARYLAND

ELSEVIER
ACADEMIC
PRESS

AMSTERDAM • BOSTON • HEIDELBERG • LONDON
NEW YORK • OXFORD • PARIS • SAN DIEGO
SAN FRANCISCO • SINGAPORE • SYDNEY • TOKYO

Academic Press is an imprint of Elsevier

Elsevier Academic Press
525 B Street, Suite 1900, San Diego, California 92101-4495, USA
84 Theobald's Road, London WC1X 8RR, UK

This book is printed on acid-free paper. ∞

Permissions may be sought directly from Elsevier's Science & Technology Rights Department in
Oxford, UK: phone: (+44) 1865 843830, fax: (+44) 1865 853333,
e-mail: permissions@elsevier.com.uk. You may also complete your request on-line
via the Elsevier homepage (http://elsevier.com), by selecting "Customer
Support" and then "Obtaining Permissions."

For all information on all Academic Press Publications
visit our Web site at www.academicpress.com

ISBN: 0-12-182780-1

PRINTED IN THE UNITED STATES OF AMERICA
04 05 06 07 08   9 8 7 6 5 4 3 2 1

Table of Contents

Section I. Biophysics, Structural Biology, and Enzymology of Chromatin Proteins

Section II. Immunochemical Assays of Chromatin Functions

Contributors to Volume 376

Article numbers are in parentheses and following the names of contributors.
Affiliations listed are current.

RHODA M. ALANI (12), *Department of Oncology, Johns Hopkins University School of Medicine, Baltimore, Maryland 21218*

FRANCISCO ASTURIAS (4), *Department of Cell Biology, The Scripps Research Institute, La Jolla, California 92037*

ANDREW J. BANNISTER (18), *Wellcome Trust/Cancer Research, United Kingdom Institute and Department of Pathology, University of Cambridge, Cambridge CB2 1QR, United Kingdom*

P. B. BECKER (1), *Adolf Butenandt Institut, Lehrstuhl für Molekularbiologie, Schillerstr. 44, 80336 Munich, Germany*

MARTIN L. BENNINK (6), *Biophysical Techniques Group and MESA+ Research Institute, Department of Science Technology, University of Twente, 7500 AE Enschede, The Netherlands*

BRADLEY E. BERNSTEIN (23), *Department of Chemistry and Chemical Biology, Harvard University, Cambridge, Massachusetts 02138*

MARGIE T. BORRA (11), *Department of Biochemistry and Molecular Biology, Oregon Health and Science University, Portland, Oregon 97239*

BRENT BROWER-TOLAND (5), *Biology Department, Washington University in St. Louis, St. Louis, Missouri 63130*

MICHAEL BUSTIN (14), *Protein Section, National Cancer Institute, National Institutes of Health, Bethesda, Maryland 20892*

JULIANA CALLAGHAN (10), *Department of Biochemistry, University of Cambridge, Cambridge CB2 1GA, United Kingdom*

MAREK CEBRAT (12), *Department of Pharmacology and Molecular Sciences, Johns Hopkins University School of Medicine, Baltimore, Maryland 21218*

JULIE CHAUMEIL (27), *Mammalian Developmental Epigenetics Group, UMR 218-Nuclear Dynamics and Genome Plasticity, Curie Institute-Research Section, 75248 Paris, Cedex 05-France*

DINA CHAYA (24), *Cell and Developmental Biology Program, Fox Chase Cancer Center, Philadelphia, Pennsylvania 19111*

PETER CHEUNG (15), *Department of Medical Biophysics, University of Toronto, Ontario Cancer Institute, Toronto, Ontario M5G 2M9, Canada*

J. CHIN (1), *Department of Biochemistry, Northwestern University, Molecular Biology and Cell Biology, Evanston, Illinois 60208-3500*

DAVID N. CICCONE (22), *Department of Molecular Biology, Massachusetts General Hospital, Boston, Massachusetts 02114*

PHILIP A. COLE (12), *Department of Pharmacology and Molecular Sciences, Johns Hopkins University School of Medicine, Baltimore, Maryland 21218*

CARLOS CORDON-CARDO (13), *Division of Molecular Pathology, Memorial Sloan Kettering Cancer Center, New York, New York 10021*

CAROLYN A. CRAIG (25), *Biology Department, Washington University in St. Louis, St. Louis, Missouri 63130*

JOHN M. DENU (11), *Department of Biochemistry and Molecular Biology, Oregon Health and Science University, Portland, Oregon 97239*

MEGHANN K. DEVLIN (12), *Department of Oncology, Johns Hopkins University School of Medicine, Baltimore, Maryland 21218*

MARIJA DROBNJAK (13), *Division of Molecular Pathology, Memorial Sloan Kettering Cancer Center, New York, New York 10021*

BRIAN DYNLACHT (20), *Department of Pathology, New York University School of Medicine, New York, New York 10016*

SARAH C. R. ELGIN (25), *Biology Department, Washington University in St. Louis, St. Louis, Missouri 63130*

CHUKWUDI EZEOKONKWO (4), *Department of Cell Biology, The Scripps Research Institute, La Jolla, California 92037*

PEGGY FARNHAM (21), *McArdle Laboratory for Cancer Research, University of Wisconsin, Madison, Wisconsin 53706*

WOLFGANG FISCHLE (9), *Department of Biochemistry and Molecular Genetics, University of Virginia, Charlottesville, Virginia 22908*

FRED K. FRIEDMAN (14), *Laboratory of Metabolism, National Cancer Institute, National Institutes of Health, Bethesda, Maryland 20892*

PHILIPPE T. GEORGEL (2), *Department of Biological Sciences, Marshall University, Huntington, West Virginia 25755*

MICHAEL GRUNSTEIN (19), *Department of Biological Chemistry, School of Medicine and Molecular Biology Institute, University of California, Los Angeles, Los Angeles, California 90095*

JEFFREY C. HANSEN (2), *Department of Biochemistry and Molecular Biology, Colorado State University, Fort Collins, Colorado 80523*

EDITH HEARD (27), *Mammalian Developmental Epigenetics Group, UMR 218-Nuclear Dynamics and Genome Plasticity, Curie Institute-Research Section, 75248 Paris, Cedex 05, France*

RACHEL A. HOROWITZ-SCHERER (3), *Department of Biology, University of Massachusetts, Amherst, Massachusetts 01003*

EMILY L. HUMPHREY (23), *Department of Chemistry and Chemical Biology, Harvard University, Cambridge, Massachusetts 02138*

STEVEN A. JACOBS (9), *Department of Biochemistry and Molecular Genetics, University of Virginia, Charlottesville, Virginia 22908*

THOMAS JENUWEIN (16), *Research Institute of Molecular Pathology (IMP), The Vienna Biocenter, Vienna, A-1030, Austria*

MONIKA KAUER (16), *Research Institute of Molecular Pathology (IMP), The Vienna Biocenter, Vienna, A-1030, Austria*

W. KEVIN KELLY (13), *Genitourinary Oncology Service and Department of Medicine, Memorial Sloan Kettering Cancer Center, New York, New York 10021*

SEPIDEH KHORASANIZADEH (9), *Department of Biochemistry and Molecular Genetics, University of Virginia, Charlottesville, Virginia 22908*

ROGER D. KORNBERG (4), *Department of Structural Biology, Stanford University School of Medicine, Stanford, California 94305*

TONY KOUZARIDES (18), *Wellcome Trust/Cancer Research, United Kingdom Institute, University of Cambridge, Cambridge CB2 1QR, United Kingdom*

SIAVASH K. KURDISTANI (19), *Department of Biological Chemistry, University of California, Los Angeles School of Medicine and Molecular Biology Institute, Los Angeles, California 90095*

G. LÄNGST (1), *Adolf Butenandt Institut, Lehrstuhl für Molekularbiologie, Schillerstr. 44, 80336 Munich, Germany*

ERNEST LAUE (10), *Department of Biochemistry, University of Cambridge, Cambridge CB2 1GA, United Kingdom*

SANFORD H. LEUBA (6), *Department of Cell Biology and Physiology, University of Pittsburgh School of Medicine, Hillman Cancer Center, UPCI Research Pavilion, Pittsburgh, Pennsylvania 15213-1863*

YUHONG LI (25), *University of Iowa, Department of Biochemistry, Iowa City, Iowa 52242*

JOHN LIS (26), *Cornell University, Ithaca, New York 14853*

CHIH LONG LIU (23), *Department of Chemistry and Chemical Biology, Harvard University, Cambridge, Massachusetts 02138*

YAHLI LORCH (4), *Department of Structural Biology, Stanford University School of Medicine, Stanford, California 94305*

PAUL A. MARKS (13), *Cell Biology Program, Memorial Sloan-Kettering Cancer Center, New York, New York 10021*

RONEN MARMORSTEIN (7), *Structural Biology Program, The Wistar Institute, Philadelphia, Pennsylvania 19104-4268*

KARL MECHTLER (16), *Research Institute of Molecular Pathology (IMP), The Vienna Biocenter, Vienna, A-1030, Austria*

KATRINA B. MORSHEAD (22), *Massachusetts General Hospital, Department of Molecular Biology, Boston, Massachusetts 02114*

SHIRAZ MUJTABA (8), *Department of Physiology and Biophysics, Structural Biology Program, Mt. Sinai School of Medicine, New York University, New York, New York 10029*

ALEXEY G. MURZIN (10), *MRC Centre for Protein Engineering, Cambridge, CB2 2QH United Kingdom*

NATALIA V. MURZINA (10), *Department of Biochemistry, University of Cambridge, Cambridge CB2 1GA, United Kingdom*

PETER R. NIELSEN (10), *Department of Biochemistry, University of Cambridge, Cambridge CB2 1GA, United Kingdom*

KENICHI NISHIOKA (17), *Department of Developmental Genetics, National Institute of Genetics, Shizuoka, Japan, 411-8540*

MATTHEW J. OBERLEY (21), *McArdle Laboratory for Cancer Research, University of Wisconsin, Madison, Wisconsin 53706*

MARJORIE A. OETTINGER (22), *Department of Molecular Biology, Massachusetts General Hospital, Boston, Massachusetts 02114*

IKUHIRO OKAMOTO (27), *Mammalian Developmental Epigenetics Group, UMR 218 – Nuclear Dynamics and Genome Plasticity, Curie Institute-Research Section, 75248 Paris, Cedex 05, France*

SUSANNE OPRAVIL (16), *Research Institute of Molecular Pathology (IMP), The Vienna Biocenter, Vienna, A-1030, Austria*

BARBARA PANNING (28), *Department of Biochemistry and Biophysics, University of California, San Francisco, San Francisco, California 94143-0448*

LAURA PEREZ-BURGOS (16), *Research Institute of Molecular Pathology (IMP), The Vienna Biocenter, Vienna, A-1030, Austria*

ANTOINE H. F. M. PETERS (16), *Research Institute of Molecular Pathology (IMP), The Vienna Biocenter, Vienna, A-1030 Austria*

DANNY REINBERG (17), *Department of Biology, Howard Hughes Medical Institute, University of Medicine and Dentistry of New Jersey, Piscataway, NJ 08854-5635*

BING REN (20), *San Diego Branch and Department of Cellular and Molecular Medicine, Ludwig Institute for Cancer Research, University of California, San Diego School of Medicine, La Jolla, California 92093-0653*

VICTORIA M. RICHON (13), *Discovery Biology, Aton Pharma, Inc., Tarrytown, New York 10591*

RICHARD C. ROBINSON (14), *Laboratory of Metabolism, National Institutes of Health, National Cancer Institute, Bethesda, Maryland 20892*

DANIEL ROBYR (19), *Department of Biological Chemistry, University of California, Los Angeles, School of Medicine and Molecular Biology Institute, Los Angeles, California 90095*

KAVITHA SARMA (17), *Department of Biology, Howard Hughes Medical Institute, University of Medicine and Dentistry of New Jersey, Piscataway, NJ 08854-5635*

STUART SCHREIBER (23), *Department of Chemistry and Chemical Biology, Harvard University, Cambridge, Massachusetts 02138*

BRIAN E. SCHWARTZ (26), *Cornell University, Ithaca, New York 14853*

J. PAUL SECRIST (13), *Discovery Biology, Aton Pharma, Inc., Tarrytown, New York 10591*

GENA E. STEPHENS (25), *Biology Department, Washington University in St. Louis, St. Louis, Missouri 63130*

PAUL R. THOMPSON (12), *Department of Pharmacology and Molecular Sciences, Johns Hopkins University School of Medicine, Baltimore, Maryland 21218*

JULISSA TSAO (21), *Microarray Centre, University Health Network, Toronto, Ontario M5G 2C4, Canada*

LORI L. WALLRATH (25), *Department of Biochemistry, University of Iowa, Iowa City, Iowa 52242*

MICHELLE D. WANG (5), *Department of Physics, Laboratory of Atomic and Solid State Physics, Cornell University, Ithaca, New York 14853*

LING WANG (12), *Department of Pharmacology and Molecular Sciences, Johns Hopkins University School of Medicine, Baltimore, Maryland 21218*

JANIS K. WERNER (26), *Cornell University, Ithaca, New York 14853*

JON WIDOM (1), *Northwestern University, Department of Biochemistry, Molecular Biology and Cell Biology, Evanston, Illinois 60208-3500*

CHRISTOPHER L. WOODCOCK (3), *Department of Biology, University of Massachusetts, Amherst, Massachusetts 01003*

PATRICK YAU (21), *Microarray Centre, University Health Network, Toronto, Ontario M5G 2C4, Canada*

KEN ZARET (24), *Cell and Developmental Biology Program, W. W. Smith Chair in Cancer Research, Fox Chase Cancer Center, Philadelphia, Pennsylvania 19111*

YUJUN ZHENG (12), *Department of Pharmacology and Molecular Sciences, Johns Hopkins University School of Medicine, Baltimore, Maryland 21218*

MING-MING ZHOU (8), *Structural Biology Program, Department of Physiology and Biophysics, Mt. Sinai School of Medicine, New York University, New York, New York 10029-6574*

XIANBO ZHOU (13), *Discovery Biology, Aton Pharma, Inc., Tarrytown, New York 10591*

JORDANKA ZLATANOVA (6), *Department of Chemical and Biological Sciences and Engineering, Polytechnic University, Brooklyn, New York 11201*

Preface

A central challenge of the post-genomic era is to understand how the 30,000 to 40,000 unique genes in the human genome are selectively expressed or silenced to coordinate cellular growth and differentiation. The packaging of eukaryotic genomes in a complex of DNA, histones, and nonhistone proteins called chromatin provides a surprisingly sophisticated system that plays a critical role in controlling the flow of genetic information. This packaging system has evolved to index our genomes such that certain genes become readily accessible to the transcription machinery, while other genes are reversibly silenced. Moreover, chromatin-based mechanisms of gene regulation, often involving domains of covalent modifications of DNA and histones, can be inherited from one generation to the next. The heritability of chromatin states in the absence of DNA mutation has contributed greatly to the current excitement in the field of epigenetics.

The past 5 years have witnessed an explosion of new research on chromatin biology and biochemistry. Chromatin structure and function are now widely recognized as being critical to regulating gene expression, maintaining genomic stability, and ensuring faithful chromosome transmission. Moreover, links between chromatin metabolism and disease are beginning to emerge. The identification of altered DNA methylation and histone acetylase activity in human cancers, the use of histone deacetylase inhibitors in the treatment of leukemia, and the tumor suppressor activities of ATP-dependent chromatin remodeling enzymes are examples that likely represent just the tip of the iceberg.

As such, the field is attracting new investigators who enter with little firsthand experience with the standard assays used to dissect chromatin structure and function. In addition, even seasoned veterans are overwhelmed by the rapid introduction of new chromatin technologies. Accordingly, we sought to bring together a useful "go-to" set of chromatin-based methods that would update and complement two previous publications in this series, Volume 170 (Nucleosomes) and Volume 304 (Chromatin). While many of the classic protocols in those volumes remain as timely now as when they were written, it is our hope the present series will fill in the gaps for the next several years.

This 3-volume set of *Methods in Enzymology* provides nearly one hundred procedures covering the full range of tools—bioinformatics, structural biology, biophysics, biochemistry, genetics, and cell biology—employed in chromatin research. Volume 375 includes a histone database, methods for preparation of histones, histone variants, modified histones and defined chromatin segments, protocols for nucleosome reconstitution and analysis, and cytological methods for imaging chromatin functions *in vivo*. Volume 376 includes electron microscopy and biophysical protocols for visualizing chromatin and detecting chromatin interactions, enzymological assays for histone modifying enzymes, and immunochemical protocols for the *in situ* detection of histone modifications and chromatin proteins. Volume 377 includes genetic assays of histones and chromatin regulators, methods for the preparation and analysis of histone modifying and ATP-dependent chromatin remodeling enzymes, and assays for transcription and DNA repair on chromatin templates. We are exceedingly grateful to the very large number of colleagues representing the field's leading laboratories, who have taken the time and effort to make their technical expertise available in this series.

Finally, we wish to take the opportunity to remember Vincent Allfrey, Andrei Mirzabekov, Harold Weintraub, Abraham Worcel, and especially Alan Wolffe, co-editor of Volume 304 (Chromatin). All of these individuals had key roles in shaping the chromatin field into what it is today.

C. DAVID ALLIS
CARL WU

Editors' Note: Additional methods can be found in Methods in Enzymology, Vol. 371 (RNA Polymerases and Associated Factors, Part D) Section III Chromatin, *Sankar L. Adhya and Susan Garges, Editors.*

METHODS IN ENZYMOLOGY

VOLUME XXXV. Lipids (Part B)
Edited by JOHN M. LOWENSTEIN

VOLUME XXXVI. Hormone Action (Part A: Steroid Hormones)
Edited by BERT W. O'MALLEY AND JOEL G. HARDMAN

VOLUME XXXVII. Hormone Action (Part B: Peptide Hormones)
Edited by BERT W. O'MALLEY AND JOEL G. HARDMAN

VOLUME XXXVIII. Hormone Action (Part C: Cyclic Nucleotides)
Edited by JOEL G. HARDMAN AND BERT W. O'MALLEY

VOLUME XXXIX. Hormone Action (Part D: Isolated Cells, Tissues, and Organ Systems)
Edited by JOEL G. HARDMAN AND BERT W. O'MALLEY

VOLUME XL. Hormone Action (Part E: Nuclear Structure and Function)
Edited by BERT W. O'MALLEY AND JOEL G. HARDMAN

VOLUME XLI. Carbohydrate Metabolism (Part B)
Edited by W. A. WOOD

VOLUME XLII. Carbohydrate Metabolism (Part C)
Edited by W. A. WOOD

VOLUME XLIII. Antibiotics
Edited by JOHN H. HASH

VOLUME XLIV. Immobilized Enzymes
Edited by KLAUS MOSBACH

VOLUME XLV. Proteolytic Enzymes (Part B)
Edited by LASZLO LORAND

VOLUME XLVI. Affinity Labeling
Edited by WILLIAM B. JAKOBY AND MEIR WILCHEK

VOLUME XLVII. Enzyme Structure (Part E)
Edited by C. H. W. HIRS AND SERGE N. TIMASHEFF

VOLUME XLVIII. Enzyme Structure (Part F)
Edited by C. H. W. HIRS AND SERGE N. TIMASHEFF

VOLUME XLIX. Enzyme Structure (Part G)
Edited by C. H. W. HIRS AND SERGE N. TIMASHEFF

VOLUME L. Complex Carbohydrates (Part C)
Edited by VICTOR GINSBURG

VOLUME LI. Purine and Pyrimidine Nucleotide Metabolism
Edited by PATRICIA A. HOFFEE AND MARY ELLEN JONES

VOLUME LII. Biomembranes (Part C: Biological Oxidations)
Edited by SIDNEY FLEISCHER AND LESTER PACKER

VOLUME LIII. Biomembranes (Part D: Biological Oxidations)
Edited by SIDNEY FLEISCHER AND LESTER PACKER

VOLUME 125. Biomembranes (Part M: Transport in Bacteria, Mitochondria, and Chloroplasts: General Approaches and Transport Systems)
Edited by SIDNEY FLEISCHER AND BECCA FLEISCHER

VOLUME 126. Biomembranes (Part N: Transport in Bacteria, Mitochondria, and Chloroplasts: Protonmotive Force)
Edited by SIDNEY FLEISCHER AND BECCA FLEISCHER

VOLUME 127. Biomembranes (Part O: Protons and Water: Structure and Translocation)
Edited by LESTER PACKER

VOLUME 128. Plasma Lipoproteins (Part A: Preparation, Structure, and Molecular Biology)
Edited by JERE P. SEGREST AND JOHN J. ALBERS

VOLUME 129. Plasma Lipoproteins (Part B: Characterization, Cell Biology, and Metabolism)
Edited by JOHN J. ALBERS AND JERE P. SEGREST

VOLUME 130. Enzyme Structure (Part K)
Edited by C. H. W. HIRS AND SERGE N. TIMASHEFF

VOLUME 131. Enzyme Structure (Part L)
Edited by C. H. W. HIRS AND SERGE N. TIMASHEFF

VOLUME 132. Immunochemical Techniques (Part J: Phagocytosis and Cell-Mediated Cytotoxicity)
Edited by GIOVANNI DI SABATO AND JOHANNES EVERSE

VOLUME 133. Bioluminescence and Chemiluminescence (Part B)
Edited by MARLENE DELUCA AND WILLIAM D. MCELROY

VOLUME 134. Structural and Contractile Proteins (Part C: The Contractile Apparatus and the Cytoskeleton)
Edited by RICHARD B. VALLEE

VOLUME 135. Immobilized Enzymes and Cells (Part B)
Edited by KLAUS MOSBACH

VOLUME 136. Immobilized Enzymes and Cells (Part C)
Edited by KLAUS MOSBACH

VOLUME 137. Immobilized Enzymes and Cells (Part D)
Edited by KLAUS MOSBACH

VOLUME 138. Complex Carbohydrates (Part E)
Edited by VICTOR GINSBURG

VOLUME 139. Cellular Regulators (Part A: Calcium- and Calmodulin-Binding Proteins)
Edited by ANTHONY R. MEANS AND P. MICHAEL CONN

VOLUME 158. Metalloproteins (Part A)
Edited by JAMES F. RIORDAN AND BERT L. VALLEE

VOLUME 159. Initiation and Termination of Cyclic Nucleotide Action
Edited by JACKIE D. CORBIN AND ROGER A. JOHNSON

VOLUME 160. Biomass (Part A: Cellulose and Hemicellulose)
Edited by WILLIS A. WOOD AND SCOTT T. KELLOGG

VOLUME 161. Biomass (Part B: Lignin, Pectin, and Chitin)
Edited by WILLIS A. WOOD AND SCOTT T. KELLOGG

VOLUME 162. Immunochemical Techniques (Part L: Chemotaxis and Inflammation)
Edited by GIOVANNI DI SABATO

VOLUME 163. Immunochemical Techniques (Part M: Chemotaxis and Inflammation)
Edited by GIOVANNI DI SABATO

VOLUME 164. Ribosomes
Edited by HARRY F. NOLLER, JR., AND KIVIE MOLDAVE

VOLUME 165. Microbial Toxins: Tools for Enzymology
Edited by SIDNEY HARSHMAN

VOLUME 166. Branched-Chain Amino Acids
Edited by ROBERT HARRIS AND JOHN R. SOKATCH

VOLUME 167. Cyanobacteria
Edited by LESTER PACKER AND ALEXANDER N. GLAZER

VOLUME 168. Hormone Action (Part K: Neuroendocrine Peptides)
Edited by P. MICHAEL CONN

VOLUME 169. Platelets: Receptors, Adhesion, Secretion (Part A)
Edited by JACEK HAWIGER

VOLUME 170. Nucleosomes
Edited by PAUL M. WASSARMAN AND ROGER D. KORNBERG

VOLUME 171. Biomembranes (Part R: Transport Theory: Cells and Model Membranes)
Edited by SIDNEY FLEISCHER AND BECCA FLEISCHER

VOLUME 172. Biomembranes (Part S: Transport: Membrane Isolation and Characterization)
Edited by SIDNEY FLEISCHER AND BECCA FLEISCHER

VOLUME 173. Biomembranes [Part T: Cellular and Subcellular Transport: Eukaryotic (Nonepithelial) Cells]
Edited by SIDNEY FLEISCHER AND BECCA FLEISCHER

VOLUME 174. Biomembranes [Part U: Cellular and Subcellular Transport: Eukaryotic (Nonepithelial) Cells]
Edited by SIDNEY FLEISCHER AND BECCA FLEISCHER

VOLUME 175. Cumulative Subject Index Volumes 135–139, 141–167

VOLUME 176. Nuclear Magnetic Resonance (Part A: Spectral Techniques and Dynamics)
Edited by NORMAN J. OPPENHEIMER AND THOMAS L. JAMES

VOLUME 177. Nuclear Magnetic Resonance (Part B: Structure and Mechanism)
Edited by NORMAN J. OPPENHEIMER AND THOMAS L. JAMES

VOLUME 178. Antibodies, Antigens, and Molecular Mimicry
Edited by JOHN J. LANGONE

VOLUME 179. Complex Carbohydrates (Part F)
Edited by VICTOR GINSBURG

VOLUME 180. RNA Processing (Part A: General Methods)
Edited by JAMES E. DAHLBERG AND JOHN N. ABELSON

VOLUME 181. RNA Processing (Part B: Specific Methods)
Edited by JAMES E. DAHLBERG AND JOHN N. ABELSON

VOLUME 182. Guide to Protein Purification
Edited by MURRAY P. DEUTSCHER

VOLUME 183. Molecular Evolution: Computer Analysis of Protein and Nucleic Acid Sequences
Edited by RUSSELL F. DOOLITTLE

VOLUME 184. Avidin-Biotin Technology
Edited by MEIR WILCHEK AND EDWARD A. BAYER

VOLUME 185. Gene Expression Technology
Edited by DAVID V. GOEDDEL

VOLUME 186. Oxygen Radicals in Biological Systems (Part B: Oxygen Radicals and Antioxidants)
Edited by LESTER PACKER AND ALEXANDER N. GLAZER

VOLUME 187. Arachidonate Related Lipid Mediators
Edited by ROBERT C. MURPHY AND FRANK A. FITZPATRICK

VOLUME 188. Hydrocarbons and Methylotrophy
Edited by MARY E. LIDSTROM

VOLUME 189. Retinoids (Part A: Molecular and Metabolic Aspects)
Edited by LESTER PACKER

VOLUME 190. Retinoids (Part B: Cell Differentiation and Clinical Applications)
Edited by LESTER PACKER

VOLUME 191. Biomembranes (Part V: Cellular and Subcellular Transport: Epithelial Cells)
Edited by SIDNEY FLEISCHER AND BECCA FLEISCHER

VOLUME 192. Biomembranes (Part W: Cellular and Subcellular Transport: Epithelial Cells)
Edited by SIDNEY FLEISCHER AND BECCA FLEISCHER

VOLUME 226. Metallobiochemistry (Part C: Spectroscopic and Physical Methods for Probing Metal Ion Environments in Metalloenzymes and Metalloproteins)
Edited by JAMES F. RIORDAN AND BERT L. VALLEE

VOLUME 227. Metallobiochemistry (Part D: Physical and Spectroscopic Methods for Probing Metal Ion Environments in Metalloproteins)
Edited by JAMES F. RIORDAN AND BERT L. VALLEE

VOLUME 228. Aqueous Two-Phase Systems
Edited by HARRY WALTER AND GÖTE JOHANSSON

VOLUME 229. Cumulative Subject Index Volumes 195–198, 200–227

VOLUME 230. Guide to Techniques in Glycobiology
Edited by WILLIAM J. LENNARZ AND GERALD W. HART

VOLUME 231. Hemoglobins (Part B: Biochemical and Analytical Methods)
Edited by JOHANNES EVERSE, KIM D. VANDEGRIFF, AND ROBERT M. WINSLOW

VOLUME 232. Hemoglobins (Part C: Biophysical Methods)
Edited by JOHANNES EVERSE, KIM D. VANDEGRIFF, AND ROBERT M. WINSLOW

VOLUME 233. Oxygen Radicals in Biological Systems (Part C)
Edited by LESTER PACKER

VOLUME 234. Oxygen Radicals in Biological Systems (Part D)
Edited by LESTER PACKER

VOLUME 235. Bacterial Pathogenesis (Part A: Identification and Regulation of Virulence Factors)
Edited by VIRGINIA L. CLARK AND PATRIK M. BAVOIL

VOLUME 236. Bacterial Pathogenesis (Part B: Integration of Pathogenic Bacteria with Host Cells)
Edited by VIRGINIA L. CLARK AND PATRIK M. BAVOIL

VOLUME 237. Heterotrimeric G Proteins
Edited by RAVI IYENGAR

VOLUME 238. Heterotrimeric G-Protein Effectors
Edited by RAVI IYENGAR

VOLUME 239. Nuclear Magnetic Resonance (Part C)
Edited by THOMAS L. JAMES AND NORMAN J. OPPENHEIMER

VOLUME 240. Numerical Computer Methods (Part B)
Edited by MICHAEL L. JOHNSON AND LUDWIG BRAND

VOLUME 241. Retroviral Proteases
Edited by LAWRENCE C. KUO AND JULES A. SHAFER

VOLUME 242. Neoglycoconjugates (Part A)
Edited by Y. C. LEE AND REIKO T. LEE

VOLUME 316. Vertebrate Phototransduction and the Visual Cycle (Part B)
Edited by KRZYSZTOF PALCZEWSKI

VOLUME 317. RNA–Ligand Interactions (Part A: Structural Biology Methods)
Edited by DANIEL W. CELANDER AND JOHN N. ABELSON

VOLUME 318. RNA–Ligand Interactions (Part B: Molecular Biology Methods)
Edited by DANIEL W. CELANDER AND JOHN N. ABELSON

VOLUME 319. Singlet Oxygen, UV-A, and Ozone
Edited by LESTER PACKER AND HELMUT SIES

VOLUME 320. Cumulative Subject Index Volumes 290–319

VOLUME 321. Numerical Computer Methods (Part C)
Edited by MICHAEL L. JOHNSON AND LUDWIG BRAND

VOLUME 322. Apoptosis
Edited by JOHN C. REED

VOLUME 323. Energetics of Biological Macromolecules (Part C)
Edited by MICHAEL L. JOHNSON AND GARY K. ACKERS

VOLUME 324. Branched-Chain Amino Acids (Part B)
Edited by ROBERT A. HARRIS AND JOHN R. SOKATCH

VOLUME 325. Regulators and Effectors of Small GTPases (Part D: Rho Family)
Edited by W. E. BALCH, CHANNING J. DER, AND ALAN HALL

VOLUME 326. Applications of Chimeric Genes and Hybrid Proteins (Part A: Gene Expression and Protein Purification)
Edited by JEREMY THORNER, SCOTT D. EMR, AND JOHN N. ABELSON

VOLUME 327. Applications of Chimeric Genes and Hybrid Proteins (Part B: Cell Biology and Physiology)
Edited by JEREMY THORNER, SCOTT D. EMR, AND JOHN N. ABELSON

VOLUME 328. Applications of Chimeric Genes and Hybrid Proteins (Part C: Protein–Protein Interactions and Genomics)
Edited by JEREMY THORNER, SCOTT D. EMR, AND JOHN N. ABELSON

VOLUME 329. Regulators and Effectors of Small GTPases (Part E: GTPases Involved in Vesicular Traffic)
Edited by W. E. BALCH, CHANNING J. DER, AND ALAN HALL

VOLUME 330. Hyperthermophilic Enzymes (Part A)
Edited by MICHAEL W. W. ADAMS AND ROBERT M. KELLY

VOLUME 331. Hyperthermophilic Enzymes (Part B)
Edited by MICHAEL W. W. ADAMS AND ROBERT M. KELLY

VOLUME 332. Regulators and Effectors of Small GTPases (Part F: Ras Family I)
Edited by W. E. BALCH, CHANNING J. DER, AND ALAN HALL

VOLUME 333. Regulators and Effectors of Small GTPases (Part G: Ras Family II)
Edited by W. E. BALCH, CHANNING J. DER, AND ALAN HALL

Section I

Biophysics, Structural Biology, and Enzymology
of Chromatin Proteins

[1] Fluorescence Anisotropy Assays for Analysis of ISWI-DNA and ISWI-Nucleosome Interactions

By J. CHIN, G. LÄNGST, P. B. BECKER, and J. WIDOM

Fluorescence anisotropy is a rapid, sensitive, and quantitative technique that is well suited to the analysis of protein-protein and protein-DNA interactions in solution. Fluorescence anisotropy is a measure of the depolarization of emitted fluorescence intensity obtained after excitation by a polarized light source, and depends directly on the relative rate of fluorescence emission versus the rate of tumbling in solution. The concept is simple: if a fluorescent molecule (or, more typically, a molecule to which a fluorescent probe has been attached) tumbles slowly in solution relative to the lifetime of fluorescence emission, then the light emitted in response to polarized excitation will remain highly polarized. However, if the molecules tumble rapidly in comparison to the emission lifetime, then, prior to emitting, they will have tumbled sufficiently so as to have "forgotten" their orientation at the moment of excitation, thus depolarizing (randomizing the polarization of) the emitted light.

Fluorescence anisotropy is applicable for analysis of macromolecular interactions because there is a good match between typical fluorescence lifetimes and typical macromolecular tumbling times. For approximately spherical molecules, the tumbling time scales as the molecular volume, that is, as the molecular weight. Thus, binding of an unlabeled macromolecule can make a significant change to the tumbling time of the molecule to which the fluorescent probe is attached, and hence to the measured anisotropy. For the studies described in the following, we utilize DNA molecules labeled at one end with the fluorescent dye fluorescein (these DNA molecules may be "naked DNA" or they may be incorporated into nucleosomes), and we use fluorescence anisotropy to monitor the binding of the *Drosophila* ISWI chromatin remodeling protein[1-3] to the labeled DNA or nucleosomes.

Fluorescence anisotropy is especially useful because of its high inherent sensitivity. Dyes such as fluorescein allow quantitative analysis of emission polarization from sub-nanomolar concentrations. Since dissociation constants are typically nanomolar or greater, this allows experiments to be

[1] T. Tsukiyama, C. Daniel, J. Tamkun, and C. Wu, *Cell* **83**, 1021 (1995).
[2] P. D. Varga-Weisz *et al.*, *Nature* **388**, 598 (1997).
[3] G. Längst and P. B. Becker, *J. Cell Sci.* **114**, 2561 (2001).

set up with the probe concentration $\ll K_d$; consequently the free concentration of the added macromolecule (ISWI, in our case), which is generally either difficult to measure or is completely unknown, will be approximately equal to the total concentration, which can be definitively measured, thus greatly simplifying the analysis of the binding measurements. Another important benefit of the sensitivity of the anisotropy measurement is that it preserves precious reagents. Measurements can be made in small volumes, and samples can be recovered and reused if desired.

Finally, as discussed later, the experiment can be carried out using inexpensive conventional fluorometers such as are found at most biochemical or chemical research laboratories, or, alternatively, using an inexpensive instrument specialized for the fluorescence anisotropy experiment.

Investigators planning to carry out such studies should study two particularly useful references, one on fluorescence theory and methodology in general[4] and one focused on fluorescence approaches to analysis of protein-DNA interactions in particular.[5] These references nicely define and explain the set of four fluorescence intensity measurements that go into a single measurement of fluorescence anisotropy; we will not duplicate this important topic here, but rather refer readers to these other sources.

Fluorescein-Labeled DNA

We use DNA sequences labeled with fluorescein attached at the 5'-end through a C6 linker. Relatively short sequences are purchased as a pair of complementary oligonucleotides, one containing 5'-fluorescein. These are annealed, and the resulting duplex purified away from any remaining single strand by reverse-phase HPLC on a Zorbax-10 column using a gradient of 10–20% acetonitrille in 0.1 M triethanolamine-acetate, pH 7.0, 0.1 mM EDTA, developed over 10 min at 1 ml min^{-1}. When longer sequences (e.g., nucleosome-length DNAs) are required, direct synthesis is not practical. Instead we use preparative scale PCR, with one of the two primers again containing 5'-fluorescein. The resulting PCR product is purified by gel electrophoresis in 1% agarose gels with standard TAE buffer, and extracted from the gel using Ultra-DA (Millipore) gel extraction kits.

DNA concentrations are quantified by UV absorbance.

[4] J. R. Lakowicz, "Principles of Fluorescence Spectroscopy," 2nd Ed. Kluwer Academic/ Plenum Press, New York, 1999.
[5] J. J. Hill and C. A. Royer, *Meth. Enzymol.* **278,** 390 (1997).

Preparation of Nucleosomes

Nucleosomes are formed by salt gradient dialysis using purified histone octamer and DNA, and the resulting nucleosomes are purified by sucrose gradient ultracentrifugation, as described.[6–8] We typically label a small amount of the fluorescein-labeled DNA with $[\gamma^{-32}]$ ATP (at the 5' end that does not have a fluorescein) using T4 polynucleotide kinase to facilitate following the sample throughout preparation and purification. Reconstitution reactions typically contain 300 ng of (^{32}P, fluorescein) double-labeled DNA, 15 μg of fluorescein-only labeled DNA, 3 μg of histone octamer, in a 300 μl volume of 2.5 M NaCl, 0.5 × TE (TE is 10 mM Tris, pH 8.0, 1 mM Na$_3$ EDTA) with 0.5 mM phenylmethylsulfonyl fluoride (PMSF), and 0.1 mM benzamidine (BZA) added as protease inhibitors. The reconstitution reactions are loaded onto ~12 ml 5–30% sucrose gradients in 0.5 × TE and centrifuged at 4° in an SW41 rotor (Beckman) at 41,000 rpm for 22–24 h. (We aim for substoichiometric reconstitution of histone octamer onto DNA, as this eliminates the possibility of overloading nucleosomes with excess histones[9] while providing useful diagnostics for the reconstitution and markers for the subsequent sucrose gradient purification.) Gradients are fractionated from the bottom in 0.5-ml fractions; fractions containing nucleosomes are identified by scintillation counting, pooled and exchanged into 0.5 × TE buffer on Centricon-30 concentrators, and analyzed by native polyacrylamide gel electrophoresis. Nucleosome concentrations are measured by UV absorbance at 260 nm. Reconstituted nucleosomes are stored at concentrations of 50 nM or greater, on ice in 0.5 × TE, and are used within 2 weeks.

Instrumentation and Technical Considerations

We use a conventional photon counting steady-state fluorometer (ISS PC-1, L-format) with rotatable polarizers in the excitation and emission paths. Alternatively, the Panvera Corporation markets a sensitive and relatively inexpensive instrument dedicated specifically to fluorescence anisotropy measurements. We generally increase the sensitivity of the PC-1 by removing the emission monochrometer and use instead a set of optical filters chosen to pass a desired broad band of fluorescence emission wavelengths, as described later.

[6] K. J. Polach and J. Widom, *J. Mol. Biol.* **254,** 130 (1995).
[7] P. T. Lowary and J. Widom, *J. Mol. Biol.* **276,** 19 (1998).
[8] J. D. Anderson, A. Thåström, and J. Widom, *Mol. Cell. Biol.* **22,** 7147 (2002).
[9] G. Voordouw and H. Eisenberg, *Nature* **273,** 446 (1978).

Sample Cleanliness

As with any sensitive experimental method in analytical biochemistry, care must be taken in certain matters to avoid potential pitfalls.

It goes without saying that both buffers and samples must be free from significant fluorescent contaminants. Fluorescence from the buffer alone and from unlabeled samples should be checked and shown to be negligible in comparison to the fluorescence obtained at the desired concentration of labeled sample. In our experience this has never proven to be a problem using dilute buffers supplemented with approximately physiological concentrations of salts and Mg^{2+} and small amounts of glycerol; nevertheless, it should be checked, especially in case of problems with the water or with contaminated glass- or plastic-ware.

Scattered Light

Even when samples are free of contaminants, one must take care to eliminate certain additional potential artifacts due to scattered light. Scattered light is particularly problematic in anisotropy measurements because it is generally perfectly polarized, and hence will systematically distort measurements of fluorescence anisotropy from the sample.

Two chief types of scattered light need to be considered in anisotropy measurements: elastic (Rayleigh) scattering and inelastic (chiefly Raman) scattering. Elastic scattering is a process in which excitation light is scattered in all directions, unshifted in wavelength, by interaction of the excitation light with molecules in the sample. Even pure solvents scatter light elastically. Such scattering is weak, yet may nevertheless be significant in comparison to the faint fluorescence from a very dilute fluorophore. Macromolecules in solution greatly increase the intensity of scattered light, in proportion to their concentration and molecular weight. Solutions containing a high molecular weight species such as nucleosomes can result in a scattering intensity that greatly exceeds the intensity of fluorescence emission from a dye attached to that same macromolecule.

If excitation and emission monochrometers were "perfect," then elastic scattering would present no problem: one could simply set the excitation and emission monochrometers to different wavelengths (e.g., the excitation and emission maxima, respectively), and there would be no leakage of scattered excitation light through the emission monochrometer. In fact, however, the finite resolution of the monochrometers, together with optical imperfections that allow light of colors outside the assumed bandpass to pass through, albeit at reduced intensity, are such that there can be significant excitation intensity at the color chosen for emission measurement, even though these colors may differ by 30 nm or more. In fact, this leakage can in practice be so great that, when combined with a relatively strong elastic light scattering from a macromolecular sample, the intensity of scattered excitation light reaching the emission detector may be significant in comparison to the intensity of fluorescence.

Raman (inelastic) scattering occurs when excitation light is scattered by solvent molecules (water, in biochemical applications) with concomitant vibrational excitation of the solvent molecules. The Raman scattered light is thus shifted in color toward the red relative to the excitation color by an amount corresponding to the vibrational energy change. This is a fixed amount in energy terms (3600 cm^{-1} for water), but corresponds to a varying wavelength change because of the reciprocal relationship between energy and wavelength ($E = hc/\lambda$ h = Planck's constant, c = velocity of light, λ = wavelength). For excitation at 280 nm, the Raman peak occurs at 311 nm, whereas for excitation at 490 nm (our typical choice for fluorescein), the Raman peak occurs at approximately 595 nm. The width of the Raman peak will be identical to that of the excitation light (measured on an energy axis, not on a wavelength axis). The Raman intensity is generally low relative to elastic scattering, but may nevertheless become significant when sample concentrations are low.

Excitation Path Filter

Both kinds of scattered light are readily eliminated with appropriate optical filters, with or without the use of an emission monochrometer. We use a bandpass interference filter in the excitation path, placed between the excitation monochrometer and the sample, to eliminate any remaining light at colors other than the desired excitation wavelength that happens to pass through the excitation monochrometer. This filter is chosen such that its wavelength of maximum transmission matches the excitation monochrometer wavelength setting, and the bandwidth of the filter is chosen to be comparable to that of the excitation monochrometer, so as to minimize unnecessary loss of excitation intensity.

Emission Path Filters

The combination of excitation monochrometer and bandpass filter in the excitation path together ensures that the excitation light is adequately clean. There remains, however, the possibilities that either light elastically scattered (at the excitation color) by the sample or Raman scattered excitation light may make it through the emission path and be counted by the emission detector.

We use a cut-on filter in the emission path to reject elastically scattered excitation light. Such filters absorb or reflect short wavelengths, while passing longer wavelengths with high transmittance, with a steep rise in transmittance occurring over a relatively narrow wavelength range. In general, colored glass filters work well for this purpose, and are available in a closely spaced series of cut-on wavelength ranges. One picks a filter that has essentially zero transmittance over the full bandpass of the excitation light, but that has high transmittance over much of the width of the fluorescence emission spectrum. In certain cases, if other constraints dictate that there will be only small shifts in color between excitation and emission wavelengths, it can be beneficial to use specialized multilayer dielectric filters, which can achieve much steeper cut-on characteristics.

Finally, one must remember to eliminate also the Raman scattered light, taking into account its full spectral width. Depending on the excitation wavelength and the fluorophore, the Raman scatter may be blue-shifted or red-shifted relative to the fluorescence emission. Even if the Raman scattering is superimposed on the fluorescence emission spectrum, the fluorescence emission spectrum will generally be much broader than the Raman band, so that it will be possible to choose filters that pass either the blue-side or the red-side of the fluorescence emission while rejecting the Raman.

The particular situation dictates the choice of filters to be used. When the Raman scatter is to the blue of the fluorescence emission, cut-on filter can be chosen to reject both elastic and Raman scattered light. When the Raman band is to the red of the fluorescence, one may need to supplement the cut-on filter with a cut-off filter. If an emission monochrometer is used, this itself may serve effectively as a cut-off filter (because of the lower intensity of Raman scattering compared to elastic scattering). Alternatively, or in addition, cut-off (short-pass) filters may be used. The selection of colored glass cut-off filters is much less extensive than for cut-on. In general, one will need to use specialized multilayer dielectric filters instead.

Once a filter combination has been chosen (whether or not an emission monochrometer will be used for the actual experiment), it is wise to verify that the filter combination works as planned by recording fluorescence emission spectra of buffer alone, of unlabeled sample, and of labeled sample (at the concentration that will be used for anisotropy measurement), scanning the emission monochrometer from below the excitation wavelength to above the fluorescence emission range and Raman band. Both buffer and unlabeled sample spectra should show negligible intensity, in comparison to the intensity from the labeled sample, at all wavelengths that will be monitored during the anisotropy measurement.

Filter Set for Use with Fluorescein

We find the following combination of filters to be highly effective for use with fluorescein, whether an emission monochrometer is present or not. We choose 490 nm as the center wavelength for excitation.

Excitation interference filter   Coherent/Ealing #35-3482 interference filter, center wavelength = 490.0 nm, bandpass =7.3 nm (full width at half-maximum transmission)

Emission cut-on (long-pass) filter   Coherent/Ealing #26-4333 OG-515 colored glass

Emission cut-off (bandpass)   Omega optical 535AF45 multilayer

Figure 1 shows excitation and emission spectra for fluorescein in panel A, compared to the transmission characteristics of the three filters in panel B. The excitation filter is well matched to the excitation maximum for

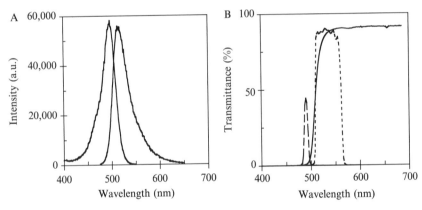

Fig. 1. Fluorescence spectra of fluorescein compared to transmission spectra of the optical filters used for measurement of fluorescein anisotropy. (A) Fluorescence excitation and emission spectra. Left-hand curve: fluorescein excitation spectrum, obtained monitoring emission at $\lambda = 520$ nM; 0.5 nM fluorescein-labeled oligonucleotide in a buffer containing 20 mM HEPES-KOH, pH 7.6, 80 mM KCl, 2 mM MgCl$_2$, 1 mM DTT, 5% glycerol. Right-hand curve: the corresponding emission spectrum, with excitation at $\lambda = 490$ nm. (B) Long-dashed curve: transmission characteristics of the 488.8 nm interference filter used in the excitation path in conjunction with the excitation monochrometer. Solid and short-dashed curves, transmission characteristics of the cut-on and bandpass filters, respectively, which are used in the emission path, typically with no emission monochrometer. The excitation filter matches the excitation maximum for fluorescein. The cut-on and leading (short wavelength) edge of the bandpass filters both strongly reject any elastically scattered excitation light; the falling (long wavelength) edge of the bandpass filter strongly rejects any Raman scattered light, which is centered at ~595 nm.

fluorescein. The colored glass cut-on filter rejects most of the excitation bandpass, while passing the majority of the emission spectrum. The specialized cut-off filter nicely rejects any Raman scatter, which is centered at 595 nm, while further strongly reducing any elastically scattered excitation light. (The cut-off [bandpass] filter also strongly rejects elastically scattered light on its own; however, the combination of the two filters gives far better blocking at the excitation bandpass than either one on its own. This improved scatter rejection is important in strongly scattering/weakly fluorescing samples.)

The performance of this filter set can be appreciated from the emission spectra in Fig. 2. Panel A shows the ability of the filter set to suppress both elastically scattered light (>10,000-fold reduction) and Raman scattering to undetectably low levels, while panel B shows that this huge reduction in background scattering intensity comes at a modest cost (2- to 3-fold) in the collectible intensity of fluorescence emission.

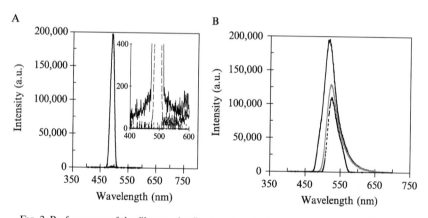

FIG. 2. Performance of the filter set for fluorescein anisotropy measurement. (A) Rejection of elastically scattered light and Raman scatter. Emission spectra recorded from a scattering solution (10 μg/ml BSA in buffer), with excitation at 490 nm using the excitation monochrometer plus the 488.8 nm interference filter. Solid line: no filters in the emission path. Long-dashed line: 515 nm cut-on filter. Short-dashed line (essentially invisible): both 515-nm cut-on plus bandpass filters. When no filters are used in the emission path, note the strong peak of scattering intensity detected when λ emission = λ excitation. Inset: same spectra plotted on a 500-fold more sensitive scale. The emission filter combination reduces the scattering signal by >10,000-fold, rendering it immeasurably low. Similarly, there is negligible remaining intensity at the Raman scattering wavelength (~595 nm). (B) Fluorescence emission from fluorescein with no filter in the emission path (solid black line), 515-nm cut-on filter only (grey line) or both cut-on and bandpass filters (dashed line). The >10,000-fold reduction in scattering intensity comes at a cost of only ~2-fold in emission intensity at the emission maximum, and only ~3-fold in total emission intensity integrated over all wavelengths (such as could be measured by a detector in the emission path if no emission monochrometer were present).

pH Sensitivity

Fluorescein has a pK_a of ~6.5, and its fluorescence intensity is dependent on the charge state. Consequently, it is important to control the pH of the solution, preferably at a pH $\geq$ 7.5, so that unanticipated small changes in pH will affect the fluorescence intensity only negligibly.

Intensity Changes

Two additional phenomena concerning the sample fluorescence intensity bear mention. First, fluorescein is sensitive to photobleaching; thus, prolonged exposure of sample to excitation light, or even room light, can significantly reduce the fluorescence intensity. Since anisotropy is an inherent property, independent of concentration, modest amounts of sample bleaching occurring during binding titrations need not be problematic, provided that there is negligible bleaching during the time required for any given anisotropy measurement. This can be checked by monitoring the total intensity during the four individual measurements that make up an anisotropy measurement.

In certain cases, however, the binding of a protein to a fluorescein-labeled DNA can affect the fluorescence intensity directly. For example, the bound protein might happen to quench (or enhance) the fluorescence quantum yield, or perhaps shift the emission color, so that the intensity measured over a given wavelength range may increase or decrease. Such effects invalidate the usual interpretation of the anisotropy changes. If the intensity changes, the likelihood is that the fluorescence lifetime too is changing; and if the lifetime changes, then the anisotropy is affected even if molecular tumbling time were to remain constant. Consequently, it will no longer be possible to assign a linear relationship between a measured anisotropy change and a probability of binding site occupancy.

Actually, such cases can be a blessing in disguise: one can often monitor the binding process simply by measuring the intensity change directly. In any event, it is necessary to pay attention to the total intensity when carrying out anisotropy measurements. Systematic changes in intensity that correlate with binding invalidate the standard interpretation of the anisotropy experiment.

Cuvettes

Samples used in fluorescence experiments may be precious, and one may wish to minimize the volume of sample used. We routinely use samples of 75–100 μl in quartz ultramicro-cuvettes (Hellma, black walled, #105.251-QS) having a sample chamber of 3 × 3 mm with a 5-mm tall aperture, that is, a 45-μl illuminated volume. These cuvettes are easy to fill and clean, and fit in the 1.25-cm square cuvette holders that are standard in most fluorometers. These cuvettes are available with the sample chamber placed at various heights above the base of the cuvette. It is important to pay attention to the height of the optical axis of the fluorometer, and to make certain that the height of the sample chamber of the cuvette matches that of the fluorometer.

Experimental Design for Binding Titrations

We use selected high-affinity nucleosome positioning sequences[7,10,11] to simultaneously provide a high degree of homogeneity in nucleosome positioning while also enhancing the stability of the nucleosomes against dissociation by mass action despite the dilute nucleosome concentrations used. ISWI is an ATP-dependent nucleosome remodeling factor that induces nucleosome sliding on nicked DNA.[12] It is expressed in *Escherichia coli* and purified by gel-filtration to near homogeneity.[13] Non-specific interactions of ISWI with DNA alone and with DNA at its entry into the nucleosome have been described qualitatively.[3] Fluorescence anisotropy measurements permit a quantitative description of these interactions.

Binding buffer for ISWI-DNA or ISWI-nucleosome binding reactions contains: 20 mM HEPES-KOH, pH 7.6, 80 mM KCl, 2 mM MgCl$_2$, 5% glycerol, 1 mM DTT, supplemented when desired with ATP, ADP, or other nucleotides or analogs. We typically make up a distinct sample for each ISWI to be investigated. This allows the DNA or nucleosomes to remain constant as the ISWI is varied over a titration. Binding reactions are 100 μl final volume, typically with 1 nM fluorescein-labeled DNA or 5 nM nucleosomes, and the desired ISWI. ISWI protein is diluted into binding buffer as appropriate, such that accurately measurable volumes are added into the 100-μl final binding reaction volumes. Samples are incubated at room temperature for 30 min prior to measurement of fluorescence anisotropy to ensure that binding reactions are well equilibrated (control studies show that binding appears to equilibrate essentially instantaneously, as judged by the absence of further changes in anisotropy, hence the 30-min equilibration time is more than sufficient). Samples are placed in quartz ultramicro-cuvettes as described earlier. We use an excitation wavelength of 490 nm, no emission monochrometer, and the

[10] A. Thåström *et al.*, *J. Mol. Biol.* **288**, 213 (1999).
[11] J. Widom, *Q. Rev. Biophys.* **34**, 269 (2001).
[12] G. Längst and P. B. Becker, *Mol. Cell* **8**, 1085 (2001).
[13] D. F. Corona *et al.*, *Mol. Cell* **3**, 239 (1999).

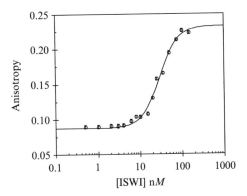

FIG. 3. Raw fluorescence anisotropy data from ISWI binding to naked DNA. Titration of a 5 nM 5′-fluorescein–labeled 35-bp long DNA with increasing concentrations of ISWI protein. (See text for buffer conditions.) The curve superimposed on the data represents a fit to a cooperative binding model.

combination of the cut-on and bandpass cut-off filter in the emission path described previously.

We always include a sample prepared with no ISWI (ISWI = 0) to establish the experimental "baseline" for each titration, and we extend the titrations to sufficiently high ISWI to allow accurate determination of the anisotropy corresponding to complete binding (complete occupancy of binding sites by bound ISWI). Note that, whatever method is used to monitor binding processes, it is important to carry titrations through the full range of the binding process. A good practice is to use "direct" plots[14] of the measured signal (anisotropy, in our case) versus the titrant concentration (ISWI, in our case) plotted on a log scale to allow representation of the wide range of titrant necessary to explore the full range from fraction bound ∼0 (no binding) to fraction bound ∼1 (∼100% binding).

We use Kalaidagraph software to fit raw binding data to desired binding models.

Results of Binding Experiments

Typical raw data resulting from such an experiment are shown in Fig. 3, for a 5′-fluorescein–labeled 35-bp long DNA, used at 5 nM. Aficionados of binding studies will recognize immediately from the raw data that the

[14] I. M. Klotz, "Ligand-Receptor Energetics: A Guide for the Perplexed." Wiley, New York, 1997.

binding curve as plotted in this manner is too "steep" to correspond to simple 1:1 binding of ISWI to DNA, implying positive cooperativity in the binding of ISWI to DNA. The titration midpoint for these given conditions (EC_{50}) is 30 nM. The concentration of fluorescein-labeled tracer is small in comparison and hence may safely be neglected (or alternatively quantitatively accounted for,[14] yielding small corrections that convert $[ISWI]_{total}$ to $[ISWI]_{free}$). If binding were simple (1:1 ISWI:DNA complex, noncooperative binding curve), then this measured EC_{50} would also be the thermodynamic K_d.

In any experimental analysis of binding processes, it is helpful to reduce the number of adjustable parameters in the curve fitting. In our anisotropy studies, we directly measure both the lower and upper "baselines" for the titrations, and hold these quantities fixed at their measured values during the curve fitting procedure. We set the anisotropy for the 0-nM ISWI baseline (the lower baseline for the curve fitting) equal to the average of several measurements on the same DNA sample lacking any ISWI, and we set the upper baseline equal to an average of the results for (replicate measurements on) the last couple or few titration points, having taken care to extend the titrations to the point at which any further increases in ISWI do not result in any further significant changes in measured anisotropy. For a given fixed assumed stoichiometry of the binding process, this leaves only one free parameter—the apparent affinity or K_d—to be determined by curve fitting. Alternatively, one may allow both the apparent affinity and the cooperativity (or "molecularity") to be simultaneously fit.

Figure 4 shows results for a negative control, in which the DNA (18-bp duplex, 5'-end labeled with fluorescein, 1 nM concentration) proves to be too short to allow high affinity binding of the ISWI. Plainly, even the raw data resulting from the anisotropy experiment can distinguish samples in which binding does occur from samples in which it does not. The EC_{50} (K_d, if 1:1 ISWI:DNA complex) for ISWI binding to this DNA is $\gg$100 nM.

Figure 5 shows the results an experiment monitoring binding to 177-bp–containing nucleosomes (again, 5'-end labeled with fluorescein, 5 nM concentration). The raw data are rescaled along the ordinate to represent the fraction of DNA with bound ISWI, simply by linearly rescaling the measured anisotropies from 0 (experimentally measured lower baseline) to 1 (measured upper baseline). The titration midpoint for this particular dataset (EC_{50}) is 15 nM, slightly lower (i.e., higher affinity) than for the 35-mer DNA of Fig. 3. An average over many datasets (data not shown) suggests that this small apparent difference in affinity between DNA and nucleosome is not statistically significant. Evidently, ISWI binds to nucleosomes with an affinity that is close to its affinity for long naked DNA.

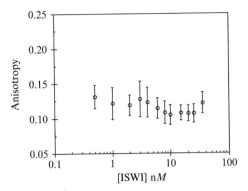

Fig. 4. Raw anisotropy data when no binding occurs. A 1 n*M* solution of a 5′-fluorescein–labeled 18-bp long (double-stranded) DNA is titrated with increasing concentrations of ISWI protein. ISWI has negligible affinity for such short DNAs.

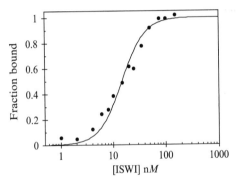

Fig. 5. ISWI binding to a nucleosomal DNA. A 5-n*M* solution of a 5′-fluorescein–labeled 177-bp DNA, assembled into nucleosomes, is titrated with increasing concentration of ISWI. Raw fluorescence anisotropy data are scaled to fraction bound (fraction of DNAs having ISWI protein bound): anisotropy data obtained in the absence of any ISWI protein establish the anisotropy for fractional occupancy = 0; the averaged anisotropy from the highest several titration points (where the signal appears to have plateaued) defines the fractional occupancy = 1. The curve represents a least-squares fit to a cooperative binding model. The DNA template used includes a 147-bp selected nucleosome positioning sequence together with an additional 30 bp of DNA extending beyond one end; a single fluorescein is attached at the 5′-end of this 30-bp extension. The DNA is assembled into nucleosomes and purified as described (see text).

Finally, in Fig. 6 we study the binding of ISWI to DNA (5′-fluorescein, 1 n*M* concentration) in the presence of 1 m*M* ADP. The titration midpoint for this dataset (EC$_{50}$) is 28 n*M*, very close to the measured 30 n*M* EC$_{50}$ for binding to naked DNA in the absence of nucleotide (Fig. 3). Evidently, the

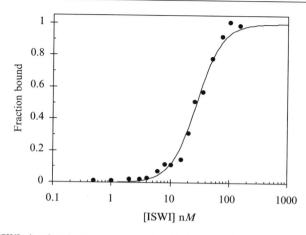

Fig. 6. ISWI titration in the presence of added nucleotide. Titration of a 1 nM 5'-fluorescein–labeled 150-bp long DNA with increasing concentrations of ISWI protein, in the presence of 1 mM ADP. The raw anisotropy data are scaled to fraction bound. Nucleotides or other cofactors or binding partners can be included in the reactions without difficulty.

affinity of ISWI for naked DNA is not influenced by the presence of high concentrations of ADP. This figure highlights the utility of anisotropy measurements to monitor binding in solutions containing other additives such as nucleotides. In contrast, it is difficult or risky to carry out such studies using a gel electrophoretic mobility shift approach, for example, since prohibitively costly amounts of nucleotide analogs might be required, and moreover these compounds may actually electrophorese. In that case, their concentrations around the complexes during a gel separation would be undefined, rendering the experiments uninterpretable.

Conclusions

Fluorescence anisotropy is well known to be useful for analysis of protein-DNA interactions, and it seems likely to be particularly useful for analysis of nucleosome remodeling factors because it is rapid, quantitative, highly sensitive (conserving precious reagents), suitable for use in the presence of cofactors such as ATP, readily measured even during rapid kinetic experiments, and very broadly applicable. It will allow analysis of the interactions of remodeling factors or their individual proteins or domains with any other species that can be specifically labeled.

[2] Biophysical Analysis of Specific Genomic Loci Assembled as Chromatin *In Vivo*

By Philippe T. Georgel and Jeffrey C. Hansen

Background

Over the last decade much progress has been made toward understanding the effects of chromatin on nuclear functions. However, virtually all previous studies of chromatin fiber organization *in vivo* have been restricted to gathering information about the locations of nucleosomes, histone post-translational modifications, regulatory DNA binding proteins, and chromatin remodeling machines relative to specific functional DNA elements, for example, promoters, origins of replications, repair sites. Despite a vastly improved understanding of the composition and configuration of functionally important genomic loci, very little is known about the higher-order organization of these chromosomal regions *in vivo*. Biophysical characterization of specific *in vivo*–assembled chromatin structures has not been possible due to technical limitations. Consequently, our knowledge of the functional effects of chromatin folding and higher-order structure has been obtained almost exclusively through use of *in vitro* model systems that mimic the solution behavior of natural chromatin.[1–4]

Recently, we have adapted the technique of agarose multigel electrophoresis (AME)[5–8] for analysis of the higher-order nucleoprotein structure of specific genomic loci that have been isolated as native chromatin in unfractionated low-salt nuclear extracts.[9,10] This approach yields analytical measurements of average macromolecular radii and surface charge density, which in turn allows one to evaluate the condensation behavior and conformational flexibility of the chromatin fragment being studied. Here

[1] J. C. Hansen and C. L. Turgeon, *Methods Mol. Biol.* **119**, 127 (1999).

[2] J. C. Hansen, *Annu. Rev. Biophys. Biomol. Struct.* **31**, 361 (2002).

[3] P. J. Horn, K. A. Crowley, L. M. Carruthers, J. C. Hansen, and C. L. Peterson, *Nat. Struct. Biol.* **9**, 167 (2002).

[4] B. Dorigo, T. Schalch, K. Bystricky, and T. J. Richmond, *J. Mol. Biol.* **327**, 85 (2003).

[5] T. M. Fletcher, P. Serwer, and J. C. Hansen, *Biochemistry* **33**, 10859 (1994).

[6] T. M. Fletcher, U. Krishnan, P. Serwer, and J. C. Hansen, *Biochemistry* **33**, 2226 (1994).

[7] L. M. Carruthers, C. Tse, K. P. Walker, III, and J. C. Hansen, *Methods Enzymol.* **304**, 19 (1999).

[8] L. M. Carruthers and J. C. Hansen, *J. Biol. Chem.* **275**, 37285 (2000).

[9] P. T. Georgel and J. C. Hansen, *Biopolymers* **68**, 557 (2003).

[10] P. T. Georgel, T. M. Fletcher, G. L. Hager, and J. C. Hansen, *Genes Dev.* **17**, 1617 (2003).

we describe how AME can be used as a biophysical method for characterizing specific *in vivo*–assembled chromatin fragments. The general con-
cepts presented in this chapter are based on our recent studies of
genomic murine mammary tumor virus (MMTV) promoters.[10]

To best describe the increasing number of specific types of "higher
order chromatin structures" observed *in vitro* and *in vivo*, Woodcock and
Dimitrov[11] have introduced a nomenclature that is analogous to that used
for proteins. Primary chromatin structure refers to a linear arrangement of
nucleosomes (i.e., beads-on-a-string), secondary chromatin structure describes condensed fiber conformations that results from intrinsic and/or
protein-mediated nucleosome-nucleosome interactions (i.e., linker his-
tone-stabilized 30 nm fiber). Tertiary structures chromatin refers to chromatin suprastructures formed through interaction of secondary structures
(i.e., long-range fiber-fiber interactions). Throughout this chapter we use
the nomenclature suggested by Woodcock and Dimitrov.[11]

Agarose Multigel Electrophoresis

Agarose gel electrophoresis generally is thought of as a preparative
technique, but it also can be used to obtain quantitative information about
macromolecular size and charge. In the presence of an applied electrical
field, the mobility, μ, of a charged macromolecule in solution is directly
proportional to its surface change density.[5,6,12] In the presence of agarose,
the "gel-free" mobility, μ_0', is reduced by the interaction of the macromolecule with the network of gel pores, P_e (pore size). Interactions with the gel
matrix are referred to as "sieving," and are dependent on the effective
radius, R_e, and conformational flexibility of the macromolecule.[5,6,12,13]

AME is an easily accessible method that is performed with the commercially available electrophoresis apparatus shown in Fig. 1A. An agarose
multigel consists of a multiple individual agarose running gels embedded
in a 1.5% agarose frame (Fig. 1B). The running gels typically range
from 0.2% to 3.0% agarose. The multigel apparatus minimizes gel to gel
variations in temperature, field strength, and buffer pH, which allows
determination of the μ, μ_0', and R_e of macromolecules with analytical precision.[5,12] The relationship between μ, μ_0', R_e, and P_e during an agarose gel
electrophoresis experiment is described in Eq. (1):

$$\mu/\mu_0' = (1 - R_e/P_e)^2. \tag{1}$$

[11] C. L. Woodcock and S. Dimitrov, *Curr. Opin. Genet. Dev.* **11**, 130 (2001).
[12] G. A. Griess, E. T. Moreno, R. A. Easom, and P. Serwer, *Biopolymers* **28**, 1475 (1989).
[13] J. C. Hansen, J. I. Kreider, B. Demeler, and T. M. Fletcher, *Methods* **12**, 62 (1997).

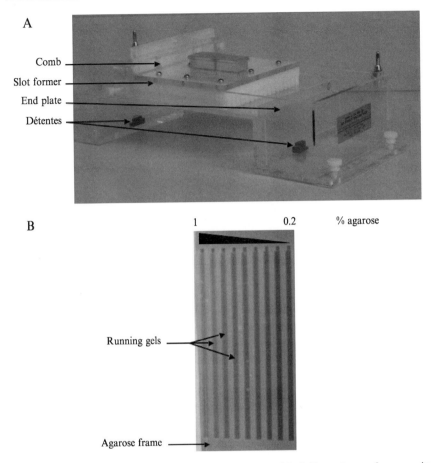

FIG. 1. (A) Multigel apparatus. (B) 9-lane agarose multigel. Percentages of agarose in running gels are in increment of 0.1%. *Note:* the figure shows only one half of an 18-lane gel (mirror image).

In an AME experiment, the sample of interest is first spiked with the spherical bacteriophage T3 ($R_e = 30.1$ nm). The T3 μ_0' is obtained by extrapolating the linear region of a plot of log μ^{T3} versus agarose percentage to 0% agarose, that is, the y-axis.[5,6,13] Using Eq. (1), the P_e of each running gel is calculated from the μ and μ_0' and known R_e (30.1 nm) of the T3 internal standard. For the unknown band(s) of interest, the μ and μ_0' are determined experimentally and the R_e in each running gel is calculated using Eq. (1) and the T3-derived values of P_e for that gel.

Extending AME to In Vivo–*Assembled Chromatin*

Previous AME studies have been performed with purified bacterio-phages[5–8,13] and defined chromatin model systems assembled *in vitro* from pure components.[14] Consequently, to apply AME for the analysis of genomic chromatin fragments isolated in low-salt nuclear extracts, it was first necessary to determine whether any components in the extracts altered the intrinsic μ, μ'_0, and R_e of the genomic chromatin bands. The results of these control experiments have been published[9] and are summarized briefly here. Naked DNA or model 12-mer nucleosomal arrays were added to nuclear extracts to mimic the release of genomic chromatin fragments into the same environment. Under these conditions super-shifted smears were observed for both the DNA and nucleosomal arrays on standard 1% agarose gels. To abolish deleterious non-specific DNA binding, herring sperm DNA (HS DNA) was added to the samples. For incubations performed with 10 μg standard unfractionated nuclear extracts[15] in a final volume of 10–20 μl, addition of up to 100-fold excess HS DNA led to single discrete bands that migrated the same as pure DNA or nucleosomal arrays alone (note that the exact amount of required excess HS-DNA is extract dependent). We have reported data for yeast, human carcinoma, *Drosophila* embryo, mouse adenocarcinoma cell extracts. Other types of extracts will require empirical optimization through titration of competitor HS DNA as described in Georgel and Hansen.[9]

AME experiments were subsequently performed with DNA and nucleosomal array samples prepared under the same conditions. In all cases, the measured R_e and μ'_0 values were the same within experimental error.[9] The quantitative data demonstrate that the intrinsic R_e and μ'_0 of nucleosomal arrays can be accurately determined in the presence of unfractionated low-salt nuclear extracts. To adapt AME for analysis of genomic chromatin fragments, it was also necessary to show that Southern blotting could be used to accurately measure the μ of specific bands in each of the running gels. This was accomplished by mixing 208-12 DNA or nucleosomal arrays in a low-salt nuclear extract as described earlier, and then measuring the band migrations by both fluorescent dyes and Southern hybridization. Again, the measured R_e and μ'_0 values were the same within experimental error.[9]

[14] J. C. Hansen, J. Ausio, V. H. Stanik, and K. E. van Holde, *Biochemistry* **28,** 9129 (1989).
[15] J. D. Dignam, R. M. Lebovitz, and R. G. Roeder, *Nucleic Acids Res.* **11,** 1475 (1983).

Methods

The only aspect of the AME approach that is not general is obtaining the specific genomic chromatin fragment of interest in a low-salt nuclear extract. Once this has occurred, the remainder of the method has been standardized. Presented in the following are the general protocols needed to:

1. Prepare a multigel.
2. Perform an AME experiment using low-salt nuclear extracts as the source of the chromatin.
3. Transfer the multigel to membrane and measure the μ in each running gel through use of Southern blotting.
4. Calculate the μ, μ'_0, and R_e of the chromatin-band(s) of interest, and the P_e of each running gel.

In addition, we describe western assays for determining the composition of the genomic chromatin fragment, and biophysical assays for formation of salt-dependent secondary chromatin structure and for the extent of conformational flexibility. To emphasize the potential of the AME approach, we conclude the article by briefly describing key aspects of our recent use of AME to characterize the active and inactive states of genomic MMTV promoters.

Equipment

Multigel units such as those shown in Fig. 1A are commercially available from Aquebogue (NY). It is important to keep all the various components of the multigel together as a single unit. A high-temperature water bath is needed to prepare the multigel. A Cole Parmer oscillating pump run at 40–50% maximum output voltage (i.e., 50–60 V for this pump) regulated through the use of a rheostat is needed to circulate the running buffer and maintain temperature control. For the reasons described later, we strongly recommend that the agarose be molecular biology grade, Low Electro Endo-Osmosis purchased from Research Organics (cat # 1170A-3). The casting and running buffer is either E buffer (40 mM Tris-HCl [pH 7.8], 0.25 mM Na$_2$ EDTA) or TAE (40 mM Tris acetate [pH 8.3], 1 mM Na$_2$EDTA). When formation of secondary chromatin structures is being assayed, E buffer is used and contains MgCl$_2$ at concentrations equal to 0.1–2.0 mM free Mg^{2+}. Do not use Tris-borate buffers in combination with low electro endo-osmosis (EEO) agarose, as this will generate anomalous undesirable electro-osmosis effects.

Multigel Preparation

Set the high-temperature water bath at 65–70°, and fill with enough water to have the entire tube of agarose (see later) submerged to about 1 cm from the lid. The gel frame is prepared with 1.5% agarose in E or TAE buffer. Weigh 1.8 g of agarose into a 250-ml flask. Tare the scale. Add 120 ml of running buffer (measure by weight not by volume; i.e., assuming a buffer density of 1.0, add 120 g of buffer) and tare again. Cover the top of the flask with a piece of plastic wrap, punch a small hole, and melt the agarose in a microwave oven. After melting, weigh the flask again. Compensate for evaporation by adding milliQ water. Cool the agarose to a temperature around 55° (i.e., comfortable to touch). Assemble the slot former and the desired comb as described in detail,[16] and as shown in Fig. 1A. Secure the end plates in the grooves with the détentes (small plastic parallelepipedic blocks). If needed, seal the plates with a small quantity of 1.5% framing agarose. Level the unit. To make the frame, pour the molten agarose into the gel bed, allow the agarose to set for 30 or 60 min (9- or 18-lane slot formers, respectively).

Begin preparing the agarose running gels. Start by labeling capped 15-ml Kimble borosilicate (threaded end, 20 × 125 for 9-lane gels and 16 × 125 for 18-lane gels) tubes with the chosen agarose concentrations (see Table I for details). Add running buffer to the tubes (see Table I for volumes) and place the tubes in the water bath at 65–70°.

After complete polymerization of the frame, carefully remove the comb and then very gently pull out the slot former. Abrupt removal will result in breaking the lanes. You may want to practice removing the slot former using standard agarose prior to using the more expensive low EEO agarose. Using a pair of small forceps, remove the small strips of agarose that often form at the edges of the wells. Burst any bubbles that may have formed in the thin layer of agarose beneath each slot. Put the comb back in its original location.

While the frame is polymerizing, prepare the agarose stock(s) for the gel dilutions. To prepare a 0.2–1% agarose multigel, start with 60 ml of molten 1% agarose in a 125-ml flask (prepared in the same way as described earlier for the 1.5% agarose stock). Immediately dilute the stock according to Table I to prepare each running gel. After adding the appropriate volume of 1% agarose to the pre-warmed running buffer, cap the tube, vortex briefly, and immediately pour the agarose into the appropriate multigel slot using disposable plastic pipettes. Prepare one agarose concentration at a time. Use a new pipette every time you change agarose concentration.

[16] J. C. Hansen, T. M. Fletcher, and J. I. Kreider, *Methods Mol. Biol.* **119**, 113 (1999).

TABLE I
DILUTIONS

Gel (%)	Buffer (ml)	Agarose stock (ml)
1% stock		
18-lane frame		
0.2	4	1
0.3	3.5	1.5
0.4	3	2
0.5	2.5	2.5
0.6	2	3
0.7	1.5	3.5
0.8	1	4
0.9	0.5	4.5
1	0	5
9-lane frame		
0.2	8	2
0.3	7	3
0.4	6	4
0.5	5	5
0.6	4	6
0.7	3	7
0.8	2	8
0.9	1	6
1	0	10
3% stock		
18-lane frame		
0.9	3.5	1.5
1	3.4	1.7
1.3	2.9	2.2
1.6	2.4	2.7
1.9	1.9	3.2
2.2	1.4	3.7
2.5	0.9	4.2
2.8	0.4	4.7
3	0	5
9-lane frame		
0.9	7	3
1	6.7	3.3
1.3	5.7	4.3
1.6	4.7	5.3
1.9	3.7	6.3
2.2	2.7	7.3
2.5	1.7	8.3
2.8	0.7	9.3
3	0	10

For a 9-lane frame, use ~7 ml of diluted agarose per lane and ~2.5 ml for an 18-lane. Note that to maintain stability of the frame, the 0.2% agarose should not be poured in the slot at the edge of the multigel frame. For a 0.2–1.0% multigel we suggest the following pouring order: 0.8, 0.9, 1.0, 0.2, 0.3, 0.4, 0.5, 0.6, 0.7. To prepare a 1.0–3.0% agarose 18-lane multigel, start with 60 ml of molten 3.0% agarose in a 125-ml flask. For each running gel, immediately dilute the stock according to Table I as described previously. For a 0.9–3.0% gel we suggest the following pouring order: 0.9, 1.0, 1.3, 1.6, 1.9, 2.2, 2.5, 2.8, 3.0.

Let the running gels set for 1 h. Pull out the comb, end plates, and détentes. Fill the multigel apparatus with running buffer. The multigels can be prepared in advance and used the following day. Make sure that the gel box is covered with plastic wrap to avoid unnecessary evaporation of buffer. Gels have been used successfully after 3 days. To prevent the gel from sliding off the gel bed during subsequent electrophoresis, place several 10-μl plastic tips in the plate grooves at opposite corners.

Gel Electrophoresis

Prior to electrophoresis, low-salt nuclear extracts (100 μg total nucleic acid) are transferred to fresh tubes containing the appropriate amount of empirically determined herring sperm DNA, 0.5 μg of bacteriophage T3, and glycerol (10% w/v final). Samples are loaded in the appropriate well and electrophoresed at 1.33 V/cm for 6 h (timed manually and precisely). Alternatively, one can perform the electrophoresis for 1 V/cm for 8 h.[13] The average buffer temperature in our experiments has been 24 ± 3° (room temperature). Using the pump system described earlier, the running buffer was continuously recirculated throughout the entire experiment at a slow and steady pace, which prevents the formation of pH or ion gradients and large temperature fluctuations.

To measure the μ of the bacteriophage T3 in each running gel, after electrophoresis a thin plastic sheet is placed under the multigel to permit transfer of the fragile gel to a tray containing either ethidium bromide or SYBR green. The tray is placed on a rotating shaker set at 60 rpm (maximum) for 30 min. A 60-min destaining period in running buffer (or distilled water) is required if ethidium bromide staining is used. The multigel is then photographed under ultraviolet light using standard positive/negative film or using a digital camera, with a UV-compatible ruler placed next to the edge of the multigel for measurement purposes.

Southern Hybridization

To detect specific genomic chromatin bands by Southern blotting (see later), the gel was subsequently treated with standard denaturing and neutralizing solutions, and transferred to Hybond N or NX membranes using 20× SSC in combination with the Schleicher and Schuell Turboblotter. The use of the Turboblotter (transferring from top to bottom) avoids having to flip the multigel, and in doing so minimizes the risk of separation of the running gels from the agarose frame. After transfer, a probe covering the desired genomic region is radioactively labeled by random priming or nick-translation. The unincorporated radioactivity is removed using gel filtration (BioSpin P30, BioRad) and the recovered probe is subsequentially phenol-chloroform extracted. The Hybond membrane is pre-hybridized at 42° for 1.5 h in hybridizing solution (6× SSC, 5× Denharts, 0.5% SDS, 50% formamide, and 0.1 mg/ml denatured salmon sperm DNA). The probe is boiled for 10 min, added to the hybridization solution, and then incubated at 42° overnight. The membranes were washed twice in 2× SSC, 0.1% SDS at 42° and 30°, respectively, and once with 0.2× SSC, 1% SDS at 27°. The membranes were then exposed to film or to a PhosporImager screen for 2–5 days.

Data Analysis

For each individual band detected by Southern blotting or fluorescent staining, the migration in centimeters is measured from the center of the well to the center of the band using NIH Image[17] or equivalent software. The digitized image of the ruler placed adjacent to the multigel (see earlier) serves as reference to set the scale (number of pixels per centimeters). Migrations (in centimeters) are subsequently converted to μ (in $cm^2/V\ s$). The gel-free mobility, μ_0' is obtained by extrapolating the linear region of a plot of log μ versus agarose percentage to the y-axis. For an ~2.5-kb chromatin fragment, the linear region falls in the range of 0.2–0.9% agarose. The correlation coefficients of the linear regressions generally are ≥0.99. Note that the linear region must be determined empirically for different sized chromatin fragments.

Using Eq. (1), the P_e of each running gel is determined from the measured μ and μ_0' and known R_e (30.1 nm) of the bacteriophage T3 internal standard. The measured P_e was compared to the range that was determined based on calculated P_e values from ~10 random multigels (see

---

[17] R. R. O'Neill, L. G. Mitchell, C. R. Merril, and W. S. Rasband, *Appl. Theor. Electrophor.* **1,** 163 (1989).

TABLE II

μ_0^{T3} AND μ_E

	μ_0^{T3} ($\times 10^{-4}$ cm^2/V s)	μ_E ($\times 10^{-5}$ cm^2/V s)
E buffer	−0.755	1.02
E buffer + 2 mM MgCl$_2$	−0.65	0.97

Table III). If the P_e values fall outside the expected ranges, the data may be suspect. The R_e of the genomic chromatin band(s) in each running gel is determined from their measured μ and μ_0' and the calculated P_e for that gel. To convert the measured μ_0' to the actual μ_0, one must correct for the effects of electroosmosis on the measured μ (μ_E). This requires empirical determination of the μ_E for each specific type of commercial agarose, which is a time-consuming and expensive process. Toward this end, we have determined the μ_E for the Research Organics low EEO agarose mentioned earlier, and found it to be constant over several lots and many years. By using this specific agarose in each multigel frame and running gel, experimental measurements of μ_E can be avoided, and the μ_0' is converted to the true μ_0 according to the following equation

$$\mu_0 = \left[(\mu_0')(\mu_0^{T3} + \mu_E)/\mu_0'^{T3}\right] - \mu_E \qquad (2)$$

where the μ_0^{T3} also is a measured constant in the same agarose (see Table II for μ_0^{T3} and μ_E values determined in E and E + 2 mM Mg^{2+} buffers).

Assay for Secondary Chromatin Structure Formation

The formation of locally folded secondary chromatin structure is assayed by determining whether the R_e of the genomic chromatin fragment decreases by ≥35% in E buffer containing 2 mM free Mg^{2+} (relative to the R_e in E buffer alone). The decrease in R_e between 0 and 2 mM Mg^{2+} also should be linear.[6,10] Importantly, when assaying for secondary chromatin structure formation, the AME experiments must be performed under conditions where the R_e is ≥10 times larger than the R_e, that is, in 0.2–1.0% low "concentration" multigels. All aspects of the AME experiment are performed as described above, the only difference being the free Mg^{2+} concentration in the casting and running buffer. For these purposes E buffer is used because of its low EDTA concentration, and the free Mg^{2+} concentration is assumed to be 0.25 mM less than the added MgCl$_2$ concentration.

TABLE III
EMPIRICALLY DETERMINED P_e RANGE (nm)

Gel (%)	P_e range (nm)
0.2	400–600
0.3	300–460
0.4	200–350
0.5	150–300
0.6	150–250
0.7	150–200
0.8	130–200
0.9	120–190
1	100–150
1.3	90–120
1.6	75–100
1.9	65–80
2.2	60–70
2.5	50–55
2.8	45–50
3	37–43

Flexibility Assay

A relative assay for the conformational flexibility of a chromatin fragment is provided by the slope of a graph of R_e versus P_e in the gel range where R_e approaches P_e, that is, in high percentage agarose gels.[5,6,12] A slope near zero is indicative of a particle whose conformational features are not deformed during electrophoresis. Such a particle is relatively inflexible, for example, unfolded nucleosomal arrays in low salt.[5] In contrast, the R_e of a more flexible particle (e.g., DNA, irregularly spaced nucleosomal arrays in low salt) decreases as the R_e approaches P_e. For a 2.5-kb chromatin fragment, this occurs between 1.2% and 3.0% agarose. Thus, relative conformational flexibility is assayed in high-concentration multigels.

Composition Analysis of the Genomic Fragments

It is very powerful to couple the biophysical measurements made using AME with western analyses of the composition of the same specific genomic fragment (see later). To accomplish this, the same low-salt nuclear extracts used in the AME experiments are loaded on duplicate agarose gels buffered with TAE. The gels are run at 5 V/cm for exactly 150 min. The first gel is probed with labeled DNA from the genomic fragments of interest to define their location in the gel. The second gel is the source of the genomic chromatin fragments used for the subsequent western analyses.

Specifically, to confirm the identity of the bands and determine the precise distance of migration, in one gel, the DNA was transferred to Hybond NX membrane and probed with the fragment-specific probe. The second gel is stained with Commassie (GelCode Blue Stain, Pierce) according to the manufacturer's specifications. Slices with 1.5 mm-thickness containing the bands of interest are excised, soaked in 1% SDS, 50 mM Tris-HCl (pH 6.8) and 1% β-mercaptoethanol for 15 min at room temperature. The slices are then transferred to a Pyrex tube containing the same buffer pre-heated to 95° for 25–35 s (or until the agarose slice becomes translucent), and the tube immediately chilled on ice. The cooled agarose slice was rapidly placed in the well of 1.5-mm thick 4–20% SDS gel and electrophoresed at 100 V for 90 min. The proteins were subsequently transferred to Immobilon-P PVDF (Amersham) membranes and immunoblotted according to standard Maniatis western blotting conditions using appropriate antibodies.

Data Interpretation

The final aspect of the AME approach is to interpret the combined biophysical and compositional data. For illustrative purposes, this section summarizes selected important observations from our work with genomic MMTV promoters under different states of transcriptional regulation.[10] For example, the inactive form of the genomic MMTV promoter in mouse 3134 cells was found to contain histone H1, and showed the same behavior in the AME assays as model H5-containing nucleosomal arrays assembled *in vitro* from pure components. Exposure of cells to dexamethasone leads to transcriptional activation *in vivo*.[18–20] The transcriptionally active genomic MMTV promoters isolated from dexamethasone-treated cells contained RNA Pol II, TBP, Octl, Brgl, and acetylated H3 tail domains, and its R_e increased to 43 nm (compared to 26 nm in the transcriptionally inactive state). Thus, the biophysical measurements support the conclusion that H1 had been replaced with several large transcription factors and multiprotein assemblages after hormone treatment. Interestingly, both the inactive and active forms of the genomic MMTV promoter were capable of forming salt-dependent secondary chromatin structure. Together, these data support a model in which transcriptional activation is associated with reorganization of secondary chromatin structures,

[18] H. Richard-Foy, F. D. Sistare, A. T. Riegel, S. S. Simons, Jr., and G. L. Hager, *Mol. Endocrinol.* **1**, 659 (1987).

[19] H. Richard-Foy and G. L. Hager, *EMBO J.* **6**, 2321 (1987).

[20] T. K. Archer, C. J. Fryer, H. L. Lee, E. Zaniewski, T. Liang, and J. S. Mymryk, *J. Steroid Biochem. Mol. Biol.* **53**, 421 (1995).

rather than decondensation of the promoter chromatin into an unfolded beads-on-a-string structure.

As with any newly introduced technical approach, the ability to accurately interpret AME data will increase in direct proportion to the number of systems that are characterized by this method in the future.

[3] Visualization and 3D Structure Determination of Defined Sequence Chromatin and Chromatin Remodeling Complexes

By Rachel A. Horowitz-Scherer and Christopher L. Woodcock

Recent progress in preparing and purifying defined sequence chromatin and chromatin remodeling complexes provides an opportunity to determine the three-dimensional (3D) changes that accompany major remodeling events, such as histone modification, the binding of transcriptional repressors and activators, and the action of ATP-dependent remodeling complexes. The compelling evidence that many of the functional changes in chromatin are intimately connected to changes in "higher-order structure" which may be epigenetically inherited, provides a powerful incentive to understand their structural basis.

The ultimate goal is to understand the different states of chromatin at an atomic level of resolution, but the range of chromatin structures accessible to X-ray crystallography is very limited. This limitation can, in principle, be overcome using lower-resolution 3D volumes generated by an electron microscope (EM) into which atomic level structures can be "docked." The range of transmission EM-based technologies available for visualizing chromatin and chromatin remodeling complexes is listed in Table I. For each method, we provide experimental guidelines based on recent work in our laboratory.

General Considerations

With any imaging technique, the information retrievable is limited by the quality of the sample, and for chromatin, the most important considerations are as follows.

Purity

The proportion of the sample constituting the structure of interest should be monitored. A higher level of contamination can be tolerated if the material under study is structurally distinct. For example, a circular

TABLE I

Method	Comments	Resources needed
Positive stain	Restricted to 2D Mostly for verifying structural integrity of chromatin in decondensed state	Transmission EM Carbon support films
Shadow casting	Quasi-3D data from shadow lengths surface relief	Transmission EM Vacuum evaporator
Negative staining	Nucleosomes, large proteins, and protein complexes resolved Partial 3D information from stereo pairs; full 3D from tomography and from single-particle reconstruction (SPR) of samples with uniform shape	Transmission EM with tilt capability Image digitization; software for tomography, SPR
Cryo-imaging	Only method that can be used with unfixed, unstained material. Tomography and SPR more technically challenging	Transmission EM, ideally with field emission gun (FEG); cryo-holder; cryo transfer device; image digitization, 3D reconstruction software

"minichromosome" is readily distinguished from contaminating linear chromatin.

Concentration

A chromatin sample containing 50 μg/ml of the material of interest will usually suffice. Concentrations much lower than this will always yield poorer images, and cryo-imaging will be all but impossible. Small volumes may be concentrated and the buffer components exchanged if needed using Minicon or Centricon (Amicon #42409, 4208) systems.

Buffer Components

Materials that alter the surface tension or otherwise interfere with adhesion or staining, such as detergents, sucrose, or glycerol, must be removed by dialysis, and the sample separated from "carrier" proteins such as BSA. If the sample is to be fixed (see later), Tris-type buffers must be avoided as they react with aldehyde fixatives. We prefer HEPES or PIPES, at the lowest concentrations that provide adequate buffering. For chromatin, the concentrations of monovalent and divalent salts have a strong influence on compaction, and will be dictated by the experiment. Protein complexes that require high salt (over ~200 mM monovalent ions) for stability are best fixed at the optimum salt, after which dilution or dialysis can be used to achieve the appropriate ionic strength.

Fixation

With the exception of cryo-imaging, all techniques require the adhesion of the sample to a support, a process that exposes chromatin to denaturing surface forces. Fixation in 0.1% EM-grade glutaraldehyde for 4 h followed by its removal by dialysis is optimal for most chromatin samples. This treatment preserves the sedimentation constant of chromatin after a post-fixation change in salt concentration. Longer fixation may result in an irreversible increase in compaction. For small volumes, minidialysers (Pierce #69570) are very convenient for this procedure.

Adhesion to Support Films

Most of the techniques discussed here require the sample to be deposited on a thin carbon film. Proper adhesion requires that the film be made hydrophilic and, for chromatin samples, a monovalent salt concentration of 50 mM is recommended.[1]

In our laboratory, carbon films are created by electron beam evaporation in an oil-free, turbopumped vacuum of 10^{-6} Torr (Baltec BAE80 system), onto freshly cleaved mica sheets, floated onto a double distilled H_2O (ddH2O) surface, and then deposited onto the rhodium side of 400-mesh copper-rhodium specimen support grids. Film stability is improved if the carbon is deposited in 3–4 bursts, separated by a few minutes. The carbon surface is rendered hydrophilic by a 1-min exposure to a glow-discharge in air at a pressure of ~50 mTorr in a Haskins PDC-3XG plasma cleaner (setting 3). Grids should be used within 1 h of glow-discharge. A properly treated carbon film will retain a thin layer of liquid when a drop of buffer is added and then wicked away with filter paper.

Positive Staining

This method is primarily useful for verifying the quality of decondensed chromatin samples, allowing the investigator to check nucleosome number and distribution, as well as the overall configuration of the array (circular or linear). A simple but reproducible procedure is as follows:

1. On a freshly exposed sheet of Parafilm, place a drop of sample (5–10 μl), 3 drops (~100 μl) of water, 3 drops of 2% aqueous uranyl acetate (UA), and a final drop of water.
2. Place a glow-discharged carbon film on the sample.

[1] C. L. Woodcock, L. Y. Frado, G. R. Green, and L. Einck, *J. Microsc.* **121,** 211 (1981).

3. After 5 min, remove the grid, blot from one side with a flag of filter paper (verifying visually that the surface remains fully wetted) and place on the first drop of water for 10 s. Remove, blot, and place on the next drop, and continue at 10 s intervals through the 3 drops of stain to the fourth drop of water, which serves to remove unbound stain. After the grid has air-dried, it is ready for EM observation.

This method provides a faint positive staining, primarily of DNA, and the microscope should be set up to maximize contrast. Alternatively, the EM may be set up for tilted beam dark field, which produces excellent reversed contrast.

A positive staining technique that provides significantly more contrast and stains protein and DNA but, in our hands, is less reproducible, is to use ethanolic phosphotungstic acid (PTA). This method is described in detail elsewhere.[2]

Shadow Casting after Drying from Glycerol (Fig. 1A)

This method is adapted from the technique described by Tyler and Branton for filamentous proteins.[3a] It has the advantage of providing quasi-3D information about chromatin conformation and tolerates salts and other buffer components that would normally have to be removed for EM preparation. The fixed chromatin sample is mixed with glycerol, and sprayed with an atomizer onto freshly cleaved mica. The mica is placed in a vacuum evaporator, whereupon the glycerol gradually evaporates, leaving chromatin particles behind as the meniscus withdraws. The 3D shape of the sample is protected from air-drying collapse, and salts and other contaminants are swept into a small area at the center of the original drop. Without breaking the vacuum, the mica is shadowed with Pt, and then coated with a layer of carbon. Finally, the mica is removed from the chamber, and the carbon/platinum replica floated off and transferred to EM grids for examination. Detailed steps are as follows:

1. The fixed sample (~50 μg/ml) is made 70% in glycerol (mix thoroughly).
2. About 20 μl is applied to the intake tube of a standard drugstore atomizer, from which the rubber bulb has been removed. The output from a "Dust-off" or similar can of compressed gas is connected to the air input orifice of the atomizer.

[2] C. L. Woodcock and R. A. Horowitz, Methods Cell Biol. 53, 187 (1998).
[3a] J. M. Tyler and D. Branton, J. Ultrastruct. Res. 71, 95 (1980).

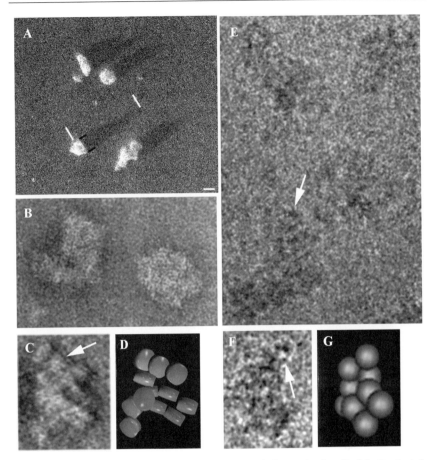

FIG. 1. Examples of the options for EM imaging of chromatin described in the text. In this case, the methylated DNA-binding protein MECP2 was bound to defined 12-mer nucleosome arrays (in collaboration with P. Georgel, P. Wade, and J. Hansen) and fixed with glutaraldehyde. (A) Shadowing after vacuum drying from glycerol preserves the 3D shape of the complex and permits measurement of height (between white lines) and diameter (between black lines). Bar = 30 nm. (B) Negative staining with uranyl acetate gives an overall sense of size and shape, but requires tomographic reconstruction to reveal details of internal substructure (C, D). A central slice from a reconstruction (C) reveals the outline of individual nucleosomes (arrow). Slice-by-slice examination of the reconstructed volume allows the location and orientation of individual nucleosomes in the array to be identified, and construction of a solid model (D) of the array, in which nucleosomes are represented by 5 × 10 nm disks. (E) ECM imaging of unstained particles yields low-contrast images of the solution conformation, without flattening or drying. Individual nucleosomes can be identified (arrow). Stereo-pair reconstruction[3b] of one particle (F) yields the 3D coordinates of individual nucleosomes (arrow), and a solid model (G) can be built from those coordinates. A stereo pair does not give orientation information, so nucleosomes are represented as 10 nm diameter spheres. Bar = 10 nm.

[3b] A. Beorchia, M. Ploton, M. Menager, M. Thiry, and N. Bonnet, *J. Microsc.* **163,** 221 (1991).

3. A 2 × 2 cm "target" of freshly cleaved mica is placed about 25 cm from the atomizer, and a 1-s burst of gas used to propel the sample as a fine spray. Trial runs should be used to establish the optimum volume of sample and distance of the mica to produce a range of individual drop sizes that are closely spaced but do not coalesce.
4. The mica is quickly transferred to a vacuum evaporator, and a high vacuum maintained for several hours. Platinum is then evaporated at an angle of ∼10°, monitoring the thickness with a film thickness monitor. Trial runs are needed to calibrate the system for an optimum shadow thickness.
5. To provide accurate size measurements as well as a complete surface profile of the specimen, the mica is mounted on a rotating platform, which is kept stationary for the first one-third of the evaporation, and rotated for the remaining two-thirds.
6. The tilt angle is set to 90°, and a self-supporting layer of carbon is evaporated onto the mica.
7. The mica is removed from the evaporator, the carbon scored into squares, and the replicas floated off onto water. 400 mesh grids are then used to pick up the squares from below.
8. In the transmission EM, the outline of the original droplets are marked by faint circular "watermarks," and the drop centers by heavy accumulations of salt and other buffer components. Shadowed sample will be found between these delimiters.
9. Excellent images are obtained from very lightly shadowed samples recorded in dark field image mode (Fig. 1A).

Negative Staining (Fig. 1B)

Selection of the appropriate negative stain requires a certain amount of trial and error; different samples inevitably demand modifications to the basic procedure. The standard stain for chromatin and many proteins is 1–2% aqueous unbuffered UA, while for the yeast SWI/SNF complex (a case discussed in detail later) and several other chromatin-associated regulatory proteins, 1.5% PTA buffered to pH 7.0 and with 0.015% glucose has provided better preservation of fine detail. Other possible stains are ammonium molybdate, vanadate, and alternate uranyl salts. It is advisable to evaluate the sample in several stains, as the appearance[4] and subsequent reconstructions[5] may differ depending on the amount of support the

[4] C. L. Woodcock, H. Woodcock, and R. A. Horowitz, *J. Cell Sci.* **99,** 99 (1991).
[5] C. L. Woodcock and W. Baumeister, *Eur. J. Cell Biol.* **51,** 45 (1990).

stain provides, and the features it accentuates. A detailed discussion of the appearance of chromatin in various negative stains may be found in Woodcock et al. (1991).[4] We use the following conditions for staining:

1. On a freshly exposed sheet of Parafilm, place a drop of sample (5–10 μl), 3 drops (∼50 μl) of water, and 3 drops of stain. Place a glow-discharged carbon film on the sample. (If diluting the sample to 50 mM NaCl, place the appropriate volume of NaCl solution on the Parafilm first, then place the grid on the drop and finally "inject" the protein sample into the drop under the grid with a fine-tipped Eppendorf pipet.)

2. After 5 min, remove the grid, blot from one side with a flag of filter paper (verifying visually that the surface remains fully wetted), and place on the first drop of water for 10 s. Remove, blot, and place on the next drop, and continue at 10-s intervals through the third drop of stain. The grid surface must not dry out during this sequence.

3. Pick the grid up off the last drop with self-closing forceps, and wick the stain off the grid. Allow to dry in a dust-free area, then store in a secure grid box in a desiccator.

For unfixed protein-only samples, an alternate method is recommend. This method takes some practice to avoid wetting the back of the grid; if this occurs the grid must be discarded:

1. Fill a Pasteur pipet with negative stain (in this case, usually 1–2% UA).

2. Apply 3–5 μl of sample to a glow-discharged carbon-coated grid clamped in self-closing forceps. Allow to adsorb for 10–60 s.

3. Hold the grid at a ∼45° angle, and allow about 10 drops of the stain to flow dropwise over the face of the grid into a waste container. The flow should be rapid enough to provide the needed rinsing action to remove excess salts.

4. Blot lightly as in the previous list and allow to air-dry.

There are several parameters to balance in the negative staining process: the stain must be thick enough to support the specimen without artifactual flattening during drying but thin enough for good contrast in the images. Altering the speed at which the grid dries provides a certain amount of control; grids can be left to dry in a chamber with up to 65% RH. Negative staining requires a threshold protein concentration to achieve an even spreading of the stain, but the specimen must be dispersed enough to allow clear discrimination of individual particles and provide a useable separation between particles for the image processing to follow.

Electron Cryomicroscopy (Fig. 1E–G)

Electron cryomicroscopy (ECM) permits the observation of specimens in their solution conformation, without stains, shadowing, or substrate adhesion and the accompanying flattening and drying artifacts. In most instances, chemical fixation is unnecessary. A drop of the sample is applied to an EM grid covered with a fenestrated substrate and allowed to spread over the holes, then blotted to form a thin film of 20–100 nm. The EM grid is then rapidly plunged into a cryogen to convert the thin layer of specimen in solution to vitreous ice. The frozen-hydrated specimen is transferred to the electron microscope and observed at temperatures that maintain the vitreous layer.

Detailed descriptions of the application of this technique to chromatin are available.[6,7] Some technical advances since those publications appeared are noteworthy:

- One of the most time-consuming and difficult steps in ECM is the production of reproducible fenestrated substrates or "holey films." It is now possible to purchase prepared grids with regularly sized (selection from 1–3 μm) and spaced holes (www.quantifoil.com).
- Manipulation of the variables of blotting, plunging, and humidity can now be computer controlled through a plunging device, the Vitrobot, originally developed at the University of Maastricht (sold by FEI Company, www.vitrobot.com).
- The possibility of performing ECM tomography has been facilitated by the development of computer interfaces to the electron microscope for control of tilting and data acquisition (www.tvips. com, www.feico.com, www.ami.scripps.edu).
- A high-tilt ($\pm 80°$) ECM specimen holder is now available (www.gatan.com).

Electron Microscope Data Collection

CCD Cameras Versus Film

The theoretical pros and cons of digital versus film recording of data have been extensively discussed[8]; here we will consider some of the practical considerations. For beam-sensitive specimens (negatively stained or frozen hydrated) the microscope must be able to record low-dose images via deflection/shuttering of the beam above the sample. For most samples

[6] J. Bednar and C. L. Woodcock, *Meth. Enz.* **304**, 191 (1999).
[7] C. L. Woodcock and R. A. Horowitz, *Methods* **12**, 84 (1997).
[8] J. Hesse, H. Hebert, and P. J. Koeck, *Microsc. Res. Tech.* **49**, 292 (2000).

in negative stain, this is in the range of 10–15 e/A^2 per exposure. Many samples can tolerate up to 25 e/A^2. Some proteins and DNA-protein complexes, dependent on a poorly understood combination of the preparative method and the negative stain, are extremely sensitive to damage. The migration of stain molecules, producing granulation, is usually obvious to the eye, but damage to the biological material is subtler. If the microscope is equipped with a CCD camera for digital data recording, it is useful to collect a series of extremely low-dose images of a single field, until the cumulative dose is ~25 e/A^2. The images can then be analyzed by calculating difference images (arithmetic pixel-by-pixel subtraction of one exposure from the next) to determine the dose at which damage becomes detectable.

Other factors that must be considered are the target resolution in the reconstruction, the size and distribution of the protein on the support film, and the area of the CCD chip. Film collection permits hundreds or thousands of particles to be selected from a single exposure, whereas the number of particles in a single CCD image is far fewer. However, film collection necessitates scanning the negatives into digitally accessible data at the appropriate resolution. The pixel size should be 3–4× smaller than the desired resolution. It is possible with either scanners or CCD imaging to greatly oversample; this should be avoided as artifacts can be introduced when data management requires interpolation to larger pixel sizes. In negative stain, the final resolution in a reconstruction of an asymmetric particle is unlikely be better than 25A, necessitating a pixel size in the raw data of 6–8A. For the yeast SWI/SNF complex, this was achieved using a 2048 × 2048 CCD camera, each CCD image containing 50–100 particles. Without convenient access to a high-resolution film scanner, the time invested in collecting larger numbers of CCD images may be offset by the time savings of eliminating darkroom work and the digitizing of negatives.

Microscope Parameters

With the target resolution determining the final pixel size, and therefore the magnification at which to record the images, other microscope parameters need be determined. While the best resolution will be achieved at intermediate voltages in field emission gun (FEG)-equipped microscopes, perfectly acceptable data can be acquired with care with LaB6 filament–equipped microscopes at 80–120 kV. The highest resolution will be obtained at the highest available voltage. The illumination conditions (small condenser aperture, coordination of spot size with spread of beam) should be optimized to provide the most coherent beam. The defocus for the exposures should be selected so that the first zero of the contrast transfer function (CTF) falls beyond the target resolution.

Single-Particle Reconstruction from Negatively Stained Images of Asymmetric Particles

Single-particle reconstruction (SPR) is the method of choice for calculating 3D reconstructions of individual proteins, or protein complexes that adopt a consistent 3D shape and have a size range from several hundred kilodaltons to several megadaltons. With "ideal" images, the best possible resolution in negatively stained samples will be ~25 Å and will require a minimum of 2000 imaged particles. The lack of perfect images usually means that 5000–10,000 images are required. Other reconstruction strategies are available for proteins that assemble into helices or 2D crystals.[9] Large cellular components and macromolecular assemblies that have highly variable individual conformations are best analyzed by electron tomography.

Specific SPR methods exist for icosahedral viruses, for complexes that exhibit rotational or other symmetries, and for particles that present a subset of preferred orientations when prepared for electron microscopy.[10] Here, we concentrate on the methods employed in this laboratory for yeast SWI/SNF, a large (1.14 MDa) complex of 12 proteins, which did not assume a preferred orientation in negative stain, and which does not possess any symmetry.

Specimen Preparation

Many protein complexes, including ySWI/SNF, are unstable in the conditions required for successful EM and must be chemically cross-linked (see earlier). For this sample, we used glutaraldehyde fixation following removal by dialysis of the glycerol and high salt required for long-term stability, and PTA proved to be the most suitable negative stain.

Overview of the SPR Process

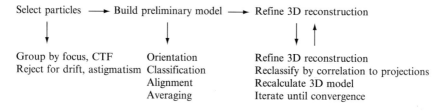

Data Preparation

The first step in calculating a single-particle reconstruction is to select the individual particles. First, each source image (negative or CCD image) must be evaluated for quality: images with drift or astigmatism are rejected and groups of images with like defocus/CTF grouped together. Then the individual particles are excised from the parent image and centered in individually "boxed" fields. In order for subsequent programs to reorient the selected particles, the box size should be 25–50% larger than the particle itself.

Most of the image processing packages that include SPR: SPIDER (www.wadsworth.org/spider), EMAN (ncmi.bcm.tmc.edu), IMAGIC (www.ImageScience.de), and SUPRIM (ami.Scripps.edu), offer the option of automating the particle selection process. This can be done by cross-correlation with projections of a previously calculated model if one exists. A semi-automated method uses a subset of user-identified particles, which are rotationally averaged and then used as a template for cross-correlation with the particles in the raw image. This was used successfully for ySWI/SNF; with sufficiently relaxed criteria, the program will "overpick" and then the operator needs only to delete improper selections, a considerably shorter task than hand-selecting each particle.

Alignment and Generation of Class Averages

Each particle must be centered in its box with x, y translations and on-plane rotations necessary to have the entire collection of particles in alignment with respect to each other. One way to align the particles is to compare them to projections of an existing model. This model can come from other data sources, or be constructed from geometric primitives based on the user's preliminary assessment of the particle shape. Caution must be maintained at this step, as the reference model can introduce a strong bias.

For ySWI/SNF, an asymmetric particle for which no a priori structural information was available,[11] a reference-free alignment was performed in EMAN.[12] This method employs algorithms that group particles by similarity in their appearance, then averages the grouped images, with the goal of having one group for each unique view of the particle. The user then examines the group averages and the members of the group to determine if a unique view is adequately represented. From these groups, the user chooses a subset that represents the best possible spread of views of the particle for the calculation of an initial model. The Fourier common lines method is employed to determine the relative orientations of the selected groups in order to build the initial model.[13] This initial model is then

[9] J. Ruprecht and J. Nield, *Prog. Biophys. Mol. Biol.* **75,** 121 (2001).

[10] P. A. Thuman-Commike, *FEBS Lett.* **505,** 199 (2001).

[11] C. L. Smith, R. A. Horowitz-Scherer, J. F. Flanagan, C. L. Woodcock, and C. L. Peterson, *Nat. Struct. Biol.* **10,** 141 (2003).

[12] S. J. Ludtke, P. R. Baldwin, and W. Chiu, *J. Struct. Biol.* **128,** 82 (1999).

[13] M. van Heel, *Ultramicroscopy* **21,** 111 (1987).

reprojected, and the members of the initial classes are then compared to this set of projections and reassigned to new classes based on their fit to the projections. The new classes are averaged, and an iterative process of model construction, reprojection, and reclassification is begun.

Final Reconstruction Parameters

Within the first few iterations, several parameters should be assessed. One is to determine whether or not the particles are in a preferred orientation. The set of calculated Euler angles for the class averages should be examined to see if any sets of views are excluded (Fig. 2A). Side-by-side comparison of the class averages and their corresponding projections from the progressing iterations provides a visual evaluation of the improving correlation between them (Fig. 2B).

Another important assessment is whether the classes consist of very small numbers of particles. This can be an indicator of significant heterogeneity in the sample. It is possible that a distinct subset of classes will be evident, indicating an alternate conformation rather than random dissociation of a complex. In this case, it may be possible to segregate the data and compute independent reconstructions of the alternate state. The iterative refinements are allowed to run until changes between successive

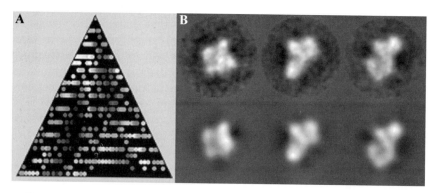

FIG. 2. Class averages: angular distribution and comparison to projections. (A) The dots in the triangle represent a well-spread distribution of orientations the particles have been classed into. The apex of the triangle is the view down the z-axis (top) of the particle, and the base represents views along the x-axis (edge). The brighter the dot, the more particles have been assigned to that orientation. If the dots appear clustered with gaps in the distribution, the particle has probably assumed a preferred orientation on the EM support film. (B) A good indicator of convergence is obtained by noting a positive match when comparing the class averages (top row) with projections of the refined model at the same orientation (bottom row). If an arithmetic subtraction is performed between the two, the resulting difference image should consist of undifferentiated noise.

iterations are minimized, and are monitored by examining the Fourier shell correlation (FSC) between rounds (Fig. 3A and B).

The resolution of the reconstruction can be estimated by dividing the data into halves, for instance, by separating then averaging all the even and odd numbered images in all the classes then independently reconstructing from the even/odd averages. A FSC between the two reconstructions is examined (Fig. 3C), and the resolution at the correlation value of 0.5 provides a reasonable estimate of the resolution of the reconstruction.[14] If the number of images comprising each class is very small, indicating either sample or staining heterogeneity, the half-averages will be very noisy, and the resolution may be underestimated.

Accuracy of the Reconstruction

The final question becomes how to determine if the reconstruction is correct. For a wholly asymmetric object, the reconstruction algorithms might be producing only one of many possible solutions and there is no quantitative way to determine if the one generated is the correct one. The most direct approach is to calculate reconstructions from more than one unique data set and comparing the results qualitatively (see section on Visualization,), and quantitatively, by calculating the FSC between the wholly independent reconstructions.[11]

Visualization

The 3D model can be studied by slicing through the volume and examining contour maps of the densities, and by creating isocontour surface maps to view the outer shape of the molecule (Fig. 4). Selecting the threshold at which to contour a surface is done by most software packages from the molecular weight of the sample (assuming a protein density of 1.35 g/cc). The "true" surface may be up to ±30% of this value. In negative stain, it is quite difficult to establish the correct cutoff. For instance, in UA stained objects, there is always the possibility of partial positive staining of the complex, especially when nucleic acids are present. These will then have an opposite density from stain-excluding regions and may be misinterpreted as "holes." Surfaces with deep pockets that accumulate stain may also be difficult to delineate accurately. There are a variety of software packages designed to perform visualization tasks. Chimera (www.cgl.ucsf.edu/chimera) and Vis5D (www.ssec.wisc.edu/~billh/vis5d.html) are freeware packages we have used; AVS (www.avs.com) and Amira

[14] I. I. Serysheva, M. Schatz, M. van Heel, W. Chiu, and S. L. Hamilton, *Biophys. J.* **77,** 1936 (1999).

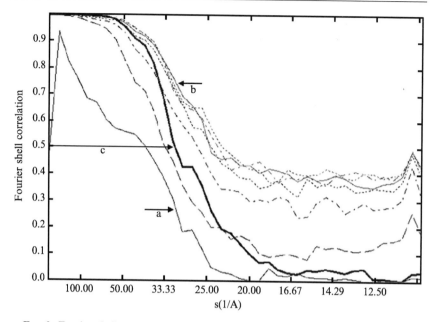

FIG. 3. Fourier shell correlation (FSC): convergence and estimation of resolution. In this convergence plot, the thin lines represent the FSC between subsequent iterations of a refinement. As refinement proceeds, these become more consistent. The thin solid lines (a, b) are plots between the (a) first and second and the (b) sixth and seventh rounds. Convergence is observed to begin after the fifth round. The bold solid line (c) is the FSC between two reconstructions each calculated from one half of the total data. A reliable estimate of resolution is the radius at 0.5, in this plot equal to ~30 A.

(www.tgs.com) are commercial packages widely used by structural biologists and currently in use in our laboratory.

SPR from ECM Images

ECM data offer the potential for considerably higher resolution than negative stain. While it is advisable to collect ECM data on an FEG-equipped microscope, the choice of film or CCD acquisition is subject to the same concerns as for conventional EM (see the previous section). The basic flow of image processing for SPR is the same, but consideration of the CTF becomes more important. Images must be collected at several defocus values to fill in missing information at the nodes of the CTF[15] and the image data must be corrected for phase reversals.[12] Space limitations

[15] J. F. Conway and A. C. Steven, *J. Struct. Biol.* **128,** 106 (1999).

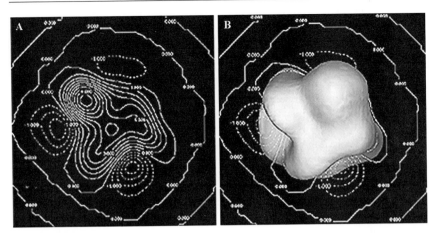

FIG. 4. Visualization. The 3D reconstruction can be examined as 2D-slices with contour lines (A) to convey information about protein density (solid lines). The dotted lines are contours of negative density, resulting from accumulation of stain at pockets in the structure. The structure can also be viewed as a solid model, with the surface drawn from a selected isoline. In (B) the surface for the 3D model has been constructed at the contour = 1.0, which was calculated to enclose a volume equal to the mass of the particle. Density and volume information from frozen-hydrated samples is much more reliable than from negative stain.

preclude a full discussion of this aspect of the image processing in the context of this article; for an excellent review, see Tao and Zhang (2000).[16]

Electron Tomography (Fig. 1C and D)

While many chromatin remodeling enzymes and other regulators of transcription may be amenable to SPR, their substrate—chromatin—is not. Although comprised of similar "subunits," the nucleosomes, chromatin itself is highly variable in conformation. While an enzyme such as SWI/SNF may have a unique binding site on the nucleosome, in EM preparations, no two arrays of SWI/SNF-bound chromatin will appear the same, and the image averaging that is key to SPR is not possible. To determine the 3D structure of, for example, a chromatin-bound transcriptional regulatory complex requires the use of EM tomography.

For tomography, the EM data are collected as a series of images as the specimen is tilted between −70° and +70° in a holder designed for that purpose. The most common approach is to use a single tilt axis, as discussed later. An alternative is to collect data for two tilt axes. This requires a

[16] Y. Tao and W. Zhang, *Curr. Opin. Struct. Biol.* **10**, 616 (2000).

tilt/rotate specimen holder and adds to the complexity of data acquisition and processing, but provides increased structural information.[17] The tilt images are aligned with respect to one another, and then a 3D volume is calculated via back-projection of the tilt data.[18]

EM tomography can be performed on either negatively stained or ECM preparations. The advantage of ECM is the retention of the solution conformation of the sample. The significant investment of resources required to create the infrastructure to perform tomographic data collection (computer controlled stage, high-tilt specimen holder, CCD camera) is further increased for frozen-hydrated samples.

Specimen Preparation for Tomography

Whether negative stain or frozen-hydrated samples are used, the support film should be mounted on thin-bar, hexagonal 600-mesh grids. These optimize mechanical support and open area to accommodate tilting. Prior to use, colloidal gold beads are applied to the back surface of the carbon. While tilt series can be aligned by cross-correlation of the imaged particles, in tilt series of negative stain and frozen-hydrated samples, the dose/image is so low that the specimen may be barely discernible. The high-density gold beads serve as fiducial markers for the alignment process. For ECM tomography, it is also possible to mix the gold beads with the sample.

Grid preparation:

1. Immediately before use, apply a drop of 0.02% polylysine, pH 8.0 to the back side of the grid for 30 s.
2. Wick off with filter paper.
3. Wash with 5 drops of ddH20, leaving the last drop on the grid for 30 s before wicking and air drying.
4. 3 μl of colloidal gold solution (bead size 5–10 nm) is applied to the same side, wicked, and rinsed with 5 drops ddH2O and air-dried.
5. The sample is applied to the front surface of the grid and negatively stained as described above.

The concentration of gold solution will need to be experimentally determined so that least 5 beads are present in the recorded image field. For tomography, the sample should be somewhat more widely spaced than for conventional imaging, to preclude particles outside the area of interest entering the field during tilting.

---

[17] P. Penczek, M. Marko, K. Buttle, and J. Frank, *Ultramicroscopy* **60**, 393 (1995).
[18] M. Radermacher, *in* "Electron Tomography" (J. Frank, ed.), p. 91. Plenum, New York, 1992.

Data Collection

Most specimen holders can only be tilted through $\pm 60°$. High-tilt holders for tomography, which range from $\pm 70°$ to $\pm 80°$ are available for room temperature or ECM work (Gatan, www.gatan.com, Fischione, www.fischione.com). There are also holders designed for double-tilt data collection, in which after the initial tilt series is collected, the sample is rotated $90°$ in the holder and a second series collected.

For acquiring tilt series of negatively stained or frozen-hydrated specimens, the ability to record extremely low-dose images is required. For this, a high-sensitivity CCD is invaluable, as images with electron doses of <1 e/A^2 can be recorded. The dose required for a single image with an optimal damage should be distributed signal/noise ratio (S/N) without specimen among all the images of a tilt series, resulting in an equivalent S/N in the back-projected reconstruction.[19] With this approach, it is possible to obtain high-quality reconstructions from images in which the sample is barely discernible.[20]

The number of tilt images needed is determined by the desired resolution in the reconstruction, but may be constrained by the beam sensitivity of the sample. A guideline for the number of tilts required is given by $n = \pi t/r$, where t is the thickness of the sample, r is the desired resolution, and n is the number of equally spaced tilts required.[21] A series of low-dose images taken sequentially over the same field can supply information on the maximum dose the sample can withstand. The low S/N of images recorded at ~ 1 e/A^2 makes it highly desirable to use automated data collection, where computer control of the microscope enables successive fields to be accurately positioned via cross-correlation, and auto-focusing is available.[22]

We employ a strategy for data collection which permits the collection of at least one full $5°$ tilt series, then a $2.5°$ interleaved series, and finally, if desired, a $1.25°$ series. An alternative to using equidistantly spaced angles in the tilt series is to collect at a more finely spaced increment in the high tilts.[21] This method may be advantageous for very thin, flat samples to increase resolution in the z-direction of the reconstruction.[23]

Recording images:

1. In Search mode, select an area to record, preferably in the center 1/3 of the grid.

[19] B. F. McEwen, K. H. Downing, and R. M. Glaeser, *Ultramicroscopy* **60,** 357 (1995).
[20] R. A. Horowitz, A. J. Koster, J. Walz, and C. L. Woodcock, *J. Struct. Biol.* **120,** 353 (1997).
[21] R. A. Crowther, D. J. DeRosier, and A. Klug, *Proc. Royal Soc. London A* **317,** 319 (1970).
[22] A. J. Koster, H. Chen, J. W. Sedat, and D. A. Agard, *Ultramicroscopy* **46,** 207 (1992).
[23] W. O. Saxton, W. Baumeister, and M. Hahn, *Ultramicroscopy* **13,** 57 (1984).

2. If the object being studied is asymmetric, the longest axis should be aligned to within $15°$ of the tilt axis as resolution is degraded perpendicular to the tilt axis.
3. Adjust eucentric height to the highest possible precision.
4. Observed at maximum tilts $\pm$ to be sure neither grid bars nor other objects in the sample obscure the area of interest.
5. Ascertain that the focus area has high-contrast features or gold beads in an easily recognizable pattern to facilitate repositioning after tilting.
6. Record the first $0°$ image. Confirm that there are five or more gold beads in the field to be reconstructed.
7. Go to maximum positive tilt.
8. Record images at $5°$ intervals through $0°$.
9. Go to maximum negative tilt.
10. Record images at $5°$ intervals through $0°$.
11. Take second $0°$ image.
12. Go to maximum positive tilt, decrease by $2.5°$.
13. Record images at $5°$ intervals through $0°$.
14. Go to maximum negative tilt, decrease by $2.5°$.
15. Record images at $5°$ intervals through $0°$.
16. Take third $0°$ image.
17. Repeat from maximum tilts offset by $1.25°$.

The purpose of starting at high tilt each half-pass is to acquire the maximum amount of high-tilt information before the specimen has accumulated much dose. The series of $0°$ images is used to assess beam damage with cumulative dose.

Alignment and Reconstruction

Alignment of the tilt series can be accomplished by cross-correlation, least squares fitting of fiducial markers, or a combination both methods. Software packages are available to accomplish either or both of these tasks (see http://3dem.ucsd.edu/#sw). Alignment proceeds until only sub-pixel corrections are necessary. The human eye is an exquisitely sensitive detector for the quality of an alignment; viewing the aligned tilt images as a movie allows one to identify any deficiencies in alignment. This strategy also provides initial information about the shape of the structure of interest, and whether excessive flattening or other preparative artifacts are present.

Reconstruction by weighted back-projection is the most common algorithm employed; other methods (e.g., maximum entropy, algebraic reconstruction "ART" and simultaneous iterative reconstruction "SIRT") are also in use. Software environments such as SPIDER and SUPRIM have extensive support for tomographic reconstruction, as do dedicated packages such as IMOD (http://bio3d.colorado.edu/imod). Several national resources (see later section) have significant tomography development projects underway, and can provide expert consultation, software, and access to state-of-the-art equipment.

Determination of Resolution in the Reconstruction

Several factors affect the resolution of tomographic reconstructions. The "missing wedge" of data from lack of images recorded between 70° and 90° creates smearing or exclusion of features in the z-direction. Also, the number of tilts and the angular increment determine the "worst case" resolution (r) as $r = \pi t/n$, the "worst case" being when the volume is densely packed with information.[21,24] Although reconstructed volumes may be low-pass filtered to exclude specimen details smaller than r, our experience with chromatin is that this level of filtering may obscure valuable information. For instance, linker DNA is frequently seen in tomographic reconstructions of chromatin where the theoretical resolution limit would preclude its visualization.[25,20]

For a qualitative evaluation of the tomographic reconstruction, one compares the 0° projection of the final reconstruction with the first 0° raw image. It is useful to do the same for several tilts also. A careful examination of a 1-pixel thick central slice of the reconstructed volume will provide an indication of what detail is likely to be available in the volume.

Visualization and Analysis

The most accessible way of evaluating the data is to slice the volume into 2D xy plane slices (Fig. 1C), to create a z-axis series. It is often highly informative to view this z-series as a movie, to study transitions between structures in the interior of the volume. The volume can be sliced to 1-pixel thickness, to generate highly detailed views of substructure not visible in projections of the whole reconstructed volume.

The most difficult aspect of tomography is understanding the 3D structure in the volumes. For chromatin, the use of isosurface contouring has proven unsatisfactory. The most useful information is accessed by viewing the entire volume as its original greyscale data, rendered with a transparency gradient to allow seeing into the volume. This task remains difficult

[24] C. L. Woodcock, in "Electron Tomography" (J. Frank, ed.), p. 313. Plenum, New York, 1992.
[25] R. A. Horowitz, D. A. Agard, J. W. Sedat, and C. L. Woodcock, J. Cell Biol. 125, 1 (1994).

computationally with the typically large data sets involved. An ideal computational visualization environment would permit access to the greyscale data while having a transparent volume to interactively move on arbitrary as well as orthogonal axes, and incorporate 3D measurement tools. While freeware volume visualization packages are available, currently each falls short in some aspect. Our laboratory currently uses the commercial package, Amira (www.tgs.com).

National Resources available in the United States.

Boulder Laboratory for 3-Dimensional Fine Structure Tomography	bio3d.Colorado.edu
National Center for Microscopy and Imaging Research Tomography	www.ncmir.ucsd.edu
National Center for Macromolecular Imaging SPR	ncmi.bcm.tmc.edu/ncmi
Resource for Visualization of Biological Complexity SPR and Tomography	www.wadsworth.org/rvbc

[4] Electron Microscopic Analysis of the RSC Chromatin Remodeling Complex

By Francisco J. Asturias, Chukwudi Ezeokonkwo, Roger D. Kornberg, and Yahli Lorch

Introduction: Use of Electron Microscopy for Analysis of Large Macromolecular Complexes

Macromolecular electron microscopy (EM) is an ideal technique for structural characterization of large (>300,000 Da) macromolecular complexes, which may be too difficult to purify to homogeneity in the large quantities required for more conventional techniques, such as X-ray crystallography. Single-molecule–thick 2D crystals suitable for EM crystallographic analysis can be prepared using the lipid layer crystallization technique,[1,2] which requires two to three orders of magnitude less material (typically 10–100 μg of protein) than would be required for preparation of 3D crystals. Even smaller amounts of material (1–5 μg) are required for the preparation of samples from which single-particle images can be obtained

---

[1] F. J. Asturias and R. D. Kornberg, *J. Biol. Chem.* **274**, 6813 (1999).
[2] E. E. Uzgiris and R. D. Kornberg, *Nature* **301**, 125 (1983).

and later analyzed to generate structures at a resolution of 10–30 Å.[3] We have used single-particle image analysis techniques to characterize the structure of RSC, an abundant, essential chromatin remodeling complex from the yeast, *Saccharomyces cerevisiae*.[4] A 3D reconstruction of RSC obtained from images of negatively stained particles suggests a possible mode of interaction between the complex and a nucleosome core particle.

Sample Preparation for Electron Microscopy of Single Particles Preserved in Stain

Preservation in a heavy metal stain offers a quick and simple method to preserve biological macromolecules for imaging in the electron microscope. A widely used stain is uranyl acetate, which works well with many molecular complexes. The electron-dense stain is excluded from the volume occupied by the specimen, and what is actually recorded is therefore an imprint of the macromolecule in the stain. The advantages of using a heavy metal salt stain are the ease and speed of specimen preparation and imaging. However, the specimen is visualized only indirectly and at low resolution, and in most cases is partially deformed as a consequence of dehydration and interaction with the substrate to which the macromolecules are adsorbed. Despite its shortcomings, preservation in stain is the best alternative for initial characterization of a macromolecule, and can provide useful structural information. In addition, a structure calculated from specimens preserved in stain can in many instances be used as a starting point for more precise structural characterization based on images of unstained specimens preserved in amorphous ice.

Specimens are typically prepared using carbon-coated grids. To prepare the grids, a thin (~10 nm) layer of amorphous carbon is evaporated onto a freshly cleaved mica surface. The carbon layer is released by slowly pushing the carbon-coated mica, carbon-coated side facing upward, at an angle (~45°) through an air-water interface. The grids to be carbon coated are previously arranged on a support below the water surface. After release from the mica, the carbon film is carefully manipulated to position it on top of the grids, and is deposited on them by slowly lowering the water level until the carbon film comes into contact with the grids.

[3] J. Frank, "Three-Dimensional Electron Microscopy of Macromolecular Assemblies." Academic Press, San Diego (1996).
[4] F. J. Asturias, W. H. Chung, R. D. Kornberg, and Y. Lorch, *Proc. Natl. Acad. Sci. USA* **99**, 13477 (2002).

Specimens are prepared using a protein solution in a low-salt (<100 mM) buffer, as the presence of high-salt concentration would interfere with staining and create problems for adequate imaging of the specimen. A protein solution with a concentration in the range of 10–100 μg protein/ml usually results in an adequate density of particles in the samples. To facilitate adsorption of the molecules to the carbon support film the grids are glow-discharged at a residual pressure of $\sim$2 $\times$ 10^{-1} mbar for 30–60 s. The nature of the gas used during glow discharge (air, water vapor, and amyl amine are typical choices) can be varied to affect the properties of the carbon film after glow discharge. Two to three microliters of the protein solution are placed on the surface of a freshly glow-discharged grid (the effect of glow discharge on the carbon surface only lasts for a few minutes), which is held using microforceps. The protein solution is left on the grid for 10–60 s to give the particles a chance to adsorb onto the support film, and then blotted away by touching the side of the grid to a piece of filter paper. The grid is then washed with two to three drops of 1–2% uranyl acetate. The final drop of stain is left on the grid for 30–60 s to ensure proper staining of the sample. The sample preparation process is completed by depositing a second carbon layer on top of the stained molecules. This is accomplished by floating a small piece ($\sim$5 $\times$ 5 mm) of carbon on the surface of a pool ($\sim$5 ml) of stain using the same procedure employed to release the carbon film used to coat the grids. The stained sample is then immersed in the stain pool and withdrawn through the interface to pick up the second carbon layer.[5,6] Excess stain is removed by bringing a piece of filter paper into contact with the backside of the grid.

Preliminary Specimen Evaluation and Data Collection

While specimens preserved in negative stain are relatively resistant to radiation damage, it is nonetheless important to minimize the electron dose received by the specimen prior to imaging. Initial evaluation of the specimens entails examining the grids at low ($\sim$2000$\times$) magnification. If the concentration of the protein solution used to prepare the specimens was adequate, properly stained specimens will have a "grainy" appearance, and examination at higher magnification ($\sim$35,000$\times$) will reveal nicely stained particles. Properly stained particles should be clearly distinguishable from the background and show internal detail. If the amount of protein adsorbed on the carbon support surface was too low, very little stain

[5] G. W. Tischendorf, H. Zeichhardt, and G. Stoffler, *Mol. Gen. Genet.* **134**, 187 (1974).
[6] G. Stoffler and M. Stoffler-Meilicke, *in* "Modern Methods in Protein Chemistry" (H. Tesche, ed.), p. 409. De Gruyter, Berlin (1983).

will be retained and the specimen will resemble a clean carbon film. Finally, an excessive amount of material adsorbed to the carbon support film will result in samples where staining is evident, but no individual particles can be distinguished when an area is examined at higher (e.g., 35,000×) magnification. After determining that the staining and protein concentration in the samples are adequate, the next step is to collect images to evaluate the preservation of the particles and the distribution of particle orientations in the samples. Since individual particles must be selected for image analysis, it is important to optimize their distribution by carefully adjusting the protein concentration used during sample preparation. Ideally, particles should be evenly distributed on the support film, and the separation between individual particles should correspond to 2–3 particle diameters. This will result in a large number of particles being included in recorded images, while ensuring that it will be possible to select individual particles without interference from neighboring ones.

Most of the data for single particle analysis are still collected on film, but the use of electronic detectors is increasing, and the trend will undoubtedly continue. The arguments that follow apply to either type of data, as the distinction between them largely disappears after film data has been digitized. Zero-tilt images of the specimens are necessary to assess particle preservation and orientation. Images for data analysis are typically recorded at magnifications in the range 35,000–65,000×. In the case of film data, the images should be scanned using a pixel size equivalent to about one fourth of the desired resolution. The number of pixels available for alignment of individual particle images is often the more critical parameter for determining how images should be digitized.

Analysis and Classification of 2D images

The RSC images used in this study were collected at a magnification of 60,000×, and scanned using a pixel size corresponding to 2.54 Å in the specimen. Once the images were digitized, individual particles were cut out using a window size such that the side of the window corresponded roughly to twice the largest dimension of an RSC particle. Reference-free alignment of the particle images followed by multivariate statistical analysis and hierarchical ascendant classification[3] provides information about the distribution of particle orientations, and about potential variability in particle conformation. RSC particles were found to occur mostly on a single preferred orientation, but to display a significant amount of variability in conformation. At low resolution and in projection, the RSC particle appears as formed by four globular domains arranged around a central cavity. The results of multivariate statistical analysis and hierarchical

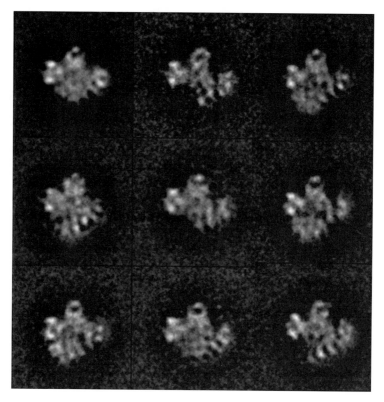

Fig. 1. Structure of RSC in projection. Images of RSC particles (5880) preserved in uranyl acetate were computationally aligned. All particles seemed to be in a similar orientation but differed in conformation and were sorted into homogeneous classes using multivariate statistical analysis and hierarchical ascendant classification.[3] A central area of lower density is apparent in several of the class averages shown, which include ~45% of all particles. Most of the variation in RSC conformation is related to a domain forming the bottom part of the structure, which is either missing (~35% of particles), or collapsed against the top of the structure (~22% of particles).

ascendant classification of RSC particle images are shown in Fig. 1. While three of the four domains remain fairly constant in their relative arrangement and orientation, the fourth one (forming the lower part of the RSC structure as shown in Fig. 1) varies in position, and in some instances appears to be either absent, or collapsed against the top part of the RSC structure. Given the large degree of variability shown by the RSC particles, 3D reconstructions were calculated for homogeneous subsets of particles defined by the hierarchical ascendant classification analysis.

Calculation of Initial 3-D Reconstructions

From a set of particles that correspond to a single conformation and a single orientation, an initial 3-D reconstruction can be easily calculated using the random conical tilt method,[7,8] which is implemented by combining information from untilted and tilted images of the same specimen area. Briefly, particles in a given orientation are selected from the untilted images, along with their corresponding tilted pairs. In-plane rotational alignment parameters for the untilted particle images, the tilt angle for the corresponding tilted particle images, and information about the relative orientation between the tilted and untilted images in a tilt pair are used to calculate a 3D structure from the tilted particle images. If additional homogeneous (same conformation and orientation) particle sets can be identified, the procedure is repeated. The resulting reconstructions are then combined to make up for the incomplete information in each individual reconstruction resulting from physical limitations in the maximum value of the tilt angle used for data collection (missing cone problem).

We were able to calculate several reconstructions of the RSC complex from homogeneous particle subsets identified by statistical analysis. Some of those reconstructions are shown in Fig. 2. Comparison of the structures confirms that the differences among them arise mostly from rearrangement of a domain in the lower part of the structure that moves as a solid module. The structures could not be combined because they correspond to different conformations of the RSC complex. An additional limitation of the reconstructions relates to particle deformation induced by preservation in stain. This phenomenon is known to be significant, as surface tension forces and dehydration cause particles preserved in stain to be flattened along the direction perpendicular to the carbon support surface. The problem can be corrected when information from different views can be combined to calculate an improved reconstruction (different reconstructions are deformed in different directions and the effect of such deformation is alleviated in the final combined volume), but the presence of different conformations and a single predominant orientation prevented us from using that approach to improve the quality of the RSC reconstruction.

Interpretation of the RSC Reconstruction

As explained earlier, examination of the different 3D reconstructions of the RSC complex shown in Fig. 2 led us to conclude that the variability observed was most likely related to a change in the position of the domain

[7] M. Radermacher, *J. Electron Microsc. Tech.* **9,** 359 (1988).
[8] M. Radermacher, T. Wagenknecht, A. Verschoor, and J. Frank, *J. Microsc.* **146,** 113 (1987).

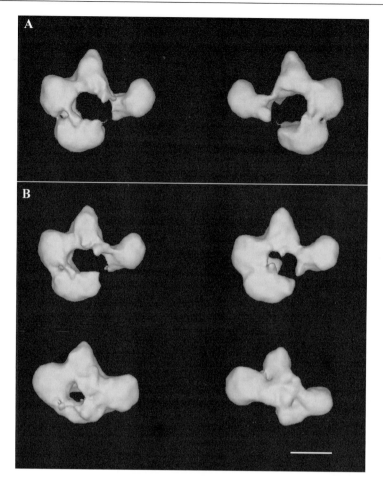

FIG. 2. Three-dimensional reconstruction of RSC. (A) RSC consists of four modules that define a central cavity. Two views of the structure (front and back) are shown. (B) The most significant variation in RSC conformation was due to the collapse (top) or absence (bottom) of a module that forms the lower part of the RSC structure. The scale bar corresponds to approximately 100 Å.

that defines the bottom portion of the central cavity. This domain seems to partially collapse under the sample preservation conditions used in the study. However, a majority of the RSC particles (~45%) show a clearly defined cavity, and that appears to be the most representative conformation of the complex. Variability in the position of the lower domain may actually have functional significance, as it could control access to the central cavity, which would be more accessible when the bottom domain moves away from the rest of the structure. The shape and size of the central cavity in the RSC complex matches closely the shape and size of a nucleosome core particle. Possible binding of a nucleosome in the cavity was tested by calculating a reconstruction of the RSC/nucleosome complex.

Studying the RSC/Nucleosome Interaction Under EM Sample Preparation Conditions

Determining the structure of a complex formed by RSC and the nucleosome core particle (histone core plus ~150 pb of double-stranded DNA) would likely help to determine the mechanism by which RSC alters the interaction between the histone core and nucleosomal DNA and renders the nucleic acid more accessible to components of the transcription apparatus. RSC interacts with the nucleosome core particle with very high affinity ($K_d \sim 10^9$), and therefore formation of the complex for examination by electron microscopy should be a straightforward matter. A rough method for quantitating formation of the RSC/nucleosome complex under the conditions required for sample preparation was implemented as follows.

First, samples of RSC alone were prepared, using a protein concentration that resulted in an adequate particle density. Under the same conditions (same grids, same buffer conditions), samples of nucleosomes alone were prepared, using a nucleosome concentration that resulted in a density of nucleosome particles similar to that of the RSC particles. Once both types of samples had been prepared, the number of particles in an arbitrary area (the area corresponding to a frame collected with an electronic CCD camera detector at 66,000× magnification) was determined for both the RSC and nucleosome samples. Having determined such numbers and checked that the desired ratio of RSC/nucleosome particles had been obtained, samples containing both RSC and nucleosomes were prepared in such a way that the final concentration of both RSC and nucleosomes in the solution used to prepare the mixed samples (as well as the buffer conditions, etc.) were identical to those previously used to prepare samples of the individual components. The RSC/nucleosome samples were then examined in the microscope, and the number of RSC and nucleosomes in the previously specified unit area were determined. A decrease in the number of nucleosomes with respect to that observed in the nucleosome (no RSC) samples could be considered as indirect evidence for formation of the RSC/nucleosome complex. The experiment was carried out by using a solution of RSC containing 160 μg protein/ml in a buffer containing 50 mM potassium acetate, 3 mM magnesium chloride, and 15 mM HEPES, pH 7.5. The nucleosome samples were prepared using a solution

TABLE I
THE NUMBER OF PARTICLES IN AN ARBITRARY AREA IS LISTED FOR RSC (FIRST COLUMN), NUCLEOSOMES (SECOND COLUMN), RSC IN THE RSC/NUCLEOSOME SAMPLES (THIRD COLUMN), AND NUCLEOSOMES IN THE RSC/NUCLEOSOME SAMPLES. THE LAST ROW SHOWS THE AVERAGE OF THE VALUES FOR EACH COLUMN[a]

RSC	Nucleosomes	RSC/nucleosomes (RSC count)	RSC/nucleosomes (nucleosome count)
72	70	35	18
65	55	43	21
67	73	41	17
67	57	28	19
72	71	35	27
69	65	36	20

[a] Analysis of these numbers indicates that there is an overall decrease in the number of particles per unit area after mixing of RSC and nucleosomes, probably as a result of mutual charge neutralization that decreases the affinity of the particles for the amorphous carbon film support. It also reveals a 40–50% decrease in the expected number of nucleosomes in the RSC/nucleosome particles, suggesting that about half of the RSC particles have a bound nucleosome.

with a concentration of 50 μg/ml in the same buffer. Formation of the RSC/ nucleosome complex was carried out by mixing RSC and nucleosome solutions resulting in the same final concentration of both components as used for preparation of single-component samples, and incubating the resulting mix for 30 min at ∼30°. The results, summarized in Table I, indicate that the number of free nucleosomes diminished by about 40–50% when the nucleosomes were incubated with RSC. The conclusion from this analysis is that the RSC/nucleosome complex seems to be at least partially stable under the acidic conditions (pH ∼ 4) of the uranyl acetate solution used to preserve the particles.

To further evaluate formation of the RSC/nucleosome complex, images of particles selected from the RSC/nucleosome samples were aligned and subject to statistical analysis as described before. While a relatively large (∼12) number of distinguishable conformations were detected in RSC samples (see Fig. 1), variability in the structure of RSC seems to be significantly reduced in the presence of nucleosomes. As shown in Fig. 3, only four distinct conformations of RSC were detected when nucleosomes were present. The decrease in the number of free nucleosomes in the presence of RSC, and the clear stabilization of the RSC conformation in the presence of nucleosomes, point to the stability of the RSC/nucleosome complex under the conditions of our EM analysis. A reconstruction calculated from the largest set of RSC/nucleosome images is shown in Fig. 4.

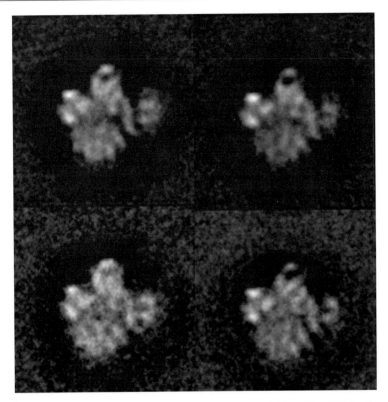

FIG. 3. Structure of the RSC/nucleosome complex in projection. The RSC/nucleosome complex was formed by incubating RSC and nucleosomes in a buffer containing 50 mM potassium acetate, 3 mM magnesium chloride, and 15 mM HEPES, pH 7.5. A total of ~9260 images of particles preserved in uranyl acetate were computationally aligned and sorted into homogeneous classes using multivariate statistical analysis and hierarchical ascendant classification.[3] As observed for RSC alone, most particles appeared to be in the same orientation. However, interaction with the nucleosome seems to stabilize the structure of the RSC complex, as evidenced by the much smaller number of homogeneous groups generated by statistical analysis.

Additional density is present in the central cavity, in agreement with our suggestion that the nucleosome might bind in that location. However, the amount of additional density observed is less than that expected on the basis of the molecular weight of the nucleosome core particle. Such discrepancy is most likely due to inclusion of unoccupied RSC particles in the RSC/nucleosome image data set, as indicated by the data summarized in Table I. Statistical analysis of the untilted RSC/nucleosome images (where the central cavity is clearly visible and does not overlap significantly

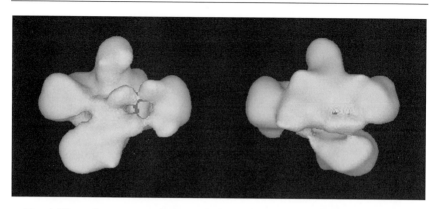

FIG. 4. Three-dimensional reconstruction of the RSC/nucleosome complex. The overall structure of the RSC complex remains constant after incubation with nucleosomes, but additional density is apparent in the central cavity. However, the additional density only corresponds to about half of that expected based on the molecular weight of the nucleosome core particle. Partial detection of the nucleosome is most likely due to the presence of unoccupied RSC particles in the data set, and to problems with staining caused by charged surfaces in the nucleosome and the RSC central cavity.

with the surrounding density that defines it) corresponding to the tilted particle images used to calculate the RSC/nucleosome reconstruction was carried out to attempt to separate empty and occupied RSC particles. The results from such analysis (not shown) were inconclusive, perhaps because the area corresponding to the central cavity includes only a relatively small number of pixels. Incomplete formation of the RSC/nucleosome complex detected by the particle count study (Table I), along with staining artifacts produced by the highly charged nature of the histone core and DNA surfaces, must contribute to the detection of an artificially low density in the central RSC cavity.

The structure of the RSC/nucleosome complex strongly suggests that interaction of a nucleosome with RSC occurs by binding of the nucleosome to the central cavity in the RSC complex. The size and shape of the central cavity match closely the size and shape of a nucleosome core particle. This is illustrated in Fig. 5, which shows our proposed model for RSC/nucleosome interaction. The RSC complex is shown in its prevalent conformation, and a low-resolution ($\sim$25 Å) structure of the nucleosome core particle (obtained by low-pass filtering of the atomic-resolution X-ray structure of the complex[9]) has been placed in the central RSC cavity.

[9] K. Luger, A. W. Mader, R. K. Richmond, D. F. Sargent, and T. J. Richmond, *Nature* **389**, 251 (1997).

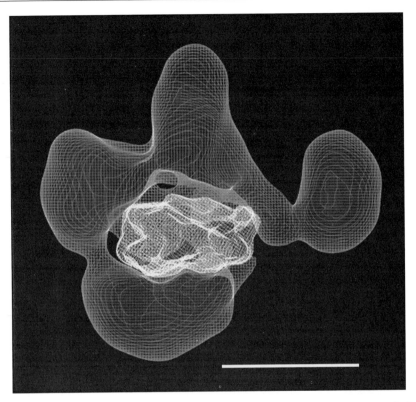

Fig. 5. Possible mode of RSC-nucleosome interaction. An X-ray structure of the nucleosome[9] was filtered to 25 Å and manually fitted in the central cavity of the RSC structure [using the program O (T. A. Jones, J. Y. Zou, S. W. Cowan, and M. Kjeldgaard, *Acta Crystallogr. A* **47** (Pt 2), 110 (1991).)]. The close fit between the nucleosome and the RSC cavity is apparent. The scale bar corresponds to approximately 100 Å. (See color insert.)

The figure illustrates the close fit of the nucleosome in the cavity. The model shown in Fig. 5 is supported by results from nuclease digestion studies of nucleosomal DNA in an RSC nucleosome complex, which indicate that digestion of nuclesomal DNA occurs preferentially at positions that would be more exposed in the model.[4]

Examination of Unstained RSC and RSC/Nucleosome Particles

While preservation of single particles in a heavy metal stain provides a simple and convenient way of obtaining structural information about a macromolecular complex, the results suffer from limitations in resolution and in the fidelity of the resulting volumes. These problems can be avoided by examining unstained particles preserved in amorphous ice (cryo-electron microscopy), which makes possible direct imaging of molecular features, and has been shown to preserve molecular structure to near atomic resolution.[10–12] Preparation of frozen-hydrated samples has been extensively described[13] and will not be repeated here. Instead, some of the issues encountered in applying the technique to the study of the RSC complex will be presented.

The first factor to consider is the type of grids to be used for sample preparation. Unstained particles can be adsorbed to a thin amorphous carbon support film, as in the case of stained samples. In fact, some of the highest resolution structures obtained by single particle analysis methods have actually been calculated from images of particles adsorbed to a support film.[14] The presence of a support film also makes it possible to determine more accurately the defocus value of the images, a parameter that is critical for calculation of a meaningful reconstruction.[15] Particles can also be preserved in a thin, unsupported film of amorphous ice, which is usually formed across holes in a relatively thick perforated carbon support film.[16] Particles preserved in an unsupported amorphous ice film are not disturbed by interaction with a substrate, and there is no contribution to the background signal from a continuous support film. However, under those conditions particles are often found to preferentially interact with the air/water interface, and this can interfere with their preservation. Also, sample preparation usually requires the use of a protein solution with a much higher (>10-fold higher) concentration, as the particles often tend to adsorb to the edges of the perforated carbon support.

Frozen-hydrated RSC samples were prepared using grids with a continuous carbon support film. The experiments were complicated by apparent interaction of the particles with the air/water interface, even in the presence of the support film. Nonetheless, a number of RSC images suitable for analysis were obtained. There are two main differences between images of RSC particles preserved in amorphous ice and those of RSC particles preserved in stain. First, the frozen-hydrates particles no longer show a preferred orientation on the substrate film. Second, the conformation of

[10] A. Miyazawa, Y. Fujiyoshi, M. Stowell, and N. Unwin, *J. Mol. Biol.* **288,** 765 (1999).
[11] G. Ren, A. Cheng, V. Reddy, P. Melnyk, and A. K. Mitra, *J. Mol. Biol.* **301,** 369 (2000).
[12] B. Bottcher, S. A. Wynne, and R. A. Crowther, *Nature* **386,** 88 (1997).
[13] J. Dubochet, M. Adrian, J. J. Chang, J. C. Homo, J. Lepault, A. W. McDowall, and P. Schultz, *Q. Rev. Biophys.* **21,** 129 (1988).
[14] J. Frank, P. Penczek, R. K. Agrawal, R. A. Grassucci, and A. B. Heagle, *Methods Enzymol.* **317,** 276 (2000).
[15] P. Penczek, M. Radermacher, and J. Frank, *Ultramicroscopy* **40,** 33 (1992).
[16] P. Penczek, R. A. Grassucci, and J. Frank, *Ultramicroscopy* **53,** 251 (1994).

the particles appears essentially homogeneous (i.e., there is no variation in the position of the lower module of the RSC structure). Images of particles preserved in amorphous ice were classified into homogeneous groups by aligning them to a set of reference projections of the RSC 3D reconstruction calculated from images of particles preserved in stain. To eliminate any possible reference-related bias, reference-free alignment was used to examine the homogeneity and alignment of each one of the particle groups defined by the classification. A gallery of the resulting reference-bias–free averages is shown in Fig. 6. The final step in obtaining a reconstruction

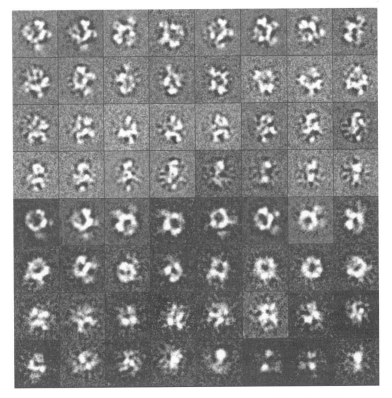

Fig. 6. Structure of unstained RSC in projection. Images of unstained, frozen-hydrated RSC particles were separated into groups based on their similarity to projections of the RSC reconstruction calculated from particles preserved in negative stain. The members of each homogeneous group were then realigned without use of a reference to avoid possible bias. Projections of the reference structure calculated from stained particles are shown in the four top rows. The corresponding group averages from unstained particles are shown in the bottom four rows. Improving the homogeneity of the groups, and determining the relative orientation of the resulting group averages, will make possible calculation of an improved RSC reconstruction.

of the RSC complex from the unstained particle images entails combining the information present in this set of projections. This requires precise knowledge of their relative orientations. Evidently, approximate information is provided by comparison with the available stain reconstruction. However, preliminary analysis indicates that the deformation induced by preservation in stain is significant, and this prevents direct use of the stain reconstruction as a meaningful reference for calculation of an improved reconstruction from the unstained particle images. We are currently working to obtain a suitable reference volume, which will allow us to make full use of the information contained in images of unstained RSC particles.

[5] Use of Optical Trapping Techniques to Study Single-Nucleosome Dynamics

By Brent Brower-Toland and Michelle D. Wang

Introduction

Over the past decade, optical trapping techniques have become a standard part of the repertoire of tools available for the study of biological molecules.[1–5] More recently, optical trapping techniques have been applied to the study of chromatin structure and even details of the structure of individual nucleosomes in a chromatin array.[6–8]

The general experimental design of optical trapping experiments with chromatin involves immobilization of one end of a linear DNA molecule on a surface, while the other end of the molecule is attached to a polystyrene microsphere (bead). The microsphere can then be used as a microscopic "handle" which can be captured and manipulated by the optical trap (Fig. 1). The optical trap can be used to exert and measure piconewton-scale

[1] K. Svoboda, C. F. Schmidt, B. J. Schnapp, and S. M. Block, *Nature* **21,** 365 (1993).

[2] H. Yin, M. D. Wang, K. Svoboda, R. Landick, J. Gelles, and S. M. Block, *Science* **270,** 1653 (1995).

[3] M. D. Wang, M. J. Schnitzer, H. Yin, R. Landick, J. Gelles, and S. M. Block, *Science* **282,** 902 (1998).

[4] J. Liphardt, B. Onoa, S. B. Smith, I. Tinoco, and C. Bustamante, *Science* **292,** 733 (2001).

[5] S. J. Koch, A. Shundrovsky, B. C. Jantzen, and M. D. Wang, *Biophys. J.* **83,** 1098 (2002).

[6] Y. Cui and C. Bustamante, *Proc. Natl. Acad. Sci. USA* **97,** 127 (2000).

[7] M. L. Bennink, S. H. Leuba, G. H. Leno, J. Zlatanova, B. G. de Grooth, and J. Greve, *Nat. Struct. Biol.* **8,** 606 (2001).

[8] B. Brower-Toland, R. C. Yeh, C. Smith, C. L. Peterson, J. T. Lis, and M. D. Wang, *Proc. Natl. Acad. Sci. USA* **99,** 1960 (2002).

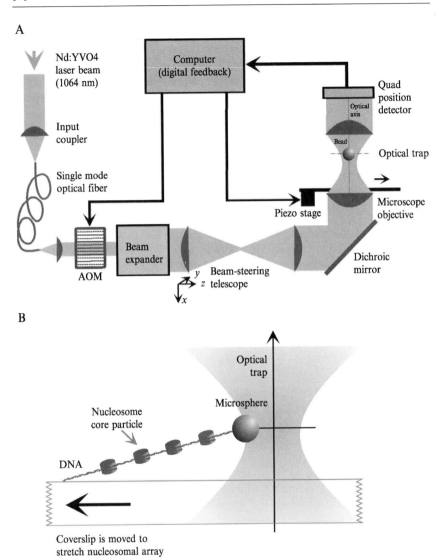

FIG. 1. Instrument and experimental configuration (not to scale). (A) Overall design of the instrument. The drawing is modified from Wang et al.[9] (B) Experimental configuration. Under feedback control, a nucleosomal array is stretched between the surface of a microscope coverslip and an optically trapped microsphere.

forces and nanometer-scale displacements on a nucleosomal DNA molecule.

Optical trapping techniques are single-molecule techniques that allow mechanical manipulation of a nucleosomal DNA molecule under physiological solution conditions. Because the sample is immobilized, solution conditions and sample components can actually be varied during the course of experimentation. The experiments can be nondestructive, permitting repeated sampling of the same nucleosomal arrays. An added advantage of the use of an optical trap is the freedom to consider individual nucleosomal structure in the context of a nucleosomal array, rather than on isolated mononucleosomes. These features distinguish optical trapping techniques from others in the repertoire of tools available for the study of chromatin.

In this chapter, we describe the optical trapping system and experimental sample preparation techniques necessary to carry out dynamic structural analysis of individual nucleosomes in nucleosomal arrays. We anticipate that optical trapping experiments will prove valuable in answering questions about chromatin structure that are difficult to access with traditional techniques.

Chromatin Sample Preparation

Preparation of Histones

Owing to the sensitivity of single molecule studies of chromatin, the use of highly purified biochemical components is critical to the success of these experiments. Large quantities of highly purified histone proteins can be prepared from various chromatin samples using standard hydroxyapatite (HAP) chromatography. Starting with washed nuclei from the tissue source of choice, their chromatin content is fragmented by mild MNase digest, and bound to HAP (BioGel HTP, BioRad Laboratories, Hercules, CA) by virtue of their nucleic acid component as previously detailed.[10,11] Linker histone and nonhistone proteins are removed by washing the HAP bed at moderate ionic strengths. Finally, core histone proteins are eluted from the HAP:DNA complex at high ionic strength.

Alternatively, highly purified recombinant histone proteins can be obtained by expression in bacteria, permitting choice and manipulation of primary sequence.[12] This flexibility permits the design of experiments

---

[9] M. D. Wang, H. Yin, R. Landick, J. Gelles, and S. M. Block, *Biophys. J.* **72**, 1335 (1997).

[10] A. P. Wolffe and K. Ura, *Methods* **12**, 10 (1997).

[11] G. Schnitzler, *in* "Current Protocols in Molecular Biology" (F. A. Ausubel *et al.*, eds.), **3**, p. 21.5.7. Wiley, New York, 2001.

[12] K. Luger, T. J. Rechsteiner, and T. J. Richmond, *Methods Enzymol.* **304**, 3 (1999).

involving minor histone variant proteins, critical structural mutants, and the introduction of cross-linking moieties.[13]

DNA Labeling and Attachment Methods

Selection of a DNA template for chromatin assembly and subsequent single-molecule analysis involves two major criteria: template length and sequence. With the experimental configuration previously described by Brower-Toland *et al.*,[8] a DNA template of around 4000 bp is optimal. A shorter DNA template will lead to greater uncertainty in the determination of the DNA extension and the tension in the DNA.[9] This is especially a concern for a saturated nucleosomal DNA due to its shortened DNA tether. A longer DNA template will introduce more Brownian motions of the trapped microsphere. DNA sequence choice is governed entirely by the goal of the experiment. We have previously utilized a DNA molecule composed of 17 repetitions of the sea urchin 5S nucleosomal positioning element in order to produce nucleosomal arrays containing, as far as possible, identical nucleosomal units.[8] However, nucleosomes assembled by salt dialysis on a nonrepetitive sequence with no bias for nucleosome positioning produce single-molecule data of sufficient quality for analysis.[14] As expected, nucleosomes on such arrays exhibit a broader range of binding energies than those assembled on naturally occurring positioning elements such as the 5S sequence.

Labeling and purification of these DNA molecules have been accomplished by standard enzymatic manipulation and chemical fractionation methods. Nonrepetitive sequences can be asymmetrically end-labeled using the polymerase chain reaction (PCR) with differentially end-labeled primers. Repetitive sequences not amenable to PCR amplification have been labeled by Klenow fill-in reaction with the appropriate labeled NTPs. Experiments performed in our lab have utilized biotin and digoxigenin-labeled nucleic acids successfully. These two labels are especially convenient because of the wide commerical availability of avidin-coated microspheres, and the existence of sufficiently high-affinity anti-digoxigenin antibodies (Roche Applied Science, Indianapolis, IN). Removal of residual unincorporated label is critical to the success of these experiments.

---

[13] A. Flaus, K. Luger, S. Tan, and T. J. Richmond, *Proc. Natl. Acad. Sci. USA* **93,** 1370 (1996).
[14] B. Brower-Toland and M. D. Wang, unpublished data.

Chromatin Assembly

Assembly of chromatin arrays for analysis from highly purified DNA and histone components can be achieved by chemical (gradient salt dialysis)[15] or by enzymatic means.[16] The salt dialysis method of assembly has the advantage of preserving sample purity and minimizing the amount of post-assembly purification required before experimentation.

Enzymatic assembly of chromatin has the pitfall of introducing a large number of protein impurities if an assembly extract is utilized. The recent development by the Kadonaga lab of a completely recombinant assembly system minimizes this complication.[17] Chromatin assembly by this method is more rapid than by salt dialysis. Moreover, it has the advantage of producing extremely regular arrays of nucleosomes in a sequence-independent fashion, without introducing artifactual structures such as dinucleosomes. With either technique, optimization of assembly conditions by post-assembly electrophoretic analysis is necessary prior to single-molecule experimentation, both to avoid artifactual structures and to ascertain the quality of array formation.

Preparation of Experimental Samples

For single-molecule studies, samples are prepared by sequential infusion of solutions into sample chambers for microscopic observation. Chamber volume is approximately 10 μl. Fluid flow through the chamber is by capillary action produced by placing an absorbent wick at one end of the chamber while delivering solutions by micropipette at the other end. All procedures are carried out at room temperature, and incubations are performed in a humid chamber to avoid evaporation. Samples prepared in this way are exquisitely sensitive to changes in physical and chemical conditions, so that consistency in all aspects of sample preparation, especially temperature, is critical. Once prepared, chromatin samples have a useful experimental lifetime of about 2 h at room temperature.

Buffer solutions:

1. PBS: 140 mM NaCl, 2.7 mM KCl, 10 mM Na$_2$HPO$_4$, 1.8 mM NaH$_2$PO$_4$, pH 7.3.
2. Blocking buffer (BB): PBS + 0.2% purified non-fat milk protein.
3. Chromatin dilution buffer (CDB): 10 mM Tris-HCl (pH 8.0), 1 mM Na$_2$EDTA, 150 mM NaCl.

[15] K.-M. Lee and G. Narlikar, in "Current Protocols in Molecular Biology" (F. A. Ausubel et al., eds.), **3**, p. 21.6.3. Wiley, New York, 2001.

[16] M. Bulger and J. T. Kadonaga, *Methods Mol. Genet.* **5**, 241 (1994).

[17] M. E. Levenstein and J. T. Kadonaga, *J. Biol. Chem.* **277**, 8749 (2002).

4. Experimental sample buffer (ESB): 10 mM Tris-HCl (pH 8.0), 1 mM Na$_2$EDTA, 100 mM NaCl, 1.5 mM MgCl$_2$, 0.02% (v/v) Tween-20, and 0.01% (w/v) milk protein.

The blocking agent that has worked best in our hands is a purified milk protein powder (Biorad Laboratories, Hercules, CA). For attachment of digoxigenin-labeled DNA samples to sample chamber surfaces we have used polyclonal sheep anti-digoxigenin (Roche Applied Science, Indianapolis, IN) with relatively uniform results. Chromatin assembled by the salt dialysis method referred to earlier are stored at concentrations ≥ 100 ng/μl at 4°, and diluted to 0.2–1 ng/μl in CDB immediately before application to sample chamber for experimentation.

1. Rinse sample chamber with 5 chamber volumes of PBS.
2. Immediately infuse 1 volume of anti-digoxigenin solution (25 ng/μl in PBS). Incubate 10 min.
3. Rinse with 5 volumes of BB. Incubate with residual blocker, 5 min.
4. Rinse with 5 volumes of PBS.
5. Immediately infuse 1 volume of diluted chromatin sample in CDB. Incubate 7.5 min.
6. Rinse with 5 volumes of CDB.
7. Infuse 1 volume of avidin-coated beads (10 pM in CDB + 0.01% (w/v) purified non-fat milk protein). Incubate 5 min.
8. Rinse with 5 volumes of ESB. Seal chamber with nail polish if no additional solutions will be infused.

Instrumentation

Mechanical measurements of a single nucleosomal array can be obtained by using a single-beam optical trapping microscope.[5,8] Here, we provide a brief overview of the instrument. The reader should refer to Wang et al.[9] and Koch et al.[5] for more detailed descriptions of the design, construction, and calibration of the optical trapping setup.

Optical Trapping System

A schematic of our optical setup is shown in Fig. 1A. The trapping laser has a wavelength of 1064 nm (Spectra-Physics Lasers, Mountain View, CA). The laser beam passes through a single-mode optical fiber (Oz Optics, Carp, ON) and an acousto-optic deflector (NEOS Technologies, Melbourne, FL), and is focused onto the sample plane by a 100×, 1.4-NA, oil immersion objective on an Eclipse TE200 DIC microscope (Nikon USA, Melville, NY). The focus of the laser serves as the center of the trap for a micron-sized microsphere. The laser light is collected by a 1.4-NA oil immersion condenser and projected onto a quadrant photodiode (Hamamatsu, Bridgewater, NJ). The photocurrents from each quadrant of the photodiode are amplified and converted to voltage signals using a position detection amplifier (On-Trak Photonics, Lake Forest, CA). The position of the optical trap relative to the sample can be adjusted with a servo-controlled 1-D piezoelectric stage (Physik Instrumente GmbH & Co., Waldbronn, Germany). Analog voltage signals from the position detector and stage position sensor are anti-alias filtered at 5 kHz (Krohn-Hite, Avon, MA) and digitized at 7–13 kHz for each channel using a multiplexed analog to digital conversion PCI board (National Instruments Corporation, Austin, TX).

Calibration of the Optical Trapping System

The instrument calibration methods may follow those of Wang *et al.*[9] In brief, the first step of the calibration determines the position of the trap center relative to the beam waist and the height of the trap center relative to the coverslip. The second step of the calibration determines the position detector sensitivity and trap stiffness. The third step of the calibration locates the anchor position of the DNA tether on the coverslip, and is performed prior to each measurement by stretching the DNA at low load (<5 pN). These calibrations are subsequently used to convert data into force and extension.

Experimental Control and Data Acquisition

To disrupt nucleosomes as shown in Fig. 1B, the coverslip is moved relative to the trapped microsphere with a piezoelectric stage to stretch the DNA. Once a surface-tethered microsphere is optically trapped, the coverslip is then moved with a piezoelectric stage to stretch the nucleosomal DNA with either a velocity clamp or a force clamp.[5,8] Both of these clamps may be implemented with digital feedback algorithms, with an average rate for a complete feedback cycle of 7–13 kHz. In the velocity clamp mode, the coverslip is moved at a constant velocity relative to the microsphere, whose position is kept constant by modulating the light intensity (trap stiffness) of the trapping laser. A disruption event, during which DNA is released from a histone octamer, is observed as a sudden reduction in the tension of the DNA. In the force-clamp mode, the position of the coverslip is modulated so that the trapping force on the microsphere is held constant by keeping its position fixed in a trap of constant stiffness. In this mode, a disruption event is observed as a step in the coverslip position.

Determination of the DNA Elasticity

Determination of the force-extension relation of a naked DNA is essential for the conversion of force and extension to number of base pairs of naked DNA (see section on Data Analysis). Marko and Siggia proposed the Worm-Like-Chain (WLC) model, which accounts for the entropic elasticity and well describes the force-extension relation in the low-force region (<5 pN).[18] Wang *et al.* extended this model to also include the enthalpic elasticity in the high-force region (>5 pN),[9] and referred to this modified form as the Marko-Siggia (MMS) model. This MMS model incorporates both enthalpic and entropic contributions to stiffness and fits the experimental results extremely well for forces up to 50 pN. In the MMS model, the elastic stiffness of DNA is parameterized by its contour length under zero tension, L_0, its persistence length, L_P, and its elastic modulus, K_0. The force (F) and extension (ξ) are simply related:

$$F = \left(\frac{k_B T}{L_p}\right)\left[\frac{1}{4(1 - \xi/L_0 + F/K_0)^2} - \frac{1}{4} + \frac{\xi}{L_0} - \frac{F}{K_0}\right].$$

The elastic parameters of dsDNA can be obtained following Wang *et al.*[9] Under our experimental conditions, L_0 per base is 0.338 nm, L_P is 43.1 nm, and K_0 is 1205 pN. Therefore, if both the force and extension are known, this relation can be used to obtain L_0, which can be readily converted to the number of base pairs.

Data Analysis

Nucleosomes can be disrupted in various ways. The two ways presented here are velocity-clamp stretching and force-clamp stretching. These two methods are roughly equivalent, but with some subtle differences. Velocity clamp allows disruption of all nucleosomes at a specified stretching velocity regardless of the strength of protein-DNA interactions within nucleosome. However, nucleosomes in an array are disrupted under slightly different force conditions, which depend on the number of nucleosomes remaining in the array at a specific disruption. Force clamp allows disruption of all nucleosomes under identical force conditions (i.e., the same force). However, more experimentation is required to determine a workable range of force: If the force is too small or too large, the time to disrupt the nucleosomes will be too long or too short to be experimentally accessible.

[18] J. F. Marko and E. D. Siggia, *Macromolecules* **28**, 8759 (1995).

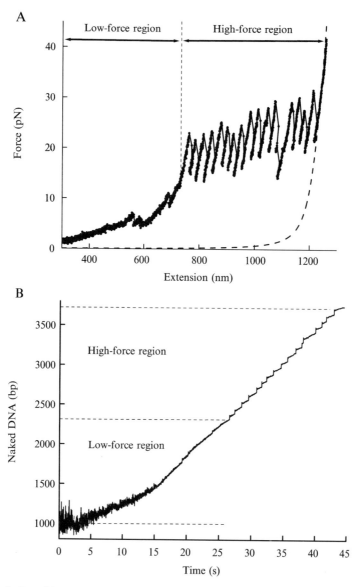

Fig. 2. Stretching a nucleosomal array with a velocity clamp at 28 nm/s. Nucleosomal arrays were prepared with avian core histones and a 3684-bp DNA fragment containing 17 direct tandem repeats of the sea urchin 5S positioning element. The biotin and digoxigenin linkers at the two ends of the DNA effectively contribute ~50 bp of DNA. (A) Force-extension curve of a fully saturated nucleosomal array. At higher force (>15 pN), a sawtooth pattern containing 17 disruption peaks was observed. Force-extension characteristics of a

Velocity Clamp

An example of data taken with a velocity clamp is shown in Fig. 2. At low force (<15 pN), the force-extension curve starts to deviate from that of the corresponding naked DNA; at higher force (>15 pN), a distinctive sawtooth pattern starts to appear; at even higher force (>40 pN), the force-extension resembles that of a naked DNA (dotted curve). Previously, Brower-Toland et al.[8] demonstrated that the high-force sawtooth pattern is indicative of a nucleosomal array, with each peak corresponding to a single nucleosome. Under the conditions used in our experiments, the spacing between adjacent peaks is ~27 nm. The observed sawtooth pattern suggests separate disruption of strong DNA-histone interactions in individual nucleosomes.

To determine the amount of DNA released from a nucleosome, the MMS model can be applied. This conversion attributes extension only to naked DNA, that is, linker DNA and DNA peeled from nucleosome core particles (NCP). This method of conversion from force-extension curve to number of base pairs of naked DNA is similar to that previously used for single-molecule studies of transcription.[3] The MMS model is only an approximation for a nucleosomal array. To achieve better precision for the conversion, a more refined model will be necessary. Conversion of the data in Fig. 2A is shown in Fig. 2B, where the amount of naked DNA is plotted as a function of time during stretching. At the beginning of stretching (0–2 s), the average amount of naked DNA is constant, indicating no DNA release from NCPs. This should correspond to the amount of linker DNA for a relaxed nucleosomal array. As force rises in the low-force region, DNA release is gradual, indicating a simultaneous release of DNA from all nucleosomes, with ~76 bp of DNA release per nucleosome. At high force, the sawtooth peaks in Fig. 2A, are converted to steps. DNA release is sudden, indicating a separate release of DNA from each nucleosome of ~80 bp.

Force Clamp

An example of data taken with a force clamp at 20.2 pN is shown in Fig. 3. Unlike the velocity clamp measurements, all the nucleosomes experienced the same force before disruption. Here, sudden disruptions of nucleosomes resulted in stepwise increases in the DNA extension, with

full-length naked DNA (dotted line) are shown for comparison. (B) Amount of naked DNA as a function of time derived from data shown in Fig. 2A. The top dotted line is a comparison with a full-length naked DNA. At higher force, the curves show 17 steps, which correspond to the 17 disruption peaks in Fig. 2A.

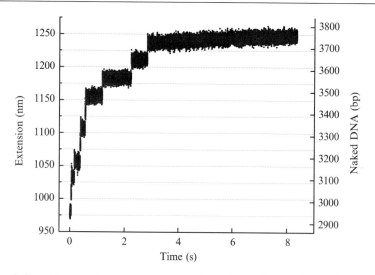

FIG. 3. Stretching a nucleosomal array with a force clamp. The graphs are plots of DNA extension (left axis) and the amount of naked DNA (right axis) versus time under constant force of 20.2 pN.

each step corresponding to one nucleosome disruption. The extension (left axis) is readily converted to number of base pairs of naked DNA (right axis). These steps (~80 bp) provide a measure of DNA release per nucleosome at high force.

Conclusion

Renewed interest in chromatin as a mediator of the structure, maintenance, and regulation of eukaryotic genomes has inspired the development of a variety of novel chromatin techniques. Optical trapping technology provides a useful addition to this repertoire of techniques. We anticipate that single-molecule optical trapping experiments on chromatin structure will complement more traditional technologies, and aid in the elucidation of the structural role in chromatin of histone and nonhistone proteins and their post-translational modifications. Likewise, optical trapping methods will be adaptable to the study of enzymatic activities such as RNA polymerases and ATP-dependent chromatin remodelers operating on chromatin structure.

Acknowledgments

We thank David A. Wacker and Dr. Robert M. Fulbright for their helpful discussions and technical assistance. Supported by grants from the NIH and the Keck Foundation's Distinguished Young Scholar Award to M.D.W.

[6] Single-Molecule Analysis of Chromatin

By Sanford H. Leuba, Martin L. Bennink, and
Jordanka Zlatanova

The importance of chromatin structure and dynamics in the functioning of eukaryotic DNA is currently widely accepted. Still, understanding the structure and how it changes in response to various signals regulating DNA transactions (transcription, replication, recombination, and repair) remains a challenging task. Recently, single-molecule techniques have been added to the arsenal of more traditional biochemical and biophysical approaches,[1] allowing the investigation of one molecule (or molecular complex) at a time, frequently in real time. This unique capability is very valuable, since it allows the elucidation of the behavior of individual molecules and an assessment of their variability. Thus, these techniques represent a significant improvement over population-average methods, in which individual differences are masked in the average parameters measured. Moreover, single-molecule techniques avoid the technically challenging task of synchronizing a population of molecules at a specific stage of a process. Even when such synchronization can be achieved, it is quickly lost soon after the synchronization block is relieved because of the stochastic nature of the individual reaction steps. Another major advantage of the single-molecule approaches is the high spatial and temporal resolution of the measurements, allowing observation of changes in the nanometer distance range and millisecond timescale.

Here, we will describe in technical detail the three single-molecule techniques that have been utilized so far to study single chromatin fibers: the atomic force microscope (AFM), optical tweezers (OT), and magnetic tweezers (MT). The AFM produces digital topographical images of samples deposited on flat surfaces by raster-scanning the surface with a sharp tip mounted on the back of a flexible cantilever[2] (Fig. 1A). Atoms on the tip interact with atoms on the surface, causing the cantilever to deflect upwards or downwards, depending on whether the tip-sample interaction is repulsive or attractive. The deflections of the cantilever are registered by a laser beam reflected off the back of the cantilever to produce a topographical image of the sample. The AFM can also be used as a force spectroscopy

---

[1] S. H. Leuba and J. Zlatanova, eds., "Biology at the Single-Molecule Level," Pergamon, Amsterdam, 2001.
[2] G. Binnig, C. F. Quate, and C. Rohrer, *Phys. Rev. Lett.* **56,** 930 (1986).

Atomic force microscope   Optical tweezers   Magnetic tweezers

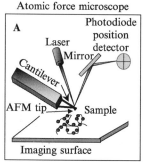
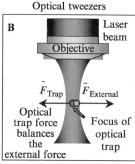
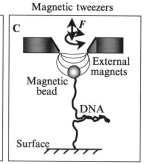

FIG. 1. Schematic of operation principles of three single-molecule instruments successfully used for chromatin research. (A) Atomic force microscope (AFM). (B) Optical tweezers (OT). (C) Magnetic tweezers (MT).

tool: in this application, the AFM tip is allowed to move only in the z-direction, up and down. If a molecule that is immobilized on the surface is also attached to the tip (either through specific linkages or through nonspecific adhesion), then the upward movement of the cantilever will stretch the molecule to produce the so-called force-extension curves.[3]

Using OT is a technique for manipulating single molecules based on the interaction of light with matter.[4] Light can exert forces on small beads of certain optical properties in such a way that the bead is kept suspended at a point close to the waist of a laser beam that is focused with an objective. If such a "trapped" bead is pulled out of this equilibrium position by some external force, a net restoring force resulting from the bead's interaction with light will effectively pull it back to this restoring position (Fig. 1B; see "Physics of Optical Trapping" and Fig. 2A–C). Thus, the trap acts as a Hookean spring, with the force on the trapped bead being proportional to its displacement from the equilibrium position in the trap (Fig. 2D). In the OT set-up, one can suspend a macromolecule between an optically trapped bead and another surface (a coverslip or another bead), which can be moved at will to apply tension to the molecular tether, and force-extension curves can be recorded.

Finally, in the MT, the macromolecule is attached between a surface and a magnetic bead (Fig. 1C). Manipulation of an external magnetic field can be used to apply stretching force to the tethered molecule, and/or to induce precisely known levels of supercoiling: stretching is achieved by changing the distance between the external magnet(s) and the cuvette, while supercoiling is introduced by rotating the external magnetic field

[3] J. Zlatanova, S. M. Lindsay, and S. H. Leuba, *Prog. Biophys. Mol. Biol.* **74**, 37 (2000).
[4] A. Ashkin, J. M. Dziedzic, J. E. Bjorkholm, and S. Chu, *Optics Lett.* **11**, 288 (1986).

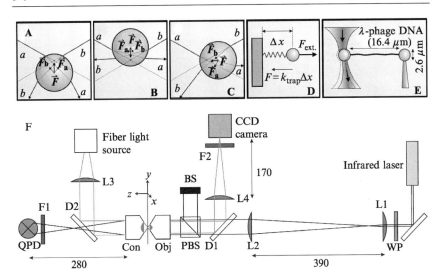

Fig. 2. Principle of optical trapping in a single-beam gradient trap. Schematic showing the propagation of two high-angle rays as they refract at the surface of a high-index spherical particle, when positioned just below the focal point (A), just above the focal point (B), or just off-axis within the focal plane (C). The net force resulting from these two rays is always pointing toward the focal point, establishing a stable trapping point at the focus. Reflection at the surface of the bead is neglected (if reflection is taken into account, the center of the trap will be just below the focal point). (D) The force exerted on the trapped bead increases linearly with the displacement of the bead from the trap center. Conceptually, this is identical to a Hookean spring with zero rest length. (E) When the trapped bead is attached to a single DNA molecule, the force exerted by the trap on the bead is equal to the tension within the molecule. The deflection of the laser beam is used to accurately measure the displacement of the bead and, therefore, the tension in the molecule. (F) Optical tweezers optical set-up. The beam of a 2-W continuous-wave infrared laser (1064 nm) is expanded using lens L1 ($f = 50$ mm) and L2 ($f = 350$ mm) (NOTE 9). The diameter of the beam slightly overfills the back aperture of the water-immersion objective (Obj), which is a necessary condition to establish a stable optical trap within the flow cell. The intensity of the laser can be tuned by turning the quarter-wavelength plate (WP) in front of the first lens. A polarizing beam splitter cube (PBS) just before the objective separates the two orthogonal polarization directions, enabling fine control of the intensity of the beam at the objective (light that is reflected by the polarizing beam splitter is stopped using a beam stop, BS). The transmitted (i.e., refracted) laser light is collected with a 0.9-NA condenser lens (Con) and projected onto a quadrant detector. Note that the quadrant detector is not positioned in a conjugate image plane (NOTE 1). This detector quantifies the deflection, which is calibrated and transformed into a force signal. A neutral density filter (F1) is used in front of the quadrant detector to reduce the intensity to measurable levels. Within this experimental set-up, the objective is also used for viewing the beads in the flow cell. The light of a fiber light source is collimated using a 40-mm lens (L3) and projected onto the back of the condenser, which results in a homogeneous illumination of the sample. The objective in combination with an additional lens L4 ($f = 170$ mm) is used to project the microscopic image onto a CCD array. An additional filter (F2) is used in front of the camera to fully block the infrared laser light. Dichroic mirrors (D1 and D2) are used to separate the laser-beam path from that of the illumination beam.

either clockwise or counterclockwise.[5–7] The magnitude of the applied force can be determined by the equipartition theorem: $F = lk_B T/<\delta x^2>$, where l is the contour length of the tether, k_B, Boltzmann's constant, T, absolute temperature, and $<\delta x^2>$, the Brownian motion of the bead in the direction perpendicular to the tether axis.

AFM imaging

Chromatin fibers are challenging to image by AFM because they are relatively soft (compared to naked DNA, for example), have features that are high above the surface, and contain histones that are notoriously sticky. Softness results in indentation of the sample by the tip, while the sticky histones may easily attach to it: both interfere with proper imaging. Distortion to the image also occurs when the vertical distance traveled by the tip is too large—which happens with samples containing topographically high features. We have recently published detailed protocols of how to image chromatin fibers with the AFM.[8] Here, we provide some further thoughts based on our prior experience with the Digital Instruments (DI, www.di.com) AFM (e.g., Zlatanova et al. (1994),[9] Leuba et al. (1994),[10] Yang et al.,[11] Zlatanova et al. (1998)[12] Leuba, Bustamante, Zlatanova, and van Holde (1998),[13] Leuba, Bustamante, van Holde, and Zlatanova (1998)[14]), and our more recent work with the Molecular Imaging (MI, www.molec.com) instrument (e.g., Karymov et al.,[15] Tomschik et al.,[16] An et al.[17]).

[5] T. R. Strick, J. F. Allemand, D. Bensimon, A. Bensimon, and V. Croquette, Science **271**, 1835 (1996).

[6] T. R. Strick, J.-F. Allemand, V. Croquette, and D. Bensimon, J. Stat. Phys. **93**, 647 (1998).

[7] T. Strick, J. Allemand, V. Croquette, and D. Bensimon, Prog. Biophys. Mol. Biol. **74**, 115 (2000).

[8] S. H. Leuba and C. Bustamante, in "Methods in Molecular Biology: Chromatin Protocols" (P. B. Becker, ed.), p. 143. Humana Press, Totowa, NJ, 1999.

[9] J. Zlatanova, S. H. Leuba, G. Yang, C. Bustamante, and K. van Holde, Proc. Natl. Acad. Sci. USA **91**, 5277 (1994).

[10] S. H. Leuba, G. Yang, C. Robert, B. Samori, K. van Holde, J. Zlatanova, and C. Bustamante, Proc. Natl. Acad. Sci. USA **91**, 11621 (1994).

[11] G. Yang, S. H. Leuba, C. Bustamante, J. Zlatanova, and K. van Holde, Nat. Struct. Biol. **1**, 761 (1994).

[12] J. Zlatanova, S. H. Leuba, and K. van Holde, Biophys. J. **74**, 2554 (1998).

[13] S. H. Leuba, C. Bustamante, J. Zlatanova, and K. van Holde, Biophys. J. **74**, 2823 (1998).

[14] S. H. Leuba, C. Bustamante, K. van Holde, and J. Zlatanova, Biophys. J. **74**, 2830 (1998).

[15] M. A. Karymov, M. Tomschik, S. H. Leuba, P. Caiafa, and J. Zlatanova, FASEB J. **15**, 2631 (2001).

[16] M. Tomschik, M. A. Karymov, J. Zlatanova, and S. H. Leuba, Struct. Fold. Des. **9**, 1201 (2001).

[17] W. An, V. B. Palhan, M. A. Karymov, S. H. Leuba, and R. G. Roeder, Mol. Cell **9**, 811 (2002).

There are two different modes of imaging with the AFM, contact mode and dynamic mode. In the contact mode, the AFM tip is in constant contact with the sample/surface during the raster-scanning. In the dynamic (so-called tapping) mode, the cantilever is oscillated at a given frequency above the surface and the effect of the sample is to dampen the amplitude of this oscillation. In tapping mode, the tip contacts the sample/surface only intermittently, thus minimizing the friction/drag forces exerted by the tip in contact mode. The different instruments achieve a controlled oscillation of the cantilever in different ways (see later).

Various commercial AFM designs have advantages and disadvantages in the ease and quality of imaging over an area of 0.5 μm × 0.5 μm. The most useful image size for chromatin fibers is 1 μm × 1 μm, 512 pixels × 512 pixels, as it gives sufficient nucleosome resolution to perform measurements of center-to-center internucleosomal distances, angles between lines connecting three contiguous nucleosomes, z-heights, and numbers of nucleosomes per 10-nm fiber length.

Digital Instruments has two designs of the instrument: the Bioscope and the Multimode. In the Bioscope, the piezo and the cantilever holder are both above the imaging surface. This scope can be mounted on inverted fluorescence microscopes: one can easily find large samples (e.g., cells or metaphase chromosomes) on the surface and then position the scanning tip directly over the region of interest. The Bioscope comes with a piezo scanner with a range of 100 μm × 100 μm. Since biological processes occur in structures in the nanometer range, scan sizes of a couple of micrometers are needed; such needs are better met by the smaller scanners (typically 16 μm × 16 μm, or 12 μm × 12 μm) of the Multimode instrument. Here, the piezo scanner is below the imaged surface, creating a risk for buffer leakage over (and destruction of) the piezo. The DI instrument uses separate piezos to drive the cantilever oscillation in tapping mode; this adversely affects the instrumental noise.

The MI AFM uses magnetically driven cantilevers to image biological samples in air and in buffers. An external solenoid drives an alternating current (AC) signal that vibrates a cantilever coated with a magnetically susceptible material (the cantilever responds to the changes in the AC magnetic field). This method of oscillation is much gentler and reduces the noise in the instrument (for more technical details, see refs. Zlatanova et al. [2000],[3] Lindsay [2001][18]). The MI liquid cell is of a simpler design than that of the DI instrument, and the piezo is placed above the imaged

---

[18] S. M. Lindsay, in "Scanning Probe Microscopy and Spectroscopy: Theory, Techniques, and Applications" (D. Bonnell, ed.), p. 289. Wiley, New York, 2001.

surface. Despite these obvious advantages of the MI instrument, it takes longer to learn how to successfully use it.

A more recent AFM company, Asylum Research (AR, www.asylumresearch.com), has come up with an x-, y-, z-dimension AFM to complement their single z-dimension puller. The puller is used in force spectroscopy experiments[3] to apply stretching forces to molecules suspended between the tip and the surface. The piezo crystal in the AR 3D imaging AFM is also placed above the imaging surface. We have no experience with this instrument.

The DI software has gone through several rounds of maturity; however, the company limits access to the original source code. The MI and AR software are more accessible (e.g., the AR AFM is run on IgorPro, Wavemetrics, Portland, OR, USA), but are less mature. The ease of use of the software, as well as the ability to archive data, and then reaccess the data in its original format is an important consideration for scientists planning to use an AFM.

Materials

- Triethanolamine, TEA (Baker, Phillipsburg, NJ, USA)
- Electron microscope (EM)-grade glutaraldehyde (EM Sciences, Fort Washington, PA, USA, Cat. No. 16120)
- Mica (Asheville-Schoonmaker Mica Company, Newport News, VA, USA)
- Paper cutter board (local office supply) to cut mica
- Scotch magic tape (3M)
- Milli-Q (Millipore) water system (Fisher Scientific, Pittsburgh, PA, USA); all buffers are prepared from 18.2 MΩ purified water
- Slick microfuge tubes (PGC Scientific, Gaithersburg, MD, USA)
- Tank of argon
- Butane professional torch (VWR, West Chester, PA, USA, Cat. No. 62379-525)

Chromatin Preparation and Dialysis; Glutaraldehyde Fixation

Chromatin fibers are isolated from cell nuclei using micrococcal nuclease-based procedures (e.g., refs. 10, 15). Extensive dialysis versus 10 mM TEA-HCl, pH 7.0, 0.1 mM EDTA is crucial to obtaining good AFM images of extended fibers. For chromatin fibers reconstituted on naked DNA using crude extracts or purified histones and chromatin-assembly proteins, it is essential to purify the chromatin fibers through a sucrose or glycerol gradient before dialysis (e.g., ref. 17).

We have used glutaraldehyde to "fix" chromatin fibers before imaging, that is, to preserve the specific morphology characteristic of specific environmental conditions. Fixation also prevents gross structural distortions that may be caused by surface interactions. A final concentration of only 0.1% glutaraldehyde seems to work fine, in accordance with other observations (Thoma *et al.* (1979)[19]; personal discussion with Jan Bednar). To lessen the loss of material to the tube walls, we use slick microfuge tubes. One microliter of glutaraldehyde from a freshly opened ampoule is added to 99 μl of a 0.1-mg/ml solution of chromatin fibers ($A_{260} = 2.0$) and the tube is left overnight on ice; rotation of the tube tends to precipitate the chromatin fibers onto the tube surface.

Preparation of Mica and Glass Surfaces, Sample Deposition, and Imaging Conditions

Square sheets of mica (cut with an office supply paper cutter) are mounted (typically with double-stick tape or superglue—the thick-gel superglue is preferable to the runny variety) onto a surface. To make a fresh, atomically flat surface (of roughness not exceeding a few angstroms), mica is peeled using Scotch magic tape. The mica surface remaining on the tape is inspected by eye to see whether it is smooth and shiny; a shiny surface is an indication that the mica surface left behind on the block is optically flat, and likely to be atomically flat when imaged by the AFM.

An alternative surface is cover glass. The glass is first rinsed with a stream of 95% ethanol from a squirt bottle, the ethanol is allowed to evaporate, and the cover glass is flamed with a torch to burn away impurities, improving its surface flatness. It is also possible to burn impurities by baking the coverslips for 5 min in a 600° oven. After flaming or baking, the cover glass is rinsed briefly with Milli-Q water and allowed to dry. When the water-rinsing step is omitted, small particles (~100 nm sized, perhaps ash) are observed in the AFM images.

Typically, 2–10 μl of sample are deposited on the surface. After 1–5 min, the sample is rinsed either with the same low-ionic strength buffer or Milli-Q water. Excess liquid is blotted with blotting paper and blown off for a few seconds with a light flow of argon or nitrogen gas.

Imaging conditions depend on the AFM make, the cantilever, the surface used and on whether the imaging is performed in air or under liquid. Typically, we have employed imaging rate of two lines per second; capturing a 512 pixels × 512 pixels image takes ~8 min. For more details see reference 8.

[19] F. Thoma, T. Koller, and A. Klug, *J. Cell Biol.* **83,** 403 (1979).

Optical Tweezers

Physics of Optical Trapping

Optical trapping and manipulation of dielectric particles was first demonstrated in 1986.[4] Radiation, when changed in direction or intensity as a result of interaction with matter (i.e., reflection and refraction at an interface), exerts force on the matter. Photons have translational momentum $p = mv = h\nu/c$, where m is mass, v, velocity, $h\nu$, the energy of the photon, and c is the speed of light. When the momentum of a photon is changed as a result of reflection and/or refraction, the matter the photon interacts with also undergoes a change in momentum (the momentum conservation law). For a continuous flow of photons, this results in a physical force exerted on the bead. The magnitude of force that radiation (with normally used light intensities) can exert on objects is rather low. However, for small particles, whose dimensions are on the order of the wavelength of the radiation, these forces are significant. Figure 2A–C illustrates the radiation forces exerted on a bead by the rays.

The optical trap is working as a Hookean spring. The force exerted on the bead is proportional to its displacement in the trap (Fig. 2D). The proportionality factor relating the displacement with the force is the trap stiffness or trap constant k (typically 0.1 pN/nm). When a DNA molecule is suspended between two beads (Fig. 2E), the force applied to the molecule through the handle is equal in magnitude and acts in opposite direction to the force experienced by the optically trapped bead. The force can be determined by measuring the displacement of the bead from the center of the trap using video analysis. However, the force can be obtained more accurately and with higher time resolution by measuring the deflection of the beam as a result of the displacement of the trapped bead (Fig. 2E). One can consider the trapped bead as a micron-sized lens, the movement of which over nanometer distances can be detected very accurately by measuring the deflection of the laser beam. This piconewton (pN) resolution in force and nanometer resolution in displacement makes the optical trap an excellent tool for measuring force-extension curves of individual macromolecules.

Materials

- Quadrant detector (SPOT-9DMI, UDT Sensors, Hawthorne, CA, USA, www.udt.com)
- Newport 3D translation stage with both manual and piezo-electric control (ULTRAlign Precision Fiber Optics Linear Stage, 561D, Newport, Irvine, CA, USA, www.newport.com)

- 2-W infrared laser (1064 nm) (CrystaLaser, Reno, NV, USA, www.crystalaser.com)
- Millennia 10-W CW laser (Spectra Physics, Mountain View, CA, USA, www.spectra-physics.com)
- High NA water-immersion objective (100×, 1.2 NA, Leica, Wetzlar, Germany, www.leica-microsystems.com)
- Horizontal vibration-isolated table (vibration-isolated Workstation 07 OWS 112, Melles Griot, Irvine, CA, USA, www.mellesgriot.com)
- Glass capillaries (Borosilicate glass capillaries, GC120TF-15, 1.2 mm OD, 0.94 mm ID, thin wall, with filament, Clark Electromedical Instruments, Reading, UK)
- Square glass capillary (ID = 50 μm, OD = 150 μm, VitroCom, Mountain Lakes, NJ, USA, www.vitrocom.com)
- Professional pipette puller (Flaming/Brown Micropipette Puller, model P-87, Sutter Instrument Company, Novato, CA, USA, www.sutter.com)
- Thin fluid line (Intramedic polyethylene tubing, non-sterile, PE 10, ID = 0.28 mm, OD = 0.61 mm, Becton Dickinson, Franklin Lakes, NJ, USA)
- 3 mm outer diameter flexible tubing (various suppliers)
- Six-way low-pressure selection valve (Upchurch Scientific, Oak Harbor, WA, USA, www.upchurch.com)
- Uniform Microspheres, CML, 2.60 μm, Bangs Laboratories, Fishers, IN, USA, www.bangslabs.com; these beads have carboxylate end-groups at the surface that are activated using EDAC (1-ethyl-2-(3-dimethylaminopropyl)carbodiimide, Sigma E7750, Sigma-Aldrich Chemie BV, Zwijndrecht, The Netherlands)
- Chemicals: λ-DNA (New England Biolabs, Beverly, MA, USA); streptavidin (Roche, Almere, The Netherlands); dNTPs (100 mM dGTP, dCTP, Sigma); bio-dATP (0.4 mM in 50 mM Tris-HCl, pH 7.5, Sigma); bio-dUTP (0.3 mM in 50 mM Tris-HCl, pH 7.5, Sigma); Klenow fragment in 50% glycerol, 100 mM KH_2PO_4, pH 7.0, 10 mM β-mercaptoethanol (Sigma)

Optics

The instrument is based on an optical microscope equipped with a CCD camera (Fig. 2F; see also Bennink [2000][20]). For the optical trap, a 2-W infrared laser is introduced. The 3-mm diameter of the emerging laser beam

[20] M. L. Bennink, Ph.D. Thesis, Department of Applied Physics, University of Twente, Enschede, 2000.

with a Gaussian-shaped intensity profile is first expanded using two lenses in order to get a beam with a diameter of just over 2 cm. This beam is projected on the back aperture of a high NA water-immersion objective. The high numerical aperture, of the objective, in combination with the slight overfilling of the back aperture, ensures that there is enough intensity in high-angle peripheral rays (having an angle of 70° with the optical axis) that mainly contribute to the creation of the gradient force within the optical trap.[21] In many optical traps, oil-immersion objectives are being used because the additional refraction of light when it enters the flow cell causes relatively more intensity in peripheral rays, which will result in a stiffer trap. However, there is a major drawback in using oil objectives. The focal point is very close to the glass surface, which limits the depth at which one can trap. Moreover, since the shape of the focus depends strongly on the distance of the focal point to the glass, the trap stiffness will depend on this distance, making calibration very complicated. We use a water-immersion objective which makes it possible to trap beads up to ∼100 μm from the cover glass surface. Moreover, since the refractive index of the immersion fluid is equal to that within the flow cell, the trap stiffness is totally independent of the distance to the cover slide.

Using the same objective that is infinity-corrected for both imaging and trapping has an advantage over using two objectives on either side of the flow cell, one to create the optical trap, and the second one to make a microscopic image. The plane in which the probe beads are captured is at a fixed position with respect to the objective lens. Furthermore, since the objective is infinity-corrected and a second lens is needed in front of the CCD array to create an image, the object plane is also fixed with respect to the objective. Now you can position the second lens, L2 ($f = 170$ mm), such that the trapped bead is in focus, ensuring that the bead is always trapped at the same position and in focus on the CCD camera.

Measurement of forces acting on the trapped bead is achieved by detecting the deflection of the transmitted light.[22,23] Small displacements of the bead from the fixed position of the trap result in deflections of the laser beam from the optical axis. This deflected light is collected by a condenser lens and projected onto a quadrant detector. In a dual-beam set-up,[24] all transmitted and refracted light is captured, allowing direct derivation of force from the measured change in momentum. In our instrument, however, the condenser lens is only collecting a small cone of the refracted

[21] A. Ashkin, *Biophys. J.* **61,** 569 (1992).
[22] L. P. Ghislain and W. W. Webb, *Opt. Lett.* **18,** 1678 (1993).
[23] L. P. Ghislain, N. A. Switz, and W. W. Webb, *Rev. Sci. Instrum.* **65,** 2762 (1994).
[24] S. B. Smith, Y. Cui, and C. Bustamante, *Methods Enzymol.* **361,** 134 (2003).

and transmitted light (NA = 0.9), and this signal therefore needs experimental calibration of the force.

The position of the quadrant detector has been widely debated within the optical tweezers community. In our set-up, the detector is positioned just behind the back focal plane (NOTE 1). The exact position of the detector behind the back focal plane is not critical, but the beam has to be within the linear regime of the detector. To determine the linear regime, you simply put the entire detector onto an x–y translation stage and determine the signal as a function of the relative position of the beam on the detector.

The entire set-up is constructed on top of a horizontal vibration-isolated table to reduce external vibrations (within the building). The beam pointing stability of the laser is crucial to the measurements (NOTE 2). Furthermore, the noise due to airflow and dust can be reduced by shielding the beam and the optical components with a Plexiglas (perspex) box. The box encloses the beam emerging from the laser all the way to the detector, leaving an opening at the flow cell to allow easy access. The side branches for the white light illumination and the CCD array are not critical and therefore not shielded.

Flow Cell Construction

For the chromatin assembly and stretching experiments, we use a flow cell that allows exchange of buffers during the course of a single experiment. The cell is constructed manually using a standard microscope glass (26 mm × 76 mm). To create an entry and an exit for the flow cell, two small holes are drilled within the glass using a diamond drill in a benchtop drill press (both available in hardware stores) at positions indicated in Fig. 3A.

The next step is to use parafilm for spacing the two glass surfaces. Fold a piece of parafilm of at least 10 cm × 10 cm into a rectangular shape of 5 cm × 10 cm and press it with your fingers such that the two layers stick together. Use a scalpel to cut a rectangle with the dimensions of the microscope slide (you can put a slide on top of the parafilm and use it as a template). Now cut a long rectangle of 4 mm × 50 mm, at the position indicated in Fig. 3B. Take out this part and put the double layer of parafilm on top of a microscope glass with holes. The positions of the entrance and exit holes are exactly at the ends of the rectangle that is removed. Place the VitroCom square capillary on top of the parafilm; the length of the capillary should slightly exceed the width of the parafilm on the top. Use a scalpel to cut it to size (wear safety glasses). Finally, take a thin microscope glass (26 mm × 76 mm) and put this on top of the assembly to seal the flow cell.

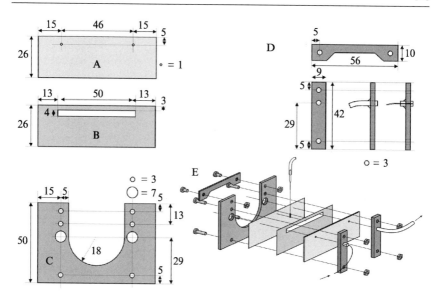

Fig. 3. The flow cell. The custom-made flow cell consists of a sandwich of a standard microscope glass with two holes (A), two layers of parafilm with a rectangular channel cut out (B), and a thin cover glass (170 μm thick) with the same dimension as the standard microscope glass. This entire sandwich is positioned onto a metal (i.e., iron) holder (C), and two plastic brackets are used to clamp the unit onto the holder (D). Inlet and outlet tubing are glued inside the holes drilled in the plastic brackets. The plastic tubing is pushed over the hole in the microscope glass and is actually clamping the sandwich on the holder (the plastic bracket is not touching the glass directly). This also ensures a leak-free connection to the flow channel. Finally, a second plastic bar (D) is attached on the top of the iron holder, which will serve as a support on which to glue the inserted micropipette. (E) Exploded view of the flow cell illustrating how the different parts should be assembled together. M3 nuts and bolts are used to clamp the brackets holding the flow cell in place and ensure a leak-free connection. The position of the small capillary between the parafilm and the thin cover slide is also indicated. The micropipette goes through the capillary into the flow cell, with its tip in the center of the flow cell. When inserted, the pipette is glued to the horizontal bar, which serves as a support.

The parafilm is not sticky enough to hold the two glasses together so you need to heat the sandwich to ~60° (NOTE 3) using a clean hot plate. Put the sandwich flat on the hotplate with the thick glass down. Put a separate microscope slide on top that allows you to push the entire sandwich gently together (if you do not use a second slide, you might break the thin cover glass). Note the transparency of the parafilm between the glass slides. It is grey and nontransparent at RT; at ~60° it starts to become transparent and sticky. Push with your finger at the point where the capillary is integrated until the glass is perfectly flat, that is, no distortion in the reflection. Take the sandwich off the plate and let it cool; the sandwich is sealed now.

To reuse the thick microscope glass with the holes, leave the sandwich in ethanol for at least a few hours. If you want to use it again right away, place it in ethanol for ~5 min, and put the point of a scalpel between the two glasses. Keeping the sandwich submerged in ethanol, gently push the scalpel edge between the two pieces of glass. Give the ethanol time to flow between the glass and the parafilm, and make sure that the thin glass slide is not bent too much. The sandwich will fall apart in a few minutes. Use a cotton-tipped applicator to clean the glass while submerged in ethanol.

For the holder you need:

- Iron plate (2 mm thick), cut as in Fig. 3C
- 4 bolts (M3 × 10) and nuts (M3)
- 2 bolts (M3 × 6) and nuts (M3)
- 2 Delrin plastic bars (3 mm thick) with three holes as indicated in Fig. 3D
- 1 Delrin plastic bar (1.5 mm thick) with two holes as indicated in Fig. 3D
- 3-mm outer diameter flexible tubing
- Thin fluid line (10–20 cm long)
- Glue (5 min-epoxy)

Place this glass sandwich in the iron holder. The U-shaped holder is cut out using a laser, but standard metal drilling and cutting tools can be used as well. Position the sandwich as indicated in Fig. 3E with the thin cover glass facing the iron. The holes created for the entrance and exit tubing should be exactly on top of the 7-mm holes within the iron holder. In between the glass and the iron, you can put one or two layers of parafilm (not shown in Fig. 3E) to prevent the glass from breaking. Cut out small holes at the position of the entrance and exit tubing holes in this spacer. This is important for aligning the tubing later.

Cut the 3-mm exit tubing with a very sharp knife such that the exit plane is exactly perpendicular to the tube axis; this will ensure a sealed connection to the glass later. The tubing is glued through the center hole in the bar. For the entrance tubing we use thin tubing for better control of the flow. This thin tubing is glued within the 3-mm tubing, which is making the connection to the glass (NOTE 4). This tubing is glued in the center hole of the second bar.

Use bolts (M3 × 10) and nuts (M3) to attach the bars over the sandwich (Fig. 3E). The sandwich is pushed against the holder by the tubing only, which is enough to keep it in place and to ensure a sealed connection. When looking from the backside through the 7-mm hole in the holder, you can see if the tubing is correctly aligned and sealed. Connect the second bar (Fig. 3D, top one) on the backside of the holder using two bolts (M3 × 6). This will be the holder for the glass micropipette.

Pulling and Placing Micropipettes in the Flow Cell; Mounting the Flow Cell into the Setup

For micropipettes, we use glass capillaries that we pull using a professional pipette puller that has four different parameters, which we optimized to produce long, tapered micropipettes with a tip diameter of 1–2 μm (the diameter of the micropipette is still less than 50 μm at ~10 mm away from the tip). Our optimized parameters are: heat, 323; pull, 100; velocity, 10; time, 240, but they may vary from one puller to the next (even of the same model), so use these numbers as a starting point and optimize them further. Start by breaking the glass tube in the middle (wrap the tube in a tissue) and clamp one piece in the instrument. After pulling, inspect the tip with a microscope, and if it looks good, gently break off the long tube, again holding it in a tissue. Then glue the micropipette to a ~20- to 30-cm long 3-mm tubing.

View the holder with the glass sandwich, the tubing and the top bar with a microscope. Use a metal support plate and small magnets to keep the flow cell in place. The tubing with the micropipette is mounted on a 3D translation stage, able to move objects with 10-μm accuracy over a range of a few centimeters. You need to move the tip of the micropipette into the square capillary that is connecting the outside world with the flow cell. Exert extreme care, since if you touch the tip head-on, it will break. Once you have the tip within the square capillary, you slowly move the pipette in until the tip is in the middle of the flow cell. Glue the tip to the plastic bar on top using two-component glue; the flow cell is ready to be mounted in the set-up.

The flow cell is positioned on a Newport 3D translation stage with both manual and piezo-electric control. An iron plate with a 4-cm diameter opening, through which the objective accesses the glass surface, is attached vertically to the stage and the flow cell is attached to it using small cylindrically shaped magnets (5 mm in height, 12 mm in diameter, available in different stores). This allows you to coarsely align the flow cell such that the tip of the micropipette is in front of the objective (NOTE 5).

Flow System

A semi-automated flow system controls the exchange of buffers within the flow cell and the flow rate (Fig. 4). The buffers are kept in containers ranging from 1 ml (for the cell extract) to 50 ml (for the normal flow buffer). All containers are screw-capped and have two holes drilled into

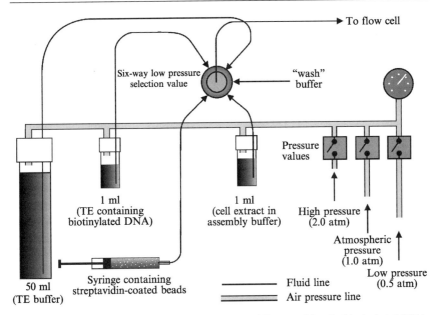

FIG. 4. The flow system used to introduce streptavidin-coated beads, biotinylated DNA, and the cell-free extract at a controlled flow rate. Different containers ranging from 1 to 50 ml are connected to a closed air-line, whose pressure is controlled using three computer-controlled pressure valves. These are connected to a high-pressure line (~2.0 atm), outside pressure (1.0 atm) and a barrel, in which low pressure is maintained (~0.5 atm). Switching these valves shortly will change the pressure in the line, and therefore the flow rate in the flow cell. Thin tubing is used to connect each container to a six-way selection valve, which specifies which container is to be used. The beads are introduced in the system using a syringe.

the cap. One hole is 3 mm in diameter, and a piece of 3-mm tubing (air-line) is inserted. A second hole, only 1 mm in diameter, provides access for a thin fluid line. The connections through the caps are sealed with 5-min epoxy. All airlines are connected to a closed air system. Increasing the pressure within the airlines will pressurize all the containers and will maintain buffer flow. The thin tubing is connected to a six-way low-pressure selection valve (NOTE 6), which enables the operator to manually select the desired container.

For containers we use transparent disposable vessels:

- 2-ml tube, screw cap, Nalgene cryovials, PP sterile, Nalgene, a subsidiary of Sybron, Rotherwas, Hereford, UK
- 12-ml tube, blue screw cap, Greiner bio-one, PS tube, sterile, diameter, 16.8 mm, height, 100 mm, Greiner BV, Alphen a/d Rijn, The Netherlands

- 50-ml tube, blue screw cap, Greiner bio-one CELLSTAR, PP tubes, sterile, skirt, diameter 30 mm, height 115 mm, Greiner BV
- 1- and 5-ml syringe, Terumo Europe NV, Leuven, Belgium
- Needles, Micro-Lance 3 (30G1/2), 0.3 mm × 13 mm, Becton Dickinson, Drogheda, Ireland

Changing the pressure within the airlines controls the flow rate and the direction of the flow. Three solenoid valves are connected to the airline; the first valve is connected to a high-pressure container (2.0 atm), the second valve is connected to a low-pressure container (0.5 atm), and the third one is left open. Opening the valves for a short time will result in a stepwise increase or decrease of the pressure within the system, thus controlling the flow rate. Opening the third valve will restore the pressure in the system to atmospheric level, stopping the flow.

The containers containing the beads, the DNA, the cell extract, and the flow buffer used to flow the DNA molecules in are connected to this system before an experiment starts. A wash buffer containing 1% Tween is used to wash the flow cell after each experiment. The beads are introduced using a syringe coupled through a needle to a fluid line connected to the selector valve (Fig. 4). A stirring magnet (5 mm long) is put into the syringe to keep the beads in suspension by moving an external magnet along the syringe. Do this just before you enter new beads into the flow cell.

Preparation of Streptavidin-Coated Beads, End-Biotinylated DNA, and Cell Extract

Streptavidin-coated beads are used as attachment surfaces for the ends of the DNA molecule. On commercial available beads, the amount of streptavidin coating the surface is too low; for this reason, we use the following protocol to obtain streptavidin-coated beads.

1. Prepare: (i) 100 ml 50 mM phosphate buffer, pH 7.4, containing 0.9% NaCl (PBS) and (ii) 100 ml 50 mM MES, pH 6.0
2. Dissolve 1 mg streptavidin in 0.5 ml MES (filtered through 0.2 μm pore-size filters) and add this to 2 ml 2% CML beads
3. Shake this mixture for 15 min at RT
4. Add 4 mg EDAC and vortex
5. Adjust the pH to 6.5 using a 0.1-M NaOH (usually you need ~50 μl)
6. Incubate for 2 h or overnight at RT (on shaker)
7. Add glycine (to a final concentration of 100 mM) to block unreacted sites

8. Centrifuge for 10 min at 10,000 rpm (Eppendorf centrifuge) and resuspend the beads in 50 mM PBS solution. Do this three times
9. Resuspend in 2 ml PBS, 1% BSA, 0.013% NaN$_3$. This suspension is kept as stock. For an experiment, add 5 μl of the stock to 1 ml of the main flow buffer (10 mM Tris-HCl, pH 7.5, 1 mM EDTA, 150 mM NaCl, 0.05% BSA, and 0.01% NaN$_3$). This final bead suspension has a concentration of 10^5 beads per milliliter and can be loaded into the syringe.

We use λ-DNA, a dsDNA fragment, 48,502 bp with 12-nucleotide ss overhangs on either end that are complementary to each other. The ends of the DNA fragment are biotinylated by filling in with Klenow polymerase.

Buffers and Solutions

- 50 mM Na$_2$HPO$_4$
- 1 M MgCl$_2$ (autoclave both solutions and mix 1 ml Na$_2$HPO$_4$ with 10 μl MgCl$_2$ to get the reaction buffer: 50 mM Na$_2$HPO$_4$, 10 mM MgCl$_2$, pH 7.5)
- Stop solution (0.1 EDTA, pH 8.0)
- TE buffer (10 mM Tris-HCl, 1 mM EDTA, pH 7.5)

Protocol

1. Incubate 100 μl (50 μg) of λ-DNA in TE buffer for 5 min at 65°.
2. Add:

   - 1 ml reaction buffer
   - 0.5 μl dNTPs (100 mM dGTP, dCTP)
   - 1.25 μl bio-dATP
   - 1.67 μl bio-dUTP
   - 1.67 μl Klenow fragment (10 units)

3. Incubate for 3 h at 37°.
4. Add 0.2 ml 0.1 M EDTA, pH 8.0.
5. Purify with Centricon concentrators (use a fixed-angle centrifuge, centrifuge for 20 min at 1000g [2200 rpm]; use a maximum of 2 ml per concentrator).
6. Wash three times with 0.8 M TE buffer.
7. Put the concentrator upside down and centrifuge once at 2200 rpm for 2 min.
8. Add 2 ml TE buffer.

The concentration of the labeled DNA in the stock solution is 25 μg/ml. The ratio A_{260}/A_{280} should be 1.8–1.9. For each experiment, 10 μl of the DNA stock solution is added to 1 ml of TE buffer. Final DNA concentration is 250 ng/ml, 4.5×10^9 molecules per milliliter.

The cell extract is isolated from *Xenopus laevis* eggs.[25] Twelve microliters of high-speed supernatant is diluted in 1 ml of assembly buffer (50 mM HEPES-KOH, pH 7.6, 50 mM KCl, 1 mM EDTA, 2 mM β-mercaptoethanol).

Attaching a Single DNA Molecule Between Beads (Fig. 5)

The main flow buffer (10 mM Tris-HCl, pH 7.5, 1 mM EDTA, 150 mM NaCl, 0.05% BSA, and 0.01% NaN$_3$, used during the attachment of the single DNA molecule to the two beads and during the stretching of the chromatin fiber) is connected to the flow system last. Pressurize the system to fill all fluid lines and remove any air in the system. Use inputs of the selection valve that are next to each other. Turning the selector to an input that is not connected is risky, since that might introduce air bubbles. Flow the main buffer for a few minutes to ensure that all beads, DNA, and proteins from the cell extract (from pre-flowing all the buffers) are removed from the flow cell.

Introduce a very small amount of the bead suspension into the flow cell by moving the piston in only 1–3 mm. (It is useful to make a small holder, e.g., out of Delrin plastic, in which to secure the syringe.) Switch back to the main buffer and put some pressure on the system ($v_{flow} = 1$–2 mm/s). At this moment you must have the trapping laser beam on; the quadrant detector will detect a relatively stable signal. As the beads enter the area within the field of view of the CCD camera, the signal on the quadrant detector displays sharp peaks, resulting from beads that move through the laser beam. Stop the flow when these peaks are visible.

Using the piezo-controlled flow cell in combination with the flow control, you can guide beads toward the position of the optical trap (Fig. 5A). Try to trap beads that are above the focal plane, that is, closer to the microscope objective. Beads below the focal plane will be pushed further down when positioned in the laser beam and thus impossible to capture. Position your objective as close to the glass as possible, increasing the chance of capturing a bead. When the bead is captured, move the flow cell such that you can transfer the bead onto the tip of the micropipette (Fig. 5B). Position the pipette ∼0.5 μm from the trapped bead and apply under-pressure using the syringe connected to the micropipette (in some cases capillary forces will make the bead jump onto the pipette). Repeat the procedure for capturing a second bead in the optical trap (Fig. 5C);

[25] G. H. Leno, *Methods Cell Biol.* **53**, 497 (1998).

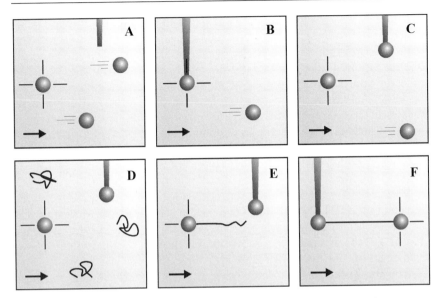

FIG. 5. Schematic presentation of how a single DNA molecule is attached between two polystyrene beads. Arrows indicate the direction of the flow inside the flow cell. The four lines forming a cross mark indicate the position of the optical trap. (A) The 2.6-μm polystyrene beads are introduced into the flow cell, and one of them is captured in the optical trap. (B) The micropipette is moved toward the trapped bead to transfer the bead onto its tip. Suction is applied to hold the bead in place. (C) A second bead is captured in the optical trap. (D) Buffer containing DNA molecules replaces the buffer containing the beads. (E) A single DNA molecule attaches to the trapped bead and stretches out in the flow direction. (F) The bead on the micropipette is moved at a distance of about 16 μm (i.e., length of the DNA) downstream from the trapped bead to connect to the free end of the DNA molecule. The presence of the invisible DNA molecule suspended between the two beads is attested to by moving the pipette away from the trap position and observing the trapped bead being pulled out of its position.

once a second bead is captured, increase the flow rate to flush the other beads out of the flow cell. To speed up this procedure, you can position the micropipette in front of the trapped bead. This will reduce the drag force on the trapped bead, thus allowing the use of higher flow rates.

When all beads have disappeared, stop the flow completely by turning the selector valve between two inputs. The signal from the quadrant detector shows a noisy pattern, representing the Brownian motion of the optically trapped bead. This signal should be centered around zero for both x and y directions (for this alignment, it is convenient to have the quadrant detector positioned onto a coarse translation table). Set the flow rate to 1 mm/s (at the position of the trap); a constant force of ~30 pN will now be detected, which is a direct result of the drag force on the trapped bead. Switch to the container with DNA molecules, let the DNA solution flow for 5-10 s, and then reverse to the main buffer. After ~1 to 1.5 min, the DNA is around the position of the trapped bead (Fig. 5D). When a DNA molecule attaches to the trapped bead, there is a jump in the force because the molecule causes a significant increase in drag force (~10 pN). If no jump is detected, repeat the procedure until you observe an attachment. During attachment, a constant flow is maintained to prevent attachment of the open DNA end to the trapped bead, resulting in a looped molecule. Only when both ends are attached (see next step) to separate beads can you stop the flow.

The last step is to attach the second bead (on the micropipette) to the free end of the DNA molecule. Move the flow cell back and forth such that the pipette is at a distance of ~16 μm downstream from the trapped bead (Fig. 5E). If the molecule becomes tethered, moving the micropipette away from the trap position will result in a force increase. Be careful not to destroy the system at this point, since it is rather easy to pull the bead out of the trap. You can move the micropipette to the other side and pull the bead out of the trap in an upstream direction (Fig. 5F).

To ensure that just one individual DNA molecule is attached between the two beads, the force-extension curve is recorded: the force is measured while the pipette is moved away from the trapped bead. Video analysis using a frame grabber and custom-written software applying a centroid method (NOTE 7) is used to determine the distance between the two beads, that is, the end-to-end distance of a DNA molecule (NOTE 8). The typical force-extension curve for a single molecule of dsDNA[26,27] exhibits a characteristic horizontal force plateau at ~65 pN. Having multiple DNA molecules attached in a parallel fashion would shift the plateau to higher forces: ~130 pN for two molecules, ~195 pN for three, etc. Other features of the force curve, such as the length of the molecule or the force development at short distances, are used to determine whether the DNA molecule is free of bound protein molecules. If this curve deviates from the expected curve, the molecule is discarded.

Assembly of Chromatin; Stretching and Relaxing of the Chromatin Fiber

To assemble chromatin fibers directly in the flow cell, we introduce the *Xenopus laevis* egg extract, a crude cellular fraction containing cell debris that is easily attracted onto the surface of the trapped bead (attempts to

---

[26] P. Cluzel, A. Lebrun, C. Heller, R. Lavery, J. L. Viovy, D. Chatenay, and F. Caron, *Science* **271,** 792 (1996).

[27] S. B. Smith, Y. Cui, and C. Bustamante, *Science* **271,** 795 (1996).

filter these out resulted in loss of activity of the cell extract). Therefore, the trap must be turned off (i.e., laser is blocked), and the beads are kept separate using a constant buffer flow. The tension within the DNA can be estimated from the flow rate using Stoke's law, but a correction has to be made for the shielding effect of the micropipette and the bead attached to it. An independent method to determine the tension within the molecule is to determine the Brownian motion of the freely suspended bead. Video analysis software is used to extract the positions of the two beads at a 25-Hz time resolution (i.e., video rate). These data are used to calculate the mean square displacement of the bead perpendicular to the direction in which the DNA molecule is pulled. This value is directly related to the force on the DNA molecule, with $F = lk_BT/<\Delta x^2>$, where l is the length of the biopolymer, k_BT the thermal energy, and $<\Delta x^2>$ the mean square displacement.

Thus, the assembly is performed under constant flow conditions and observed as shortening of the apparent end-to-end distance as a result of the formation of nucleosomal particles[28] ($\sim$240 nucleosomal particles along the length of the λ-DNA molecule). Carefully controlling the flow rate allows the observation of chromatin fiber assembly as a function of the tension within the DNA.[28]

To stretch the reconstituted chromatin fiber, the cell extract is replaced by the main flow buffer at a low flow rate so that the force does not exceed 5 pN. At this point, the trap is turned on, its intensity being slowly increased from zero to ensure that the free bead is gently caught in the trap, again avoiding large forces on the suspended chromatin fiber. Next, the micropipette is moved away at a constant velocity (typically 1 μm/s) while the force generated in the fiber is continuously monitored as its length increases. When the distance between the two beads reaches $\sim$20–22 μm (the contour length of the ds λ-DNA molecule is $\sim$16 μm, so the DNA at this point is overstretched), the direction of the micropipette movement is reversed to relax the molecule at the same speed.

The deflection signal representing the force exerted on the chromatin fiber is measured using a LabView data acquisition card at a frequency of 1.2 kHz; such a frequency allows the observation of very small steps of tens of nanometers. In the DNA stretching experiments, the end-to-end distance measurements are performed using real-time video analysis of the precise positions of the two beads, thus limiting the sampling rate to a maximum of 25 Hz. For the chromatin stretching experiments, we needed sampling at higher frequencies. Since the micropipette bead is moved away at a

[28] M. L. Bennink, L. H. Pope, S. H. Leuba, B. G. de Grooth, and J. Greve, *Single Mol.* **2,** 91 (2001).

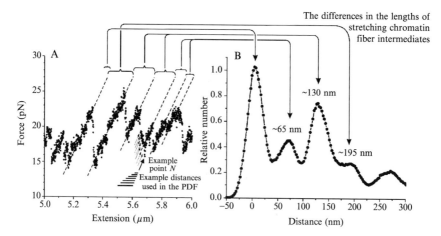

FIG. 6. Analysis of the opening events. (A) Enlargement of a portion of a force-extension curve. Portions of the curve, in which the force is continuously rising, then falling abruptly, can be clearly distinguished, indicative of opening events. The step sizes involved in the opening events can be analyzed using a pairwise distribution function. Determining the slope of the apparent linear portions (i.e., stretch modulus) and assuming that this parameter does not change significantly from one chromatin structural intermediate to the next, the contour length of the fiber at each data point can be determined. Then, the contour length data can be analyzed by a pairwise distribution function. This function determines all distances between any two points, that is, for each point N, the distances to points $N-1$, $N-2$, etc., as well as to $N+1$, $N+2$, etc., are determined; then a frequency distribution graph is plotted. The large arrow indicates a typical point N, and the smaller arrows indicate points $N-1$, $N-2$, $N-3$, etc. The horizontal lines are the distances between point N and $N-1$, $N-2$, $N-3$, etc. (B) The analysis on the data in this force-distance curve results in a multi-peak histogram, in which peaks are found at 65 nm, and multiples thereof, indicating that there is a discrete opening length during stretching; the presence of the peaks at ~130 and ~195 nm reflect two and three simultaneous opening events, respectively. The brackets above the graph in (A) indicate the major occurrences of distances that account for the major peaks in (B), and connecting arrows help show this relationship.

constant speed, an interpolation algorithm can be successfully used to determine the position of this bead at 1.2 kHz. The precise position of the second (trapped) bead is calculated from the acquired deflection signal using the trap stiffness. As depicted in Fig. 2D, the force acting on the trapped bead and its displacement from the center of the trap are coupled with the trap stiffness. The position of that bead, and thus the length of the chromatin fiber, can therefore be obtained at a 1.2-kHz frequency.

The force-extension curve of a chromatin fiber looks quite different from that of the naked DNA molecule, measured before the addition of the cell extract (Fig. 6A; see also Fig. 3 of Bennink, Leuba et al. [2001][29]).

[29] M. L. Bennink, S. H. Leuba, G. H. Leno, J. Zlatanova, B. G. de Groot, and J. Greve, Nat. Struct. Biol. 8, 606 (2001).

The tension within the chromatin fiber starts to develop already at $\sim$2 to 3 μm extension, apparently revealing a wormlike chain behavior as has been observed for the naked DNA molecule. When the tension reaches $\sim$20 pN, a sudden drop in force is observed on further extension, indicating a domain opening event within the chromatin structure. Multiple force peaks are observed on further extension, indicating multiple opening events. These structural rearrangements continue until the length of the chromatin fiber reaches the contour length of the naked DNA molecule. From that point onward the structure behaves like a DNA molecule without any bound histones, that is, the force-extension curve exhibits the characteristic 65-pN plateau. During the relaxation part of the cycle, the behavior is that of naked DNA, showing no signs of bound proteins.

Analysis of the Opening Events

To get more detailed information on the exact structural rearrangements within each opening event, the length increase involved is determined. For this, each ascending part of the force signal must be fitted to a modified wormlike chain model.[30,31] This model includes both entropic and enthalpic (i.e., intrinsic) elasticity of the polymer:

$$F = \frac{k_B T}{L_p} \left[\frac{1}{4(1 - (x/L_0) + (F/S))^2} - \frac{1}{4} + \frac{x}{L_0} - \frac{F}{S} \right],$$

where L_p is the persistence length, L_0 the contour length, x the end-to-end distance of the polymer, $k_B T$ the thermal energy (4.1 pN nm), and S is the stretch modulus, defined as the force needed to extend the polymer to twice its contour length.

However, an accurate fit to this model is impossible because data points are available only between $\sim$15 and 40 pN. Within this force range, the entropic contribution to the elasticity of the polymer is negligible, which reduces the expression to $F = (S/L_0)x - S$. This expression is used to fit the linear portion of the curve before each drop. Using stretch moduli of several discernible fiber intermediates along the stretch curve and assuming that this parameter does not change significantly on a single opening event, a contour length for each fiber intermediate can be determined.

Plotting the contour lengths as a function of time clearly reveals the stepwise increase in length, on each opening event.[29] A powerful method to determine whether the observed steps are quantized is the pairwise

[30] J. F. Marko and E. D. Siggia, *Macromolecules* **28,** 8759 (1995).
[31] M. D. Wang, H. Yin, R. Landick, J. Gelles, and S. M. Block, *Biophys. J.* **72,** 1335 (1997).

distribution function (Fig. 6). This function bins all possible differences between each pair of calculated contour lengths in the data set and plots the result as a histogram. The presence of peaks in this histogram is a clear signature of a quantized step present in the data. Figure 6B reveals a peak at 65 nm, and multiples thereof (i.e., 130 nm, 195 nm).[29] There is also a strong peak around zero, which is a direct result of the pairwise distribution function: Many data points that originally are on one rising part of the curve are reduced to about the same value for the contour length, producing differences close to zero.

Magnetic Tweezers

Materials

- Square glass cuvettes, 1 mm × 1 mm × 50 mm (VitroCom, Mountain Lakes, NJ, USA, www.vitrocom.com)
- 2.8-μm streptavidin-coated magnetic beads (Dynabeads, Dynal, Oslo, Norway, www.dynal.no)
- Steel ball bearing (Allied Industrial Technologies, Cat. No. 100KSFF, H401)
- Planetary-geared 12 V DC motor, 30 rpm (Cramer, Cat. No. 800HN-DC-3277, available from Digi-Key, Thief River Falls, MN, USA, Cat. No. CRA203-ND, www.digikey.com)
- Elenco Precision regulated power supply, model XP-620 (1.5–15 V)
- Rubber O-rings 6.7 cm × 7.3 cm × 0.3 cm (2–5/8 in × 2–7/8 in × 1/8 in) (Allied Industrial Technologies, Cat. No. 01-231 or 01-146)
- NdFeB/NIB magnets (Indigo Instruments, Waterloo, Canada, www.indigo.com, Cat. No. 33512)
- 0.8-mm inner diameter silicon tubing (BioRad, Hercules, CA, USA, Cat. No. 731–8210)
- 3-Com HomeConnect video camera (obsolete; try www.ebay.com alternative digital cameras from Orange Micro, Anaheim, CA, USA, www.orangemicro.com, and Logitech, Fremont, CA, USA, www.logitech.com)
- Six-way low-pressure bulkhead selection valve (Upchurch Scientific, Oak Harbor, WA, USA, Cat. No. V-241)
- Low-pressure injection valve (Upchurch, Cat. No. V451)
- Seiwa Optical America objective micrometer (Cat. No. RET-PCS81X, available from Fisher Scientific, Pittsburgh, PA, USA, Cat. No. #S11144)

- Videopoint software (Pasco Scientific, Roseville, CA, USA, www.pasco.com) for manual particle tracking of a bead in QuickTime movies in NIH Image
- Chemicals: 3-aminopropyltriethoxysilane (APTES) (Sigma, St. Louis, MO, USA); toluene, research grade; NP40 (Igepal) (Sigma) blue dextran (Sigma); mPEG (polyethylene glycol)-NH_2 (5000 Da) (Shearwater, now Nektar, Huntsville, AL, USA); PEG (3350 Da) (Sigma)

Instrumental Setup

In the horizontal magnetic tweezers,[32] a white light source (either a fiber optic or a 150-W bulb) is focused onto a square glass cuvette (Fig. 7A) through a condenser lens, placed roughly halfway between the light source output and the cuvette (focusing can be facilitated by using a white postcard to view the beam of light at the cuvette). The beam passes through a hollow steel bearing on which a rare earth magnet is placed. The cuvette is held by a machined piece of Delrin plastic bridge (Fig. 7B and C) bolted onto the three-axis flexure stage. We have fashioned a paperclip (not shown) to hold the cuvette tightly in place in the Delrin bridge. The focused light is collected by an objective juxtaposed to the cuvette. Most experiments were performed with 90× oil (1.2 NA) or 40× air (0.6 NA) objectives. If the objective has a fixed focal plane, typically a distance of 160 mm, then the video camera is placed 160 mm behind the objective. Inexpensive (~$100) digital cameras can be used. These cameras typically have either USB or Firewire direct connections to a PC and thus do not need an extra analog/digital computer board. Public-domain software such as NIH Image or ImageJ can be used to collect movies in AVI format.

The magnification of the image on the video screen is determined by the magnification of the objective and the distance from the camera to the objective. There are a couple of ways to calibrate the number of nanometers per pixel in the image. One can use the dimensions of the 2.8-μm magnetic bead for rough calibration. More accurate calibration can be done with an objective micrometer; the micrometer is a grid of fine lines on a glass slide typically separated by distances of 10 μm. This gridded slide is placed at the position of the cuvette, and images are captured with the video camera located at positions used during an actual experiment. It is possible to slide the camera toward the objective to increase the field of

[32] S. H. Leuba, M. A. Karymov, M. Tomschik, R. Ramjit, P. Smith, and J. Zlatanova, *Proc. Natl. Acad. Sci. USA* **100**, 495 (2003).

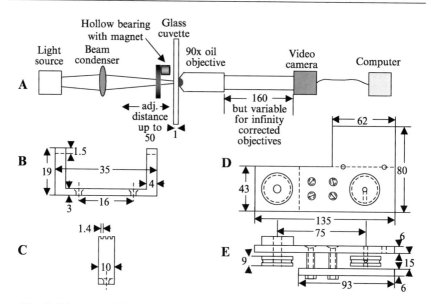

Fig. 7. Schematic of the horizontal version of the magnetic tweezers set-up and some technical parts of the magnetic tweezers instrument. This schematic in (A) indicates some important distances (in millimeters). The hollow bearing with the mounted magnet is movable up to 50 mm away from the square glass cuvette in order to adjust the force on the magnetic bead. (B) Side-view drawing of the Delrin bridge to hold the square glass cuvette. (C) Cross-section at one end of the Delrin bridge. The three 1.4-mm furrows at the top of the piece are the three possible locations of the cuvette. (D) Side-view drawing of the DC motor-driven assembly (fabricated out of aluminum plate and Teflon pulleys) for rotating the external magnet(s) on the embedded bearing. Upper right side of drawing is space on aluminum plate to mount DC motor that directly rotates the pulley on the right. The two pulleys are connected by a rubber O-ring. (E) Bird'seye view of the same assembly. In this view it is possible to see that the pulley on the left is directly connected to the hollow ball bearing that is mounted on the upper left side of the aluminum plate in (E). Dimensions are in millimeters.

view. This allows observation of a larger number of beads, finding one that is tethered, centering it on the screen, and then enlarging the image by manually moving the camera away from the objective. This is a flexibility advantage of our horizontal MT set-up over the more traditional set-up with the magnet and the cuvette positioned above the objective of an inverted fluorescence microscope. In that set-up, the distance between the eyepiece (or CCD camera) and the objective is fixed (160 mm); thus, a video image can only be zoomed in and out by changing the objective; in doing so one often loses the bead of interest.

DC Motor/Bearing Set-Up to Rotate Magnet (Fig. 7D and E) and Gravity Flow System

The planetary-geared motor runs a pulley connected to another pulley via a rubber O-ring. Switching the electrodes in the power supply runs the motor in the opposite direction, thus allowing to supercoil topologically constrained DNA molecules either positively or negatively.

Gravity-fed flow systems consistently provide the same speed and avoid undesirable vibrations of peristaltic pumps. Half a dozen buffer bottles, placed ∼70 cm above the instrument, are connected to a six-way low-pressure valve. The outflow from the valve goes to a low-pressure injection valve, which has an injection loop of ∼200 μl, that can be used to inject the proteins of interest. The length of time it takes for the injected buffer to reach the cuvette can be checked by coloring the buffer with blue dextran, and looking for the appearance of blue color on the screen.

Incorporation of Biotinylated Nucleotides into DNA

The procedure is described in the optical tweezers section. To prepare DNA fragments that can be suspended between the surface and a bead in a topologically constrained manner, we use PCR-based procedures (see Strick [1999][33]).

Surface Treatment of Square Glass Cuvette: Activation of the Glass Surface with APTES (Steps 1–3), Coating the Walls of the Cuvette with Streptavidin (Steps 4–7), and Attachment of Biotinylated λ-DNA Carrying a Magnetic Bead to the Cuvette Walls (Steps 8–10)

Surface treatment of the glass cuvette is the major critical step in performing the experiment. The use of glutaraldehyde to link streptavidin to the APTES-coated surface is beneficial in stabilizing the attachment of the tethered complexes. The high stability of the connection to the surface allows us to routinely use 2 M NaCl to remove bound proteins and utilize the same tethered DNA molecule for another experiment.

Buffers

- Phosphate-buffered saline (PBS) (150 mM NaCl, 3 mM KCl, 2 mM KH_2PO_4, 10 mM NaH_2PO_4, pH 7.2)
- Washing buffer (used to wash out the unattached beads in the cuvette after it is assembled into the instrument): PBS with 1% NP40, 0.2% PEG (3350 Da), 1 mM EDTA, 0.2% biotin

[33] T. Strick, Ph.D. Thesis, Department de Physique, Ecole Normale Superieure, Paris, 1999.

- Blue buffer: PBS with blue dextran (just enough to see the blue), 1% NP40, 0.2% PEG, 2 M NaCl (this buffer is added to the DNA/beads and used to visualize their entry into the cuvette)
- Histone-dissociation buffer (used to dissociate histones from the DNA when chromatin fibers are to be disassembled; it can also used to rinse the DNA/beads in the initial stage of the experiment): PBS with 0.2% biotin, 1% NP40, 0.2% PEG, 1 mM EDTA, 2 M NaCl

1. Place six cuvettes in 5 ml of 1% APTES in toluene in a 15-ml polypropylene tube in a rotator at RT for 5 min.
2. Transfer the cuvettes to another tube with 10 ml toluene and rotate for 10 min.
3. Transfer the cuvettes to glass Petri dish and bake at 110° for 30–45 min. At this point cuvettes can be stored in the Petri dish for several days.
4. Pipette 10% glutaraldehyde from a freshly opened ampoule into a cuvette; after 5 min non-shaking incubation, rinse the glutaraldehyde out of the cuvette with a water bottle for 1 min.
5. Mix 4 μl of streptavidin solution (5 mg/ml) with 96 μl PBS, pipette the mixture into a cuvette, and let it sit for 10–20 min (we have reused the diluted streptavidin solution 5–10 times). (If the DNA is end-labeled with digoxigenin, a much longer deposition time, 20–60 min, is needed for the antidigoxygenin antibody to adsorb stably.)
6. Rinse the cuvette once with a water bottle.
7. Place the cuvette in 10 ml PBS containing 50 mM glycine and 0.1% mPEG-NH$_2$ (both reagents block the unreacted primary amino groups on the walls of the cuvette).
8. Mix in 0.6-ml microfuge tube 2 μl of 5 M NaCl and 1 μl of 2.8-μm streptavidin-coated Dynabeads suspension. Add 0.5 μl of biotinylated λ-DNA (50 μg/ml in 10 mM Tris-HCl, 1 mM EDTA, pH 7.5). Immediately add blue buffer to 50 μl total volume. Wait for 5–10 min ($\sim$2 M NaCl neutralizes the charges on the DNA phosphate backbone and allows streptavidin to bind to the biotinylated DNA more easily). The average number of DNA molecules per bead is 1/2.
9. Put the DNA/beads into the cuvette, connect tubing, and let sit for 3–5 min (NOTE 10).
10. The beads settle on the bottom of the cuvette, so the capillary needs to be rotated 90° around its longitudinal axis to visualize the bottom through the objective. Blue buffer is used to wash out free beads/

DNA, and washing buffer brings the NaCl concentration down to 150 mM before Brownian motion analysis is performed.

Locating a Tethered Magnetic Bead

With the flow turned off and the DC motor rotating the ball bearing to which the external permanent magnet is attached, it is possible to observe single magnetic beads rotating in a circle (irregular-circular or oval paths suggest more than one DNA molecule tethering the surface to the bead). Once a tethered bead is identified, the camera is moved further from the objective so that the area covered by the rotating bead occupies roughly the area of the video screen.

Calibrating the Force; Measuring the Change in Tether Length Due to Chromatin Assembly

With the flow turned off and the external magnet at a fixed position, Brownian motion analysis can be performed to determine the force acting on the single DNA molecule. We will explain the particle-tracking analysis with Videopoint, although much more sophisticated programs are available. Using the movie function of NIH image, 1000 frames of video are collected at 1 frame per second. The movie is opened within Videopoint, and the center of the tethered bead for all 1000 frames of the movie is mouse-clicked. (Manual clicking on 1000 video frames is tedious; it is possible to use a particle-tracking macro-function in ImageJ after thresholding the images to remove the background.) The table of x and y coordinates of the bead is imported into KaleidaGraph or Excel (or other spreadsheet program). The variance of the x-data points is automatically computed, and this variance is the $<\delta x^2>$ term of the equipartition formula (see above).

The first movie is collected with the magnet at the closest possible position to the cuvette. Next, the external magnet stage is moved 0.5 mm away from the cuvette, and another 1000-frame movie is collected; this procedure is repeated until the distance to the cuvette is 0.5 cm. At each distance, we determine $<\delta x^2>$. The other parameter needed to determine the force at any given distance of the external magnet to the cuvette is the tether length. Assuming that the DNA molecule is maximally extended (i.e., approaches its contour length of 16.4 μm) when the magnet is at its closest possible position to the cuvette, a maximum force of ~22 pN is calculated using the equipartition theorem. (Using 16 μm as the initial estimate for the length of the tether at this position of the magnet is a reasonable assumption, since at forces >3 pN the extension l is within ~10% of the contour length of λ-DNA.)[6] Once a first approximation force is calculated by using this value of l and the measured $<\delta x^2>$, the actual extension at this force is estimated by interpolation from the force-extension curve (Fig. 8 in ref. 6); a second approximation force is calculated by using this interpolated value of l. After one round of iteration, the accuracy of our measurements is $\sim$10% for forces >3 pN and $\sim$20% for forces in the range of 0.5–3 pN.

Once we have the calibration of forces at each distance of the magnet to the cuvette, all we need to do in the actual chromatin assembly experiments is determine the shortening of the DNA tether on formation of nucleosomes. To understand how this is done with our set-up, let us first explain how changes in tether length are determined in the laboratory of Croquette and Bensimon (e.g., ref. 5). In that instrument, two magnets are used instead of one. These magnets, spaced at 2 mm, create a homogeneous magnetic field for the tiny magnetic bead, and keep the bead always at the same x-, y-position on the screen (apart from the low-range Brownian motions around this position). Whatever happens to the tethered DNA molecule to change its length, the only observable change (apart from the change in the Brownian motion) is a change in the z-dimension, Δz.

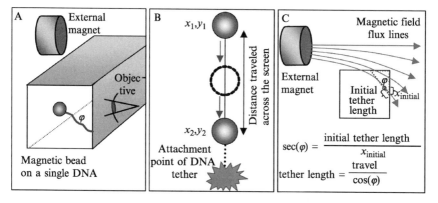

FIG. 8. Schematic to explain how the measurements of travel of the bead across the screen are done, and how the tether length is calculated. (A) Cross-section of the cuvette with DNA-tethered magnetic bead and the placement of the external magnet above the incident light; such a placement tilts the tether so that the angle φ is about 60°C. (B) Schematic of the positions of the magnetic bead in the video image on chromatin assembly when the only applied external force is the magnetic force; the only movement of the bead during assembly is in the x-dimension of the video screen. We have defined this movement of the bead across the screen as the "travel across the screen." (C) Schematic of how the angle φ is determined and how travel is converted to actual changes in tether length. The angle φ is determined by trigonometry using $x_{initial}$ and the initial length of the DNA ($\sim$16 μm). The actual tether length during the assembly/disassembly experiment is then calculated by dividing the travel (depicted in [B]) by the cosine of φ.

Following the movements of the bead in z, that is, in direction perpendicular to the focal plane is technically nontrivial and requires precalibration. Such calibration is based on observing the size of interference rings around the bead that become larger as the bead moves away from the focal plane.

We determine Δz using a different, somewhat simpler (but less accurate) approach that involves no prior calibration. We place the only external magnet not right above the sample, but to the side, so that the DNA tether is at an angle (typically $\sim 60°$) to the imaged surface (Fig. 8A). Since the movement of the bead on the screen is only in the x-direction (subject to only the magnetic force, once the flow is stopped) (Fig. 8B), the travel of the bead on the screen and knowledge of the angle are sufficient to determine Δz, using trigonometry (Fig. 8C). The external magnet pulls on the DNA tether at an angle, so it is important to keep the angle the same during the course of the experiment; this can be easily achieved by moving the external magnet in both the vertical and horizontal dimensions by two linear micrometer stages.

Assembling a Single Chromatin Fiber and Analysis of the Assembly

Core histones (final total concentration 0.1 mg/ml) and nucleosome assembly protein 1 (NAP-1; 0.2 mg/ml) in 150 mM NaCl, 10 mM Tris-HCl, pH 8.0, 1 mM EDTA are incubated at 37° for 30 min, diluted tenfold in the same buffer containing 1% NP-40, 0.2% PEG, and injected into the 200-μl loop of the injection valve. Using gravity flow, the mixture is allowed to flow into the cuvette. The position of the tethered bead is followed until the bead starts moving against the flow (typically 2–3 min), indicating that the assembly process has begun: the wrapping of the DNA around the histone octamer leads to shortening of the DNA tether. At this point the flow is turned off. The tethered bead is now subjected to external force from the permanent magnet only, and we can follow the travel of the bead across the video screen as a function of the applied external force (see above).

We can now study the rate of chromatin assembly as a function of applied force. The instrument allows to rheostatically change the force in the course of a single experiment by moving the external magnet further and further away from the cuvette in a stepwise fashion. Reducing the applied force resulted in an immediate increase in the rate of chromatin assembly.[32]

Notes

NOTE 1: In many set-ups the quadrant detector is positioned in the back focal plane of the condenser lens, or a position that is optically conjugate to this. When properly set up, the detector will be completely insensitive to lateral displacements of the optical trap and will only sense displacements of the bead from the trap center. In our set-up, the position of the trap is fixed, avoiding the need for an optical scheme that will render the deflection insensitive to trap movements. The quadrant detector is behind the focal point such that the ring-shaped illumination pattern is ~0.5 cm in diameter. This position, however, is not critical.

NOTE 2: We first used a Millennia 10-W CW laser, which has good beam pointing stability, but is expensive and large. The laser head is 44 cm × 15 cm × 15 cm and needs to be positioned on top of the optical table. The controller unit is 60 cm × 40 cm × 26 cm. Both are connected with a thick cable, which is not flexible and hard to install. Now, we use a 2-W infrared (1064 nm) laser (CrystaLaser) that is only 16 cm × 7 cm × 3 cm, with a controller box of only 15 cm × 15 cm × 5 cm. For a single trap, a 2-W laser is sufficient to exert forces of ~100 pN on a 2.6-μm polystyrene bead.

NOTE 3: To get ~60°, start with a hot plate that has not been used for a while. Put your finger on the plate and turn it on. Put the glass sandwich on the plate as well. Wait until it feels hot (you have to take your finger off) and turn off the hot plate. This is about the right temperature you need.

NOTE 4: Cut a piece of 3-mm tubing ~1 to 1.5 cm in length. Pull the thin tubing (10–20 cm in length) through this piece and fill it up with two-component glue (the thin tubing should be coming through; otherwise you risk closing the thin tubing with the glue). When solidified, use a sharp knife to cut the tubing perpendicular to its axis, creating an end that is able to make a good seal with the glass.

NOTE 5: The high NA objective used for optical trapping and for creating the microscopic image has a field of view of only 40 μm × 40 μm. This makes it tedious to find the tip of your micropipette. First align the micropipette (by eye) as well as you can in front of the objective. Then switch to the microscopic image and use the manual control on the stage to locate the edge of the flow channel. Scan along the edge until you see the capillary through which the pipette is entering the flow channel, and follow the pipette to the tip.

NOTE 6: For connecting tubing to the selector valve, there are small cone-shaped pieces, provided with the valve. These cones, however, do not fit the thin tubing that we use. For connecting the thin line, we use Micro-Lance needles. Using a very sharp scalpel, cut the needle into two pieces at the point where the thin metal tube ends in the plastic part. This will provide a smooth end-face in which the metal tubing ends in the center. Take the plastic gray screws that are on the selector valve and increase the size of the hole such that the needle is able to go through. The end-face is still slightly larger in diameter and the needle will stick out of the screw.

Now attach the thin line to the metal needle and clamp it in the selector by carefully screwing in the gray unit.

NOTE 7: For accurate position detection of the two beads, custom-written video analysis routines that use a centroid method have been developed. Beads imaged by the CCD array appear as light, circular objects surrounded by a black edge. The bead position is determined as an average of the positions of all the pixels that are within this black circular edge. For this, an algorithm starts at a pixel inside this edge and moves slowly outward in all directions, until a preset threshold level is reached (i.e., that is lower than the value of the edge). This results in 600–800 pixels that are next to each other, forming the inner part of the imaged bead. The x- and y-position of the bead is determined as the average x- and y-position of these pixels (accuracy of determination of $\sim$4 nm).

NOTE 8: The end-to-end distance is obtained by subtracting the radii of the two beads from the center-to-center distance as determined with the centroid method (see NOTE 7).

NOTE 9: In a beam-expanding system you normally use two lenses that are separated at a distance equal to the sum of their focal distances. Within this optical layout, the distance between the two lenses (L1 and L2 in Fig. 2F) is only 390 mm, instead of the expected 400. The beam coming off lens L2 is thus slightly converging. Using such a beam will create a clear image of the trapped bead, since the bead is trapped stably just below the focal point. In practice, correct aligning can be done by slightly moving lens L2 back and forth until the bead is in focus on the CCD array.

NOTE 10: Connecting the tubings in a way that avoids bubble formation requires practice. Passing bubbles appear to irreparably stick the tethered beads to the surface, making them useless. It is recommended to agitate the DNA-connected beads during the deposition process by making a couple of short flow pulses (turning on and off the flow valve).

Acknowledgments

We thank Drs. Mikhail Karymov, Miroslav Tomschik, Paul Smith, and Guoliang Yang, and Kirsten van Leijenhorst-Groener, Ravi Ramjit, and Waldemar Koscielny for help with experiments, Dr. Haocheng Zheng for assistance with figures, and Dr. Jan Bednar for discussion. This research has been supported by an NCI K22 grant and startup funds from the University of Pittsburgh School of Medicine (SHL), startup funds from Polytechnic University (JZ), and the Dutch Foundation of Fundamental Research (FOM) and MESA$^+$ Research Institute, University of Twente (MLB).

[7] Biochemical and Structural Characterization of Recombinant Histone Acetyltransferase Proteins

By RONEN MARMORSTEIN

Chromatin is modified in distinct ways to modulate gene expression.[1,2] These modifications are mediated by ATP-dependent nucleosome remodeling proteins that move the nucleosome cores along the DNA,[3,4] and histone-modifying enzymes that post-translationally modify the N-terminal histone tail regions.[5] Post-translational histone modifications have been proposed to effect gene expression through both steric/electrostatic effects and through a signal-transduction cascade that has been called the "histone code hypothesis."[6,7] These post-translational histone modifications include methylation, acetylation, phosphorylation, and ubiquitinilation (that occurs at the histone C-terminal tail).

The enzymes that regulate histone acetylation, called HATs, were one of the first to be characterized biochemically.[8] To date, over 20 different HAT proteins have been characterized and they fall into distinct families that generally have high sequence similarity within the catalytic HAT domain and related substrate specificity within families, but have divergent sequence and substrate specificity between families (Table I).[9] In addition, HAT proteins usually contain other conserved domains with autonomous transcription-related functions that also seem to be family specific, consistent with the ability of HAT proteins to coordinate other transcription-related activities.[1] *In vivo*, HAT enzymes are the catalytic subunits of multisubunit protein complexes, and distinct HAT proteins acetylate specific lysine residues within specific N-terminal histone tails.[10] There have been several reports showing that subunits of HAT complexes, among other transcription-related activities, play a modulatory role on the level of enzymatic activity and the specificity for substrate.[11–13] Nonetheless,

[1] R. Marmorstein, *Nat. Rev. Mol. Cell. Biol.* **2**, 422 (2001).
[2] Y. Zhang and D. Reinberg, *Genes Develop.* **15**, 2343 (2001).
[3] C. L. Peterson, *EMBO Reports* **3**, 319 (2002).
[4] K. E. Neely and J. L. Workman, *Biochim. Biophys. Acta* **1603**, 19 (2002).
[5] P. A. Grant, *Genome Biol.* **2**, REVIEWS0003 (2001).
[6] B. D. Strahl and C. D. Allis, *Nature* **403**, 41 (2000).
[7] S. L. Schreiber and B. E. Bernstein, *Cell* **111**, 771 (2002).
[8] J. E. Brownell, J. Zhou, T. Ranalli, R. Kobayashi, D. G. Edmondson, S. Y. Roth, and C. D. Allis, *Cell* **84**, 843 (1996).
[9] R. Marmorstein, *Cell. Mol. Life Sci.* **58**, 693 (2001).
[10] D. E. Sterner and S. L. Berger, *Microbiol. Mol. Biol. Rev.* **64**, 435 (2000).

TABLE I
HAT FAMILIES AND THEIR TRANSCRIPTION-RELATED FUNCTIONS

HAT	Organism	Function	Histone[a]
GCN5/PCAF family			*H3 (K14)*
Gcn5	Yeast to human	Coactivator (adaptor)	
PCAF	Human	Coactivator	+ Non-histone proteins
MYST family			*H4*
Sas2	Yeast	Silencing	
Sas3	Yeast	Silencing	(H3)
Esa1	Yeast	Cell cycle progression	
MOF	Fruit fly	Dosage compensation	
Tip60	Human	HIV Tat interaction	
MOZ	Human	Leukemogenesis	
HBO1	Human	Origin recognition interaction	
TAFII250 family	Yeast to human	TBP-associated factor	*H3*
CBP/p300 family	Worm to human	Global coactivator	*All + non-histone proteins*
SRC family	Mice and human	Steroid receptor coactivators	*H3/H4*
SRC-1		coactivators	
ACTR/AIB1/pCIP/TRAM-1/RAC3			
SRC-3			
TIF-2			
GRIP1			
ATF-2	Yeast to human	Sequence specific DNA-binding activator	*H4, H2B*
HAT1 family	Yeast to human	Replication-dependent chromatin assembly (cytoplasmic)	*H4*

[a] Only preferred histone substrates are indicated.

[11] R. Balasubramanian, M. G. Pray-Grant, W. Selleck, P. A. Grant, and S. Tan, *J. Biol. Chem.* **277,** 7989 (2002).

[12] L. A. Boyer, M. R. Langer, K. A. Crowley, S. Tan, J. M. Denu, and C. L. Peterson, *Mol. Cell* **10,** 935 (2002).

[13] P. A. Grant, A. Eberharter, S. John, R. G. Cook, B. M. Turner, and J. L. Workman, *J. Biol. Chem.* **274,** 5895 (1999).

considerable information on HAT function has been derived from biochemical and structural characterization of the isolated HAT domains. This chapter will focus on the biochemical and structural characterization of recombinant HAT proteins. In particular, I will focus on the methodologies used to prepare HAT proteins for such studies and highlight what we have learned from these studies. Finally, I will discuss what we have still to learn about the activity of HAT proteins.

Identification of HAT Domains for Overexpression in Bacteria

Bacterial expression systems have been effectively used to overexpress the catalytic domain of several HAT proteins. In my laboratory, the expression of intact HAT protein or proteins containing additional domains tethered to the catalytic domain, for example, the bromodomain, has resulted in the expression of protein that is largely insoluble or, if soluble, proteolytically cleaved between the HAT domain and the other associated domain. HAT proteins from the Gcn5/PCAF family were the first nuclear HAT proteins to be overexpressed for biochemical and structural studies.[14–16] One often uses sequence homology within a family of related proteins to define structural domains, and this strategy was effectively used for overexpressing HAT domains from the Gcn5/PCAF family for biochemical analysis. The high degree of sequence homology among the Gcn5/PCAF family of HAT proteins within a region that colocalized to a region that was mutationally sensitive for HAT activity and transcription[17,18] was used to define the structural catalytic domain (Fig. 1A). The HAT domain for overexpression included this region of homology and 2–3 additional flanking residues. The proteins and amino-acid boundaries chosen for overexpression were yeast Gcn5 (yGcn5, residues 99–261), human PCAF (hPCAF, residues 493–658) and *Tetrahymena* GCN5 (tGCN5, residues 48–210). In the case of *Tetrahymena,* each of the UAG codons, which code for Gln in *Tetrahymena* but code for STOP in bacteria, had to be changed to CAG, thus coding for Gln when tGcn5 was expressed in bacteria.[16]

[14] A. Clements, J. R. Rojas, R. C. Trievel, L. Wang, S. L. Berger, and R. Marmorstein, *EMBO J.* **18,** 3521 (1999).

[15] R. C. Trievel, J. R. Rojas, D. E. Sterner, R. Venkataramani, L. Wang, J. Zhou, C. D. Allis, S. L. Berger, and R. Marmorstein, *Proc. Natl. Acad. Sci. USA* **96,** 8931 (1999).

[16] J. R. Rojas, R. C. Trievel, J. Zhou, Y. Mo, X. Li, S. L. Berger, D. Allis, and R. Marmorstein, *Nature* **401,** 93 (1999).

[17] M. H. Kuo, J. X. Zhou, P. Jambeck, M. E. A. Churchill, and C. D. Allis, *Genes Develop.* **12,** 627 (1998).

[18] L. Wang, L. Liu, and S. L. Berger, *Genes Dev.* **12,** 640 (1998).

Each of the HAT domains was cloned into the PRSET plasmid vector for overexpression in *Escherichia coli* BL21 (DE3) cells. Induction at 37° with IPTG resulted in robust protein induction; however, upon cell disruption by sonication, the majority of protein was found in the insoluble protein fraction. In contrast, if the protein was induced with IPTG and induced overnight at 15°, a significant fraction of the protein (between 25% and 50%) was found in the soluble protein fraction. Although the reason for this is not clear, it has been proposed that the lower temperature allows the protein to fold more slowly, resulting in more "native like" folding and less improper folding that typically leads to protein aggregation.

My laboratory was also able to overexpress the HAT domain from members of the MYST HAT family that were suitable for biochemical study[19]; however, this has not been as straightforward as with the HAT domains from the Gcn5/PCAF family. As with the Gcn5/PCAF HAT domain, the regions chosen for overexpression relied heavily on sequence homology among the MYST HAT members (Fig. 1B). Various deletion constructs of the yeast Esa1 members of the MYST HAT family were prepared and bacterial extracts containing these deletion mutants were screened for HAT activity. This analysis localized the HAT active domain to residues 147–445 (the C-terminus of the protein). Induction of a yEsa1 protein construct at 15°, harboring residues 147–445, showed good expression of soluble protein; however, the protein showed visible degradation during purification. Inspection of the sequence conservation within the MYST family revealed that residues 147–162 and 434–445 were poorly conserved, and based on this we prepared a yEsa1 construct harboring residues 160–435 that was amenable to the overexpression of intact soluble protein that was suitable for biochemical and structural studies. Subcloning analogous regions of the human homologue of yEsa1, we have more recently been able to overexpress hMOZ and hTIP60 in bacteria (unpublished). However, in these cases the proteins were prepared as N-terminal 6 × His-fusion proteins and also contained a C-terminal KKK sequence to improve protein solubility.

Purification of HAT Proteins

Purification of the Gcn5/PCAF HAT domains took advantage of the fact that these proteins use the acetyl-Coenzyme-A (Ac-CoA) cofactor and employed a CoA agarose affinity column in the purification.[14–16] Chromatography on CoA-agarose was followed by cation exchange chromatography

---

[19] Y. Yan, N. A. Barlev, R. H. Haley, S. L. Berger, and R. Marmorstein, *Mol. Cell* **6,** 1195 (2000).

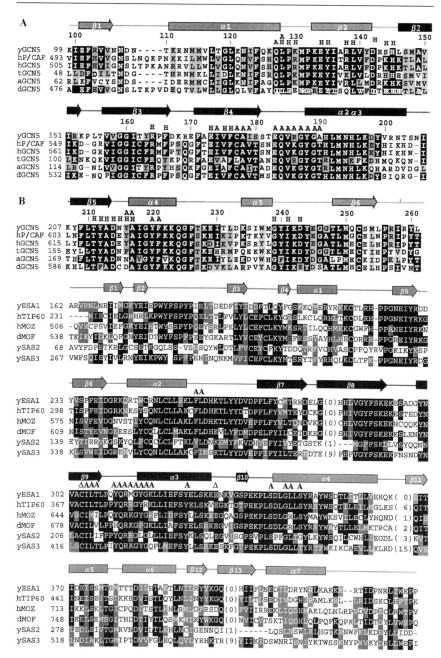

FIG. 1. Sequence homology among HAT proteins. (A) Sequence homology among the Gcn5/PCAF family of HAT proteins. Residues with black and grey shading are identical and conserved, respectively. Secondary structural elements within the enzymatic core (black) and

(SP-sepharose) and gel filtration chromatography (Superdex-75). CoA-agarose appeared to be critical for enriching for "natively folded" protein that was capable of binding CoA and monomeric protein that eluted at an appropriate position on gel filtration chromatography. We also observed that the proteins were more stable when purified and stored in a low-pH buffer, and we found Na citrate, pH 6.0, to be a suitable buffer. We presume that the low pH has two advantages. First, it is far from the pK_a of these proteins (which is above 8) and therefore increases the surface charge of the protein for increased solubility. Second, enzymatic studies reveal that the protein has low activity at this pH, thus disfavoring autoacetylation that may destabilize the protein. Following purification of the Gcn5/PCAF HAT domains, they were each concentrated to between 20 and 40 mg/ml in a buffer containing 20 mM Na citrate, 150 mM NaCl, and 1–10 mM reducing agent and stored at $-70°$ until use.

Purification of the yEsa1 HAT domain was carried out essentially as described for the Gcn5/PCAF HAT domains.[19] However, in the case of yEsa1, CoA-agarose affinity resulted in significant protein loss, and so only cation exchange and gel filtration chromatography were used. In addition, protein concentration beyond $\sim$8 mg/ml resulted in protein precipitation, so protein was generally not concentrated beyond this point.

Crystallization of HAT Proteins

The Gcn5, PCAF, and Esa1 HAT proteins were crystallized using hanging drop vapor diffusion and standard crystallization screens.[14–16] We have found that crystals of liganded forms of the HAT proteins are more easily obtained than the free proteins, whereby the CoA or Ac-CA bound forms are the most straightforward to obtain. Indeed, the nascent tGcn5 and yGcn5 crystals that we were able to obtain were not "true" nascent proteins. The nascent yGcn5 contained the C-terminus from a symmetry-related yGcn5 molecule in the histone-binding cleft[15] and nascent tGCN5 contained a HEPES buffer molecule bound to the CoA-binding region.[16] This, coupled with the fact that the PCAF HAT domain could only be crystallized in the presence of CoA or Acetyl-CoA in our laboratory, suggests that the HAT domains of these proteins become more rigid in the presence of a ligand, making them more amenable to crystallization. This is consistent with another published report showing that the cofactor-liganded forms

N- and C-terminal regions (grey) are color-coded. The "A" and "H" symbols represent residues that mediate Ac-CoA and histone contacts, respectively, and the catalytic glutamate general base is indicated with a "Δ" symbol. (B) Sequence homology among the MYST family of HAT proteins. The color-coding and symbol designation are as described in (A). The catalytic glutamate and cysteine residues are indicated with a "Δ" symbol.

of Gcn5/PCAF have greater stability in solution (as measured by protease resistance) than their unliganded counterparts.[20]

Interestingly, crystallization of a ternary complex with CoA cofactor product and a histone H3 peptide substrate could only be obtained with the tGcn5 member of the recombinant Gcn5/PCAF HAT domains. This does not appear to be related to differential affinities of histone peptides for members of the Gcn5/PCAF HAT domains, but rather to accessibility of the histone peptide binding site within the crystal lattice. Indeed, we have recently been able to co-crystallize tGcn5 with histone H4 and p53 peptides (unpublished), two substrates that have K_m/K_{cat} values 1000- to 10,000-fold lower than histone H3, respectively.[21]

Overall Structure of HAT Proteins

My laboratory has determined the structure of HAT proteins from the Gcn5/PCAF[14–16,22] and MYST subfamily[19,23] in various liganded forms, and Dutnall and co-workers have determined the structure of the cytoplasmic yeast HAT1 enzymes bound to CoA.[24] A comparison of these structures reveals structural homology within a central core domain and structural divergence in regions amino-and carboxyl-terminal to this core domain (Fig. 2). The structure of ternary complexes between *Tetrahymena* Gcn5 (tGCN5) with CoA and either an 11 (16)- or 19-residue (unpublished) histone H3 peptide cantered around the preferred lysine-14 target reveals the mode of cosubstrate binding. The protein structure contains an L-shaped cleft, with CoA bound in the short segment and the histone peptide bond in the long segment of the L-shaped cleft. A correlation of the structure with related biochemical and mutagenesis studies reveals that the central domain plays a particularly important role in Ac-CoA binding and catalysis, while the regions N- and C-terminal to the catalytic core domain play a particularly important role in histone substrate specific binding within Gcn5.

Enzymology of HAT Proteins

To study the enzymology of HAT domains, our laboratory has used a sensitive, fluorescence-based assay using the sulfhydryl-sensitive dye, CPM (7-diethylamino-3-(4'-maleimidylphenyl)-4-methylcoumarin), that reacts

[20] J. E. Herrera, M. Bergel, X. J. Yang, Y. Nakatani, and M. Bustin, *J. Biol. Chem.* **272**, 27253 (1997).

[21] R. C. Trievel, F.-Y. Li, and R. Marmorstein, *Anal. Biochem.* **287**, 319 (2000).

[22] Y. Lin, C. M. Fletcher, J. Zhou, C. D. Allis, and G. Wagner, *Nature* **400**, 86 (1999).

[23] Y. Yan, S. Harper, D. W. Speicher, and R. Marmorstein, *Nat. Struct. Biol.* **9**, 862 (2002).

[24] R. N. Dutnall, S. T. Tafrov, R. Sternglanz, and V. Ramakrishnan, *Cell* **94**, 427 (1998).

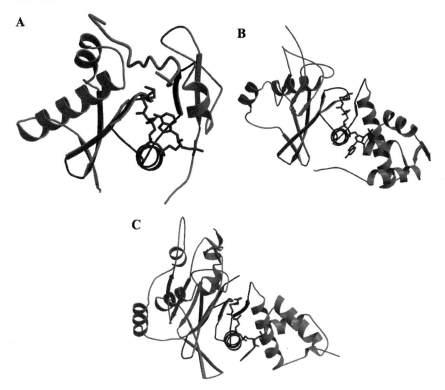

FIG. 2. Overall structure of HAT proteins. (A) Schematic structure of the Gcn5/PCAF HAT domain. tGcn5 on complex with CoA (red) and a 19-residue histone H3 peptide (green). The structurally conserved catalytic core domain is colored in blue, and the structurally variable N- and C-terminal domains are colored in aqua. (B) Schematic structure of the yEsa1 member of the MYST HAT domain in complex with CoA (red). The color coding is as indicated in (A). (C) Schematic structure of the yHAT1 HAT domain in complex with Acetyl-CoA (red). The color coding is as indicated in (A). (See color insert.)

with free CoA, that is generated upon substrate acetylation.[21] Enzymatic reactions are typically carried out in the presence of protein, Ac-CoA, and substrate and quenched by the addition of isoproponal. CPM is then added to the quenched reaction, which reacts with the free sulfhydryl of the CoA reaction product. The fluorescence of the samples is then read with emission and excitation filters of 385 and 465 nm, respectively, which is appropriate for the CoA-modified CPM chromaphor. To obtain steady-state parameters, reactions are carried out in the linear range of protein concentration and reaction time with saturating concentrations of one of the substrates and varying the concentration of the other substrate. Reactions are also typically carried out in 96-well white plates and analyzed with a fluorescence microplate reader. This assay is sensitive, reproducible, and convenient; however, a shortcoming of the assay is that the CPM chromaphore reacts avidly with all accessible free sulfhydryl groups and will therefore react with free cysteine residues within the protein or substrate. If either the protein or substrate contains surface-exposed cysteine residues, they will react with the CPM molecule and thereby increase the background of the signal. The assay cannot be carried out in real time because the reaction of CPM with free sulfhydryls on the protein may also modify the activity of the protein. Nonetheless, the fluorescence-based assay has been used successfully to obtain steady-state kinetic parameters for each of the recombinant HAT proteins that we have worked with in the laboratory, including yGcn5, tGcn5, hPCAF, and yEsa1.[19,21,23]

Other assays that have been employed to enzymatically characterize HAT proteins include radioactive assays and a spectrophotometric enzyme-linked assay. For the radioactive assays, [³H]acetyl-CoA[25] or [¹⁴C]acetyl-CoA[26] is used as a substrate to generate the labeled histone products. The amount of labeled histone is then quantitated by either blotting the labeled histone onto filter paper for scintillation counting (for [³H]) or separating the labeled histone on SDS Tris-Tricine polyacrylamide gels and quantifying by phosphorimage analysis (for [¹⁴C]). The continuous spectrophotometric enzyme-linked HAT assay that has been developed that uses a coupled enzyme system with either α-ketoglutarate dehydrogenase or pyruvate dehydrogenase.[27]

The enzymology that has been carried out on the Gcn5/PCAF family and the yEsa1 member of the MYST family reveals that they use different catalytic mechanisms. The Gcn5/PCAF HAT proteins use a sequential ordered mechanism, whereby the acetyl moiety of Ac-CoA is transferred directly from the cofactor to the Nζ nitrogen of the substrate lysine residue (Fig. 3A).[15,16,25] In contrast, the yEsa1 member of the MYST family of HAT proteins employs a ping-pong catalytic mechanism, whereby the acetyl moiety from Ac-CoA is first transferred to a cysteine residue within the active site of the protein, and in a second step this acetyl moiety is transferred to the Nζ nitrogen of the substrate lysine residue (Fig. 3B).[23] By sequence homology, it appears that all members of the MYST HAT family may use the same catalytic mechanism as yEsa1. Despite the different catalytic mechanisms between the two HAT families, both contain a conserved glutamate residue in the active site that appears to

[25] K. G. Tanner, R. C. Trievel, M.-H. Kuo, R. Howard, S. L. Berger, C. D. Allis, R. Marmorstein, and J. M. Denu, *J. Biol. Chem.* **274,** 18157 (1999).

[26] O. D. Lau, T. K. Kundu, R. E. Soccio, S. Ait-Si-Ali, E. M. Khalil, A. Vassilev, A. P. Wolffe, Y. Nakatani, R. G. Roeder, and P. A. Cole, *Mol. Cell* **5,** 539 (2000).

[27] Y. Kim, K. G. Tanner, and J. M. Denu, *Anal. Biochem.* **280,** 308 (2000).

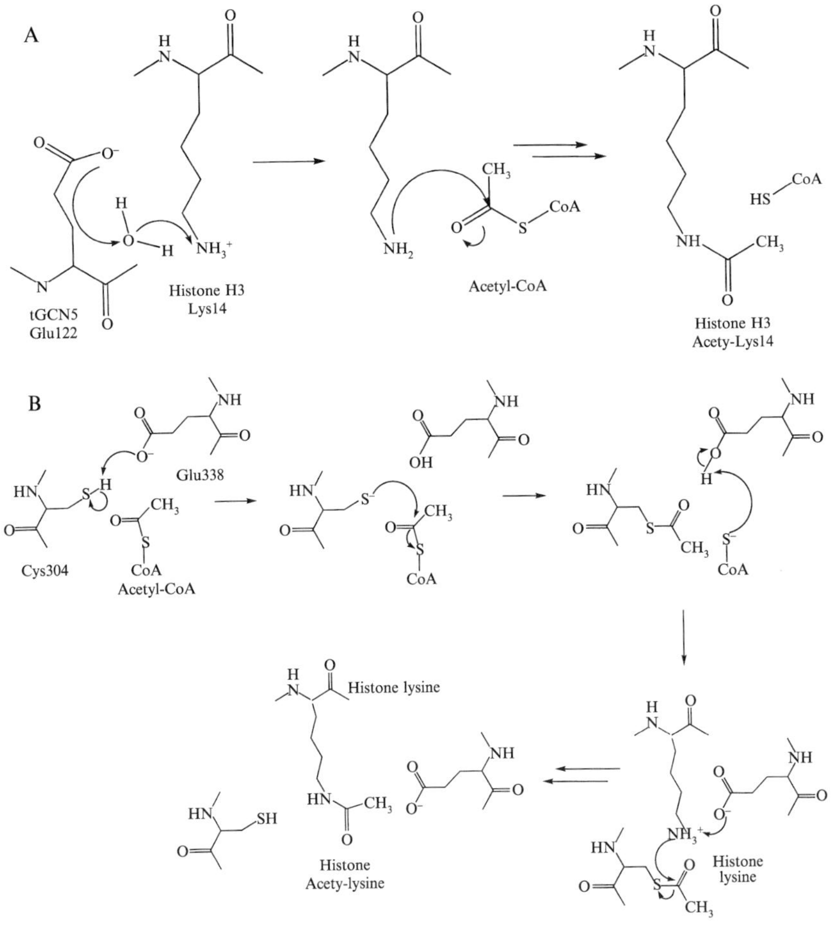

FIG. 3. Enzymatic mechanism of HAT proteins. (A) Proposed catalytic mechanism for the Gcn5/PCAF HATs. Residue numbering for tGcn5 is indicted. (B) Proposed catalytic mechanism for the MYST HATs. Residue numbering for yEsa1 is indicated.

deprotonate the Nζ nitrogen of the target lysine substrate prior to acetyl transfer to substrate.

It is very surprising that the Gcn5/PCAF and MYST family of HAT enzymes, two functionally and structurally related enzyme families, use different catalytic mechanisms. Interestingly, a recent report from the Cole group shows that the p300 HAT, a member of yet another HAT family, appears to use yet a different catalytic mechanism.[28] It is attractive to

[28] P. R. Thompson, H. Kurooka, Y. Nakatani, and P. A. Cole, *J. Biol. Chem.* **276**, 33721 (2001).

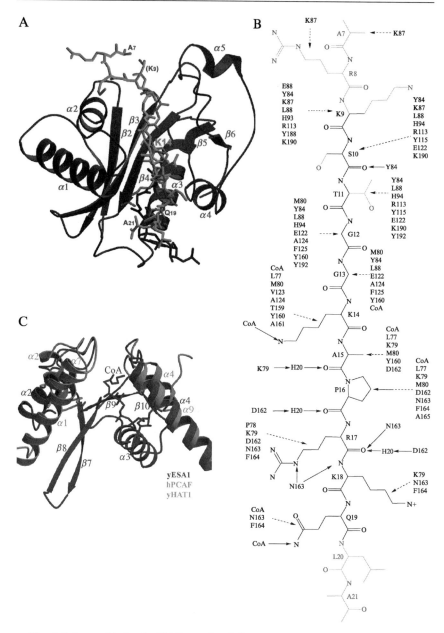

FIG. 4. Substrate binding by HAT proteins. (A) tGcn5/CoA/histone H3 complex. CoA is colored in purple and the ordered regions of the 19-residue histone H3 peptide (residues 7–21) are colored in green. An overlay of the histone H3 peptide from the ternary complex with the 11-residue histone H3 peptide (residues 9–19) is shown in grey. (B) Schematic tGcn5-histone H3 interactions in the ternary tGcn5/CoA/histone H3 complex shown in (A). Regions of the

hypothesize that the difference in catalytic mechanism between these different HAT families may be related to their different *in vivo* properties.

Histone-Specific Binding by HAT Proteins

The crystal structure of the ternary complex between tGCN5, CoA, and an 11-residue histone H3 peptide (H3p11) reveals that the histone substrate sits in a cleft formed by the central protein core and the pantetheine arm of CoA and is flanked by N- and C-terminal HAT domain segments that mediate most of the tGCN5-histone H3 contacts.[16] The participation of the pantetheine arm of CoA in histone H3 contacts is consistent with the important structural role played by CoA in histone recognition. Coupled with the increased stability of the cofactor-liganded Gcn5/PCAF enzymes, we propose that Ac-CoA binds the enzyme prior to histone substrate binding. One surprising feature of the tGCN5/CoA/H3p11 structure is that most of the peptide-protein interactions are mediated through the backbone of the peptide, in its C-terminal region, whereas the N-terminal residues 9–13 are flexible with disordered side chains in the crystal structure. This result, coupled with the biochemical studies demonstrating that GCN5/PCAF family members exhibit a high degree of specificity for lysine 14 of histone H3,[17,18,29,30] raised the possibility that there are other structural determinants for substrate binding specificity by the GCN5/PCAF HAT domain. In support of this, the PCAF HAT domain has a greater than 100-fold enhanced substrate specificity (k_{cat}/K_m) for a 19–amino acid H3-derived substrate peptide (residues 5–23; H3p19) when compared to H3p11,[21] with only a slightly elevated activity toward a 27-residue histone H3 peptide (1.9-fold relative to H3p19). Taken together, these studies suggest that the 19-residue histone H3 peptide is sufficient to mediate most, if not all, of the histone H3 interactions with the Gcn5/PCAF HAT domain.

A more recent structure of tGcn5 in ternary complex with CoA and a 19-residue histone H3 peptide shows that 15 of the 19 histone H3

[29] M. H. Kuo, J. E. Brownell, R. E. Sobel, T. A. Ranalli, R. G. Cook, D. G. Edmondson, S. Y. Roth, and C. D. Allis, *Nature* **383**, 269 (1996).
[30] R. L. Schiltz, C. A. Mizzen, A. Vassilev, R. G. Cook, C. D. Allis, and Y. Nakatani, *J. Biol. Chem.* **274**, 1189 (1999).

histone H3 peptide that are ordered in the ternary complex with the 19-residue peptide, but not ordered in the ternary complex with the 11-residue peptide, are colored in grey. (C) Superposition of putative substrate binding sites of HAT proteins. The superposition is generated by superimposing the core domains from the Gcn5/PCAF member, PCAF (pink), and the MYST member, yEsa1 (blue), with yHAT1 (green). Only the core domain and CoA of yEsa1 (blue) is shown for clarity. (See color insert.)

residues are ordered with only two disordered residues at each end of the peptide (Fig. 4A and B) (unpublished). Surprisingly, relative to the complex containing H3p11, the H3p19 complex reveals significantly more extensive protein-peptide interactions. The greatest structural alterations between the two peptides occur in residues N-terminal to lysine 14, as H3p19 makes more interactions in this region. Most strikingly, a core of 12 residues of histone H3 centered on lysine 14 is well ordered in the tGCN5/CoA/histone H3p19 complex. Each of these residues makes sequence-specific contacts to protein regions N- and C-terminal to the catalytic core of tGCN5. This comparison reveals that residues outside the core 12-residue sequence of the histone H3 peptide anchor and reposition the core of histone H3 for more optimal enzyme-histone contacts.

The protein regions N- and C-terminal to the catalytic domains of Esa1 and HAT1 show structural divergence to the corresponding regions of Gcn5/PCAF (Fig. 2). Despite this, a supposition of the respective core domains of the two protein families reveals that the N- and C-terminal regions of Gcn5 that specifically contact histone substrate have structurally overlapping counterparts in the Esa1 structure (Fig. 4C).[19] This suggests that the Gcn5/PCAF and MYST HATs, and possibly other HAT families, may have a similar structural scaffold for substrate-specific recognition and that the sequence divergence within this scaffold may contribute to substrate-specific binding.

Conclusions and Future Prospects

The methodologies that we have outlined for preparing recombinant HAT domains of the Gcn5/PCAF and MYST family of HAT proteins should be generally applicable for the preparation of HAT domains from other HAT families. The functional and structural studies carried out on the Gcn5/PCAF and MYST HAT families have provided important insights into the overall architecture of HAT proteins and the mode of catalysis and substrate binding. The architectural features of the Gcn5/PCAF and MYST family lead to the hypothesis that all HAT proteins will have a structurally conserved catalytic core domain with divergent catalytic mechanisms and structurally divergent but related N- and C-terminal domains that are correlated with substrate binding. The structure of HAT domains from other HAT families needs to be determined in order to test and extend this hypothesis. In particular, the mode of substrate binding specificity is an area that requires considerably more attention. In addition, it is clear that other protein subunits of the *in vivo* HAT complexes modify HAT functional at both the level of catalysis and substrate binding. The structural and mechanistic basis for this is an important issue that will require structural studies of relevant HAT complexes. These structures may be at the heart of understanding how HAT activity is coordinated with other histone-modifying activities to faithfully modulate transcriptional regulation.

Acknowledgments

I would like to acknowledge all past and present members of my laboratory who have contributed to the studies discussed in this chapter. In particular, I would like to thank A. Clements, M. Holbert, A. Poux, J. Rojas, T. Sikorski, R. Trievel, and Y. Yan.

[8] Use of Nuclear Magnetic Resonance Spectroscopy to Study Structure-Function of Bromodomains

By SHIRAZ MUJTABA and MING-MING ZHOU

Characterization of the evolutionarily conserved protein modular domains in signaling proteins that recognize post-translationally modified amino acids or unique sequence motifs in a protein has revolutionized our understanding of regulation of protein-protein interactions or enzyme activities in signal transduction that govern cell growth, proliferation, differentiation, and apoptosis.[1] Chromatin remodeling represents another important frontier in cell biology.[2,3] While recent studies have identified numerous conserved protein modules in many proteins and enzymes linked to chromatin remodeling,[3–5] their detailed molecular mechanisms remain elusive. Nuclear magnetic resonance (NMR) spectroscopy is a powerful tool not only for determination of high-resolution 3D structures of protein domains but also for investigation of their biochemical functions. The resulting structural and functional inferences can help us gain important insights into the molecular mechanisms underlying chromatin-mediated epigenetic control processes, including transcriptional activation and repression, as well as gene silencing. Here we describe the

[1] T. Pawson and P. Nash, *Genes Dev.* **14**, 1027 (2000).

[2] R. D. Kornberg and Y. Lorch, *Cell* **98**, 285 (1999).

[3] S. Bjorklund, G. Almouzni, I. Davidson, K. P. Nightingdale, and K. Weiss, *Cell* **96**, 759 (1999).

[4] F. Jeanmougin, J. M. Wurtz, B. L. Douarin, P. Chambon, and R. Losson, *Trends Biochem. Sci.* **22**, 151 (1997).

[5] R. Aasland, T. J. Gibson, and A. F. Stewart, *Trends Biochem. Sci.* **20**, 56 (1995).

procedures recently used to delineate structure-function relationships of the bromodomains.

The bromodomain of ~ 110 amino acids, first reported in the *Drosophila* protein brahma (hence the name),[6,7] represents an extensive family of evolutionarily conserved protein modules found in many chromatin-associated proteins and in nearly all known nuclear histone acetyltransferases (HATs).[4] It has been long implicated from yeast genetic and biochemical studies that bromodomains play an important role in chromatin remodeling[8-10] on the basis of their importance in the assembly and activity of multi-protein complexes of chromatin remodeling,[11,12] as well as by the fact that the bromodomain module is indispensable for the function of GCN5 in yeast.[13,14] For example, it has been shown that deletion of a bromodomain in human HBRM, a protein in the SWI/SNF remodeling complex, causes both decreased stability and loss of nuclear localization.[15,16] Bromodomains of Bdf1, a *Saccharomyces cerevisiae* protein, are required for sporulation and normal mitotic growth.[17] Finally, bromodomain deletion in Sth1, Rsc1, and Rsc2, three members of the nucleosome remodeling complex, can cause a conditional lethal phenotype (in Sth1)[18] or a strong phenotypic inhibition on cell growth (in Rsc1 and Rsc2).[19] Notably, the phenotypic effect observed in Rsc1 and Rsc2 results from deletion of only the second but not the first bromodomain, suggesting that these two bromodomains serve distinct functions through interactions with different

[6] J. W. Tamkun, R. Deuring, M. P. Scott, M. Kissinger, A. M. Pattatucci, T. C. Kaufman, and J. A. Kennison, *Cell* **68**, 561 (1992).

[7] S. R. Haynes, C. Dollard, F. Winston, S. Beck, J. Trowsdale, and I. B. Dawid, *Nucleic Acids Res.* **20**, 2603 (1992).

[8] J. E. Brownell, J. Zhou, T. Ranalli, R. Kobayashi, D. G. Edmondson, S. Y. Roth, and C. D. Allis, *Cell* **84**, 843 (1996).

[9] P. Filetici, C. Aranda, A. Gonzalez, and P. Ballario, *Biochem. Biophys. Res. Commun.* **242**, 84 (1998).

[10] G. A. Marcus, N. Silverman, S. L. Berger, J. Horiuchi, and L. Guarente, *EMBO J.* **13**, 4807 (1994).

[11] C. E. Brown, L. Howe, K. Sousa, S. C. Alley, M. J. Carozza, S. Tan, and J. L. Workman, *Science* **292**, 2333 (2001).

[12] D. E. Sterner, P. A. Grant, S. M. Roberts, L. J. Duggan, R. Belotserkovskaya, L. A. Pacella, F. Winston, J. L. Workman, and S. L. Berger, *Mol. Cell. Biol.* **19**, 86 (1999).

[13] T. Georgakopoulos, N. Gounalaki, and G. Thireos, *Mol. Gen. Genet.* **246**, 723 (1995).

[14] P. Syntichaki, I. Topalidou, and G. Thireos, *Nature* **404**, 414 (2000).

[15] C. Muchardt, B. Bourachot, J. C. Reyes, and M. Yaniv, *EMBO J.* **17**, 223 (1998).

[16] C. Muchardt and M. Yaniv, *Semin. Cell. Dev. Biol.* **10**, 189 (1999).

[17] P. Chua and G. S. Roeder, *Mol. Cell. Biol.* **15**, 3685 (1995).

[18] J. Du, I. Nasir, B. K. Benton, M. P. Kladde, and B. C. Laurent, *Genetics* **150**, 987 (1998).

[19] B. R. Cairns, A. Schlichter, H. Erdjument-Bromage, P. Tempst, R. D. Kornberg, and F. Winston, *Mol. Cell* **4**, 715 (1999).

biological ligands.[19] The recent NMR-based structure-function analysis of the prototypical bromodomain from the transcriptional coactivator PCAF (p300/CBP-associated factor) demonstrates that bromodomains function as acetyl lysine–binding domains,[20] which offers insights into the molecular basis of biological functions of bromodomains in a wide variety of cellular events, including chromatin remodeling and transcriptional activation.[21–24] In this chapter, we describe use of the NMR-based methods to study structure and function of bromodomains, which is generalizable for other conserved protein modular domains in chromatin remodeling.

Preparation of Protein Samples

Protein Expression and Stable Isotope Labeling

The cDNA construct that encodes the bromodomain of PCAF (residues 719–832) used in the NMR structural analysis is designed on the basis of sequence analysis, which shows that this region of PCAF contains the conserved bromodomain present in many other proteins.[4,20] The expression construct is ligated into the pET14b vector (Novagen) between *Nde*1 and *BamH*1 sires. The recombinant protein expressed in *Escherichia coli* BL21 (DE3) cells contains a hexahistidine tag at its amino terminus followed by a thrombin cleavage site. After confirming the clone by DNA sequencing, protein expression studies are conducted in *E. coli* BL21 (DE3) cells. Initial optimization of experimental conditions for expression and solubility of the protein is conducted with small- and large-scale culture in Luria-Bertani (LB) media. Subsequently, protein samples for the NMR structural study is expressed in an M9 minimal medium. Ingredients of the minimal medium consist of NaCl (0.5 g/L), NH_4Cl (1 g/L), KH_2PO_4 (3 g/L), and Na_2HPO_4-H_2O (6 g/L), which after sterilization by autoclave are added to glucose (4 g/L), vitamin B1 (0.0005%), $MgSO_4$ (1 μM), $CaCl_2$ (100 μM) and ampicillin (100 mg/L). For a typical protein expression experiment, bacterial BL21(DE3) cells transformed with pET14b-bromodomain are grown overnight at 30° in 100 ml of the M9 minimal medium containing ampicillin (100 mg/L). This starter culture is used to inoculate 1 L of fresh M9 media, which is then incubated at 30° for a few hours until OD_{600} reaches about 0.5. Protein expression is induced by adding 200–400 μM

[20] C. Dhalluin, J. E. Carlson, L. Zeng, C. He, A. K. Aggarwal, and M.-M. Zhou, *Nature* **399,** 491 (1999).
[21] M. H. Dyson, S. Rose, and L. C. Mahadevan, *Front Biosci.* **6,** 853 (2001).
[22] F. Winston and C. D. Allis, *Nat. Struct. Biol.* **6,** 601 (1999).
[23] B. D. Strahl and C. D. Allis, *Nature* **403,** 41 (2000).
[24] L. Zeng and M.-M. Zhou, *FEBS Lett.* **513,** 124 (2001).

isopropyl-β-D-thiogalactopyranoside (IPTG) to the cell culture, which is further incubated for another 10–12 h at 18°. Cells are harvested by centrifugation at 3000g for 30 min at 4°, and cell pellets are collected for protein purification (see later) or quick freezing in liquid nitrogen for storage at 80°. This expression procedure is also used to make various stable isotope (^{15}N, ^{13}C/^{15}N or ^{2}H/^{13}C/^{15}N)-labeled proteins for the NMR structural study. Particularly, the labeled proteins are prepared from bacterial cells grown in a minimal medium containing ^{15}NH$_4$Cl with or without ^{13}C$_6$-glucose in H$_2$O or ^{2}H$_2$O. ^{15}NH$_4$Cl and ^{13}C$_6$-glucose provide sole nitrogen and carbon sources for recombinant protein expressed in *E. coli*.

Protein Purification

The harvested bacterial cells are resuspended in a lysis buffer [50 mM Tris-HCl of pH 8.0, containing 10% glycerol, 1% NP-40, 300 mM NaCl, 1 mM PMSF, and EDTA free protease inhibitors (one tablet per liter of cell culture) (Roche)] and subjected to sonication. Cellular debris is removed by centrifugation at 100,000g for 20 min, and the supernatant obtained is used for subsequent protein purification.

The bromodomain protein is first purified by affinity chromatography on a nickel-IDA column (Invitrogen) using an FPLC system (Amersham Pharmacia Biosciences). The cell lysate is applied to the nickel resin column of 5 ml (for 3–4 L of cell culture) that is pre-equilibrated in a binding buffer [50 mM Tris-HCl of pH 8.0, containing 250 mM NaCl, 5 mM β-mercaptoethanol (β-ME), 1 mM PMSF, and protease inhibitors], and subsequently washed with 10–20 column volumes of the binding buffer, followed by 10 column volumes of a washing buffer [50 mM Tris-HCl of pH 8.0, containing 250 mM NaCl, 5 mM β-ME, 1 mM PMSF, and protease inhibitors, plus 20 mM imidazole]. The hexahistidine-tagged bromodomain protein is eluted from the column in 1.0-ml fractions with an elution buffer with an 20–500-mM imidazole gradient in 50 mM Tris-HCl of pH 8.0, containing 250 mM NaCl, 5 mM β-ME, 1 mM PMSF, and protease inhibitors. Fractions containing the pure protein are pooled and dialyzed to a thrombin cleavage buffer [50 mM Tris-HCl of pH 8.0, containing 250 mM NaCl, and 5 mM β-ME]. The hexahistidine tag in the recombinant protein is removed with thrombin treatment (~1 unit thrombin/mg protein) overnight at 4°. The thrombin cleavage reaction is stopped by addition of 1 mM PMSF. The protein sample is concentrated and applied to a size exclusion chromatography column for further purification using an FPLC system in a phosphate buffer [100 mM phosphate of pH 6.5, containing 0.5 mM EDTA and 1 mM DTT]. Peak fractions are collected, concentrated to ~0.5 mM, and dialyzed to the final NMR buffer in H$_2$O/^{2}H$_2$O (9:1) consisting of

100 mM phosphate, pH 6.5, containing 0.5 mM EDTA, 5 mM perdeuterated DTT. Typically, 5–10 mg of pure bromodomain protein is obtained from 1 L of cell culture.

Protein Structure Determination by NMR

Three-dimensional structure of a protein can be determined by using heteronuclear multi-dimensional NMR methods.[25] The heteronuclear NMR methods separate the proton signals of a protein in the NMR spectra according to chemical shifts of their attached heteronuclei (such as ^{15}N and ^{13}C), thus minimizing signal overlapping problems in the protein spectra. Also, NMR resonance assignments of the protein can be obtained in a sequence-specific manner, which assures the accuracy of data analysis for high-resolution structure determination. In addition, because of favorable ^{1}H, ^{13}C, and ^{15}N relaxation rates caused by partial deuteration of the protein, factional deuteration in combination with ^{13}C and ^{15}N-labeling is often employed for protein structure determination by NMR.[26,27] All the NMR data are processed and analyzed using NMR software programs of NMRPipe[28] and NMRView.[29] For sequential backbone and side chain assignments and structure determination of the protein, a set of NMR experiments is described briefly later.

Backbone Assignments. Sequence-specific backbone assignment is achieved using a suite of deuterium-decoupled 3D NMR experiments that include HNCA, HN(CO)CA, HN(CA)CB, HN(COCA)CB, HNCO, and HN(CA)CO experiments.[30] Using the triple-labeled (75% ^{2}H, ^{13}C, and ^{15}N) protein sample, we perform constant-time experiments to gain higher digital resolution and use a water flip-back scheme to minimize amide signal attenuation from water exchange.

Side-Chain Assignments. Sequential side-chain assignments are accomplished from a series of 3D NMR experiments with alternative approaches to confirm the assignments. These experiments include 3D ^{15}N-edited TOCSY-HSQC, HCCH-TOCSY, (H)C(CO)NH-TOCSY, and H(C)(CO)NH-TOCSY.[25]

[25] G. M. Clore and A. M. Gronenborn, *Meth. Enzymol.* **239,** 249 (1994).

[26] M. Sattler and S. W. Fesik, *Structure* **4,** 1245 (1996).

[27] M.-M. Zhou, K. S. Ravichandran, E. T. Olejniczak, A. P. Petros, R. P. Meadows, M. Sattler, J. E. Harlan, W. Wade, S. J. Burakoff, and S. W. Fesik, *Nature* **378,** 584 (1995).

[28] F. Delaglio, S. Grzesiek, G. W. Vuister, G. Zhu, J. Pfeifer, and A. Bax, *J. Biomol. NMR* **6,** 277 (1995).

[29] B. A. Johnson and R. A. Blevins, *J. Biomol. NMR* **4,** 603 (1994).

[30] T. Yamazaki, W. Lee, C. H. Arrowsmith, D. R. Mahandiram, and L. E. Kay, *J. Am. Chem. Soc.* **116,** 11655 (1994).

NOE Analysis/Distance Restraints. Distance restraints are obtained from analysis of [15]N- and ^{13}C-edited 3D NOESY data, which are collected with different mixing times to minimize spin diffusion problems. The nuclear Overhauser effect (NOE)-derived restraints are categorized as strong (1.8–3 Å), medium (1.8–4 Å), or weak (1.8–5 Å) based on the observed NOE intensities. We also employ the recently developed ARIA program[31] that is integrated with X-PLOR for the iterative automated NOE analysis. ARIA-assigned NOE peaks are manually checked and confirmed to ensure the success of ARIA/X-PLOR-assisted NOE analysis and structure calculations.

Slow Exchange Amides. Amide protons involved in hydrogen bonds are identified from an analysis of the amide exchange rates measured from a series of 2D ^{1}H/^{15}N-heteronuclear single quantum coherence (HSQC) spectra recorded after adding ^{2}H$_2$O to the protein sample.

Stereospecific Methyl Groups. Stereospecific assignments of methyl groups of Val and Leu residues are obtained from an analysis of carbon signal multiplet splitting using 10% ^{13}C-labeled protein sample, which can be readily prepared using 10% ^{13}C-glucose containing M9 minimal medium.[32]

Dihedral Angle Restraints. ϕ angle constraints are generated from the $^3J_{HNH\alpha}$ coupling constants measured in a 3D HNHA-*J* experiment.[33] Stereospecific assignments of β-methylene protons, which give information on $\chi 1$, angles can be obtained from HNHB[34] and ^{15}N-edited TOCSY with a short mixing time.[35]

Structure Calculations and Refinements. Structures of the protein are calculated using a distance geometry/simulated annealing protocol with the X-PLOR program.[36–38] The structure calculations employ inter-proton distance restraints obtained from [15]N- and ^{13}C-resolved NOESY spectra of the protein or protein/peptide complex. The initial structures are typically calculated with only manually assigned NOE-derived distance restrains, which are used to assist further NOE assignments and identify hydrogen bond partners for the slow exchange amide protons. The converged structures are refined with the experimental restraints of dihedral angles and

[31] M. Nilges and S. O'Donoghue, *Prog. NMR Spectrosc.* **32,** 107 (1998).

[32] D. Neri, T. Szyperski, G. Otting, H. Senn, and K. Wüthrich, *Biochemistry* **28,** 7510 (1989).

[33] G. Vuister and A. Bax, *J. Am. Chem. Soc.* **115,** 7772 (1993).

[34] J. C. Matson, O. W. Sörensen, P. Söresen, and F. M. Poulsen, *J. Biomol. NMR* **3,** 239 (1993).

[35] G. M. Clore, A. Bax, and A. M. Gronenborn, *J. Biomol. NMR* **1,** 13 (1991).

[36] A. T. Brunger, "X-PLOR Version 3.1: A System for X-Ray Crystallography and NMR." Yale University Press, New Haven, CT, 1993.

[37] J. Kuszewski, M. Nilges, and A. T. Brunger, *J. Biolmol. NMR* **2,** 33 (1992).

[38] M. Nilges, G. M. Clore, and A. M. Gronenborn, *FEBS Lett.* **229,** 317 (1988).

hydrogen bonds. Final structure calculations employ the manual and the ARIA-assisted NOE distance restraints, together with hydrogen bond distance restraints and dihedral angle restraints. The distance restraint force constant used in the calculations is typically 50 kcal/mol/$\mathring{A}^2$, and no NOE is violated by more than 0.3 $\mathring{A}$. The torsion restraint force constant is 200 kcal/mol/rad^2, and no dihedral angle restraint is violated by more than 5°. Only the covalent geometry terms, NOE, torsion, and repulsive van der Waals terms are used in the structure refinement. A large, and negative Lennard-Jones potential energy should be observed for the final structures, indicating good non-bonded geometry of the structure. Procheck[39] analysis is also performed to show the majority of the protein residues that are in preferred and allowed regions of the Ramachandran map. Finally, structures of proteins determined using NOE-derived distance restraints and dihedral angle restraints can be further refined with use of residual dipolar couplings, which can be measured in bicelle-based liquid crystalline or cross-linked polyacrylamide gel medium and implemented in the final refinement stage of the structure calculations.[40–43]

The Bromodomain Structure

The 3D structure of a prototypical bromodomain from the transcriptional coactivator PCAF determined by NMR shows that the bromodomain adopts an atypical left-handed four-helix bundle (helices α_Z, α_A, α_B, and α_C) (Fig. 1A).[20] A long intervening loop between helices α_Z and α_A (termed the ZA loop) is packed against the loop connecting helices B and C (named the BC loop) to form a surface accessible hydrophobic pocket, which is located at one end of the four-helix bundle, opposite the amino and carboxy termini of the protein. Mutagenesis studies suggest that tertiary contacts among the hydrophobic and aromatic residues between the two inter-helical loops contribute directly to the structural stability of the protein.[20] This unique structural fold is highly conserved in the bromodomain family, as supported by several more recently determined structures of bromodomains from human GCN5[44] and S. cerevisiae GCN5p[45]

[39] R. A. Laskowski, J. A. Rullmannn, M. W. MacArthur, R. Kaptein, and J. M. Thornton, J. Biomol. NMR 8, 477 (1996).

[40] J. H. Prestegard, Nat. Struct. Biol. 5 Suppl., 517 (1998).

[41] J. J. Chou, S. Li, C. B. Klee, and A. Bax, Nat. Struct. Biol. 8, 990 (2001).

[42] S. Cavagnero, H. J. Dyson, and P. E. Wright, J. Biomol. NMR 13, 387 (1999).

[43] J. J. Chou, S. Gaemers, B. Howder, J. M. Louis, and A. Bax, J. Biomol. NMR 21, 377 (2001).

[44] B. P. Hudson, M. A. Martinez-Yamout, H. J. Dyson, and P. E. Wright, J. Mol. Biol. 304, 355 (2000).

[45] D. J. Owen, P. Ornaghi, J. C. Yang, N. Lowe, P. R. Evans, P. Ballario, D. Neuhaus, P. Eiletici, and A. A. Travers, EMBO J. 19, 6141 (2000).

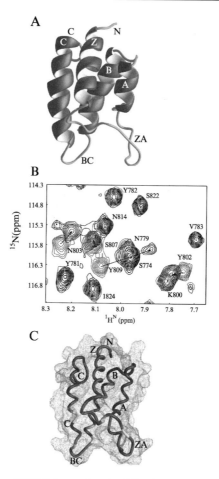

Fig. 1. Ligand binding of PCAF bromodomain. (A) Three-dimensional NMR structure of the PCAF bromodomain. (B) Superimposed region of the 2D ^{15}N-HSQC spectra of the bromodomain ($\sim$ 0.5 mM) in its free form (dark) and complexed to a histone H4 peptide containing acetylated lysine 8 (SGRGKGG-AcK-GLGK, where AcK is acetyl-lysine) (molar ratio 1:6) (light). The movement of protein resonances upon ligand binding is indicated by arrows connecting from the free to the ligand-bound forms. (C) Ribbon and dotted-surface diagram of the bromodomain depicting the location of the lysine-acetylated H4 peptide-binding site. Bromodomain residues that exhibit major chemical shift changes of the backbone amide ^{1}H and ^{15}N resonances upon binding to the AcK histone H4 peptide as observed in the 2D ^{15}N-HSQC spectra are indicated in lighter color. Note that most of these perturbed residues are located in the ZA and BC loops. [From C. Dhalluin, J. E. Carlson, L. Zeng, C. He, A. K. Aggarwal, and M.-M. Zhou, *Nature* **399**, 491 (1999).]

as well as the double bromodomain module of human $TAF_{II}250$.[46] The structural similarity among these bromodomains is very high for the four helices with pairwise root-mean-square deviations of 0.7–1.8 Å for the backbone C_α atoms. The majority of structural deviations are localized in the loop regions, particularly in the ZA and BC loops. This observation is in an agreement with relatively high sequence variations in these loops.[4] The modular structure supports the notion that bromodomains act as a functional unit for protein interactions, and multiple bromodomain modules can be placed sequentially in a protein to serve similar or distinct functions.[4,20]

Ligand Binding Study by NMR

The unique advantage of protein structural analysis by NMR is that in addition to determination of a 3D structure of a protein, resonance assignments obtained in the structural study provide a map of the entire protein at atomic details-level, which could be used in biochemical analysis of protein-ligand interactions. Because the NMR resonances of protein residues are highly sensitive to local chemical and conformational changes, binding of a ligand to a host protein could be detected by resonance perturbations of protein residues directly or indirectly involved in interactions with the ligand.[47,48] For highly specific interactions between a host protein and a ligand, protein resonances change as a function of ligand concentration until complete saturation of the protein by the ligand. NMR titration of protein and ligand binding, therefore, can be used to determine binding affinity (dissociation constant, K_D) of the complex. The ligand-binding study is most conveniently performed by using 2D ^{15}N-HSQC spectra that record backbone amide proton and nitrogen resonances of a uniformly ^{15}N-labeled protein. Because the ligand is typically not ^{15}N-labeled thus invisible in ^{15}N-HSQC spectra, protein resonance changes in the spectra are relatively easy to monitor. With the resonance assignment of a protein, using this NMR method, one can identify the location of the ligand-binding sites on the protein. The NMR titration study of the PCAF bromodomain shows that the protein can bind to acetyl-lysine–containing peptides derived from major acetylation sites on histones H3 and H4 in a highly specific manner, and the interaction is dependent on acetylation of lysine (Fig. 1B and C).[20]

[46] R. H. Jacobson, A. G. Ladurner, D. S. King, and R. Tjian, *Science* **288**, 1422 (2000).
[47] P. J. Hajduk, R. P. Measdows, and S. W. Fesik, *Q. Rev. Biophys.* **32**, 211 (1999).
[48] J. M. Moore, *Curr. Opin. Biotech.* **10**, 54 (1999).

This unique ability of NMR spectroscopy to detect relatively weak but highly specific interactions between a host protein and a ligand is also demonstrated by our recent study of the bromodomain of transcriptional coactivator CBP (CREB-binding protein) and the tumor suppressor protein p53. It is known that transcriptional activity of p53 in cell cycle arrest, senescence or apoptosis is tightly controlled by acetylation of its C-terminal lysines, that is, K320 by PCAF,[49] and K373 and K382 by CBP,[50] which results in its association with transcriptional coactivators including CBP.[51,52] Our study shows that p53/CBP association involves CBP bromodomain binding to p53 at the acetylated lysine 382 but not K373 or K320, which can be demonstrated by NMR titration with lysine-acetylated p53 peptides containing different acetylation sites (Mujtaba and Zhou, unpublished results). Notably, amino acid sequences of the acetylated K373 and K382 peptides used in the NMR-binding study are identical except that the acetylated lysine is in a different position. Remarkably, only the latter p53 peptide causes ligand concentration–dependent resonance perturbations of protein residues in 2D ^{15}N-HSQC spectra of the bromodomain, underscoring the highly selective nature of the CBP bromodomain/p53 AcK382 recognition. None of the p53 peptides showed any detectable binding to the structurally homologous bromodomain from PCAF.[20,53]

To understand the detailed structural basis of molecular recognition of acetyl-lysine–containing peptide, one can determine the 3D structure of a bromodomain in complex with a synthetic, lysine-acetylated peptide derived from the known binding site in its biological binding partner protein. Intermolecular NOE distance restraints required for structure determination of the complex can be obtained from a ^{13}C-edited (F_1) and ^{15}N, and ^{13}C-filtered (F_3) 3D NOESY spectrum collected for a sample containing isotope-labeled protein and unlabeled ligand.[54] In addition, a 2D ^{13}C/^{15}N-filtered ^{1}H-^{1}H NOESY spectrum can provide intra-molecular NOEs for the peptide molecule bound to the protein. Structure calculations of the complex using both inter- and intra-molecular NOE distance restraints together with other experimentally determined hydrogen bond distance restraints and dihedral angle restrains are similar to those described earlier

[49] L. Liu, D. M. Scolnick, R. C. Trievel, H. B. Zhang, R. Marmorstein, T. D. Halazonetis, and S. L. Berger, *Mol. Cell Biol.* **19**, 1202 (1999).

[50] W. Gu and R. G. Roeder, *Cell* **90**, 595 (1997).

[51] N. A. Barlev, L. Liu, N. H. Chehab, K. Mansfield, K. G. Harris, T. D. Halazonetis, and S. L. Berger, *Mol. Cell* **8**, 1243 (2001).

[52] C. Prives and J. L. Manley, *Cell* **107**, 815 (2001).

[53] S. Mujtaba, Y. He, L. Zeng, A. Farooq, J. E. Carlson, M. Ott, E. Verdin, and M.-M. Zhou, *Mol. Cell* **9**, 575 (2002).

[54] M. Sattler, J. Schleucher, and C. Griesinger, *Prog. NMR Spectrosc.* **34**, 93 (1999).

for the structure determination of the protein alone using XPLOR program.

Protein–Peptide Binding Assays

Recent structural studies of different bromodomains show that bromodomains share a conserved left-handed four-helix bundle fold, and acetyl-lysine–containing peptides are bound between the ZA and BC loops. Residues important for acetyl-lysine recognition in different bromodomains are conserved; however, ligand selectivity differ due to a few but important differences in bromodomain sequences. These include variations in the ZA loops, which have relatively low sequence conservation and amino acid deletion or insertion in different bromodomains; and differences in bromodomain residues that directly interact with residues surrounding acetyl-lysine in a target protein. Structure-based mutational analysis can be employed to determine the structural basis of molecular recognition of a bromodomain/ligand complex. For example, in an effort to determine the molecular determinants of the selective recognition between the PCAF bromodomain and HIV-1 Tat at the acetylated lysine 50, we performed structure-based mutational analysis of the complex in an *in vitro*–binding assay using the recombinant and purified GST-fusion bromodomain and an N-terminal biotinylated, p53 peptide containing lysine-acetylated K382 that is immobilized onto Streptavidin agarose beads.[53]

The cDNA encoding GST-fusion bromodomain of PCAF is cloned into the pGEX4T-3 vector (Amersham Biosciences). The GST bromodomain is expressed in *E. coli* BL21(DE3) codon-plus cells using a procedure similar to that described earlier for NMR protein sample preparation, except that the GST protein is prepared in LB medium. The harvested bacterial cells culture is re-suspended in lysis buffer [20 mM Tris of pH 8.0, 150 mM NaCl, 1.0% NP-40, 1 mM PMSF, 5 mM β-ME, 10% glycerol, 5 mM EDTA, 1 mM PMSF, and DNAse]. After cell lysis by sonication and removal of cell debris by centrifugation, the supernatant is purified by affinity chromatography using a glutathione Sepharose resins according to manufacturer's (Amersham Bioscience) instruction. The binding assay is performed by incubating an equal amount of the PCAF bromodomain with biotinylated HIV Tat peptide for 2 h at 22° in the binding buffer (50 mM Tris buffer of pH 7.5, containing 50 mM NaCl, 0.1% BSA, and 1 mM DTT). The protein-peptide complex is pulled down by 10 μl of Streptavidin agarose (Novagen) by further incubation of 30 min. To minimize non-specific binding, the beads are washed extensively (thrice) with a high salt- and detergent-containing washing buffer (50 mM Tris buffer of pH 7.5, containing 300 mM NaCl, 0.1% NP-40, and 1 mM DTT). Proteins eluted from the agarose beads are separated by SDS-PAGE and protein/peptide interaction is visualized by western blotting using anti-GST antibody (Sigma) and horseradish-peroxidase–conjugated goat anti-rabbit IgG.

Such a GST pull-down assay is also used to assess other bromodomains' binding or effect of site-directed mutation of PCAF bromodomain residues on protein binding to the HIV-1 Tat peptide. Furthermore, this binding assay can be used for mutational analysis of Tat peptide residues in a peptide competition assay, in which a non-biotinylated peptide carrying a mutation at a specific amino acid residue competes with the biotinylated wild-type Tat AcK50 peptide for binding to the GST-fusion PCAF bromodomain. The molar ratio of the mutant and wild-type peptides in the mixture is kept at 1:2. Because of the high sensitivity of western blotting detection, this binding study can be performed at a protein concentration (10 μM) much lower than that required for NMR study ($\sim$200 μM), thus ensuring specificity of protein-peptide interactions. Using this GST pull-down–based assay, mutational analyses of the protein and peptide residues validate the molecular interactions observed in the NMR structure of the complex, and identify key residues of the PCAF bromodomain that are important for recognition of the acetyl-lysine and its flanking residues in the HIV-1 Tat peptide.

Perspective

The NMR-based methods described here have enabled establishment of the biochemical functions of bromodomains as acetyl-lysine–binding domains.[20,44,53] This new mechanism of regulating protein-protein interactions via lysine acetylation has broad implications in a wide variety of cellular processes, including chromatin remodeling and transcriptional activation.[21–23,53,55] Such a powerful NMR structure-based approach is readily applicable to investigate the biological functions of the other evolutionarily conserved protein modular domains that play an important role in regulation of chromatin remodeling.

Acknowledgment

The work was supported by a grant from the National Institutes of Health to M.-M. Z. (CA87658).

[55] A. Dorr, V. Kiermer, A. Pedal, H.-R. Rackwitz, P. Henklein, U. Schubert, M.-M. Zhou, E. Verdin, and M. Ott, *EMBO J.* **21**, 2715 (2002).

[9]   Assays for the Determination of Structure and Dynamics of the Interaction of the Chromodomain with Histone Peptides

By STEVEN A. JACOBS, WOLFGANG FISCHLE, and
SEPIDEH KHORASANIZADEH

The histone code hypothesis predicts that multiple histone modifications, acting in a combinatorial or sequential fashion on one or multiple histone tails, specify unique downstream functions.[1] In order for this process to come to fruition, there must exist enzymatic protein modules which write the histone code by modifying specific residues of the histone tails, and protein recognition modules which read the code by binding to specific modifications. Indeed this appears to be the case, as numerous protein modules which enzymatically modify histone tails have been identified. Histone acetyltransferase domains (HAT) acetylate lysines.[2,3] Histone deacetylase domains (HDACs) remove acetyl marks from lysines,[4] SET domains methylate specific lysines,[5] histone arginine methyltransferases methylate arginine residues, and kinases phosphorylate serine and threonine residues. Recognition protein modules discovered to date include the chromodomain which interacts with methylated lysines[6–8] and the bromodomain which interacts with acetylated lysines; for review, see Marmorstein and Berger (2001).[9] To study the specificities and affinities of the interactions of chromodomains with histone tails we have successfully applied biophysical methods, including nuclear magnetic resonance (NMR) spectroscopy, X-ray crystallography, fluorescence spectroscopy, and isothermal titration calorimetry (ITC), all of which are described here.

NMR spectroscopy is best used to quickly screen peptides for interaction with a specific protein module and determine the extent of a minimally structured complex. NMR or X-ray crystallography can subsequently be used to

---

[1] B. D. Strahl and C. D. Allis, *Nature* **403,** 41 (2000).
[2] S. Y. Roth, J. M. Denu, and C. D. Allis, *Annu. Rev. Biochem.* **70,** 81 (2001).
[3] R. Marmorstein and S. Y. Roth, *Curr. Opin. Genet Dev.* **11,** 155 (2001).
[4] R. Marmorstein, *Structure (Camb)* **9,** 1127 (2001).
[5] M. Lachner and T. Jenuwein, *Curr. Opin. Cell Biol.* **14,** 286 (2002).
[6] M. Lachner, D. O'Carroll, S. Rea, K. Mechtler, and T. Jenuwein, *Nature* **410,** 116 (2001).
[7] A. J. Bannister, P. Zegerman, J. F. Partridge, E. A. Miska, J. O. Thomas, R. C. Allshire, and T. Kouzarides, *Nature* **410,** 120 (2001).
[8] S. A. Jacobs, S. D. Taverna, Y. Zhang, S. D. Briggs, J. Li, J. C. Eissenberg, C. D. Allis, and S. Khorasanizadeh, *EMBO J.* **20,** 5232 (2001).
[9] R. Marmorstein and S. L. Berger, *Gene* **272,** 1 (2001).

solve the structure of a complex formed between a protein module and a target peptide in order to gain the details of interaction at atomic resolution. Binding affinities and thermodynamics of interaction are best determined by ITC and/or fluorescence spectroscopy.

NMR Spectroscopy

In addition to the determination of high-resolution structure, this technique can be used to rapidly screen for specific protein-protein, protein-peptide, or protein-small molecule interactions, map the protein interaction surface, and calculate the binding affinity. All of these applications have been used successfully to study the interactions of chromodomain and bromodomain chromatin binding modules with histone tails. For example, NMR has been used to determine the specificity of binding of the HP1 chromodomain to methylated H3 tail peptides and map the surface of interaction.[8] It has also been used to solve the high-resolution structure of the HP1 chromodomain bound to a methylated histone H3 tail.[10]

NMR techniques do have limitations. First of all, the protein module must be able to be expressed in milligram quantities in a few liters of minimal media in order to incorporate ^{15}N and/or ^{13}C labeling at an efficient cost (every liter of minimal media requires 1 g of ^{15}N ammonium chloride at \$30/g and 4 g of ^{13}C glucose at \$200/g). In addition, the protein must be soluble and stable to >200 μM in a buffer suitable for NMR measurements at room temperature (samples of 50–100 μM may be studies when using a state-of-the-art cryoprobe in the NMR spectrometer). Phosphate buffer in the pH range 6–7.5 is ideal to ensure that backbone amide hydrogens are fully occupied for maximum sensitivity of detection.

To screen for protein-peptide interactions by NMR spectroscopy, the main tool is the two-dimensional [^1H-^{15}N]-HSQC experiment. The spectrum resulting from this experiment correlates the ^1H and ^{15}N chemical shifts, thus providing a backbone fingerprint of the protein. The value of the chemical shift of each residue is highly dependent on the local chemical environment of that residue. Thus, addition of a protein or peptide which specifically binds to the target protein produces a new spectrum with a defined number of large chemical shift perturbations. For example, the ability of protein modules to bind modified histone tail peptides can be monitored by titrating an unlabeled peptide into a ^{15}N-labeled protein solution, and collecting the [^1H-^{15}N]-HSQC spectrum of the protein. Figure 1A shows the [^1H-^{15}N]-HSQC spectra of the free chromodomain (black cross peaks)

[10] P. R. Nielsen, D. Nietlispach, H. R. Mott, J. Callaghan, A. Bannister, T. Kouzarides, A. G. Murzin, N. V. Murzina, and E. D. Laue, *Nature* **416,** 103 (2002).

A

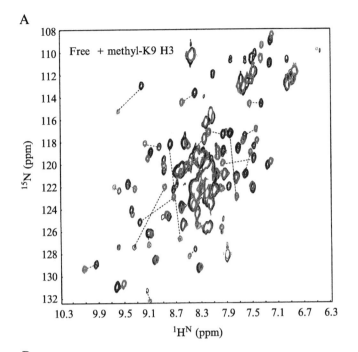

B

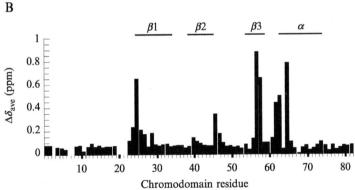

FIG. 1. NMR analysis of the HP1 chromodomain interaction with the dimethyllysine 9 histone H3 peptide. (A) The [^{1}H-^{15}N]-HSQC spectrum of the free chromodomain (black cross peaks) and after the addition of the dimethyllysine 9 histone H3 peptide (grey cross peaks). Each dashed line corresponds to the perturbation of a residue. (B) Chemical shift perturbation ($\Delta\delta_{ave}$) upon H3 peptide binding for the residues of the chromodomain.

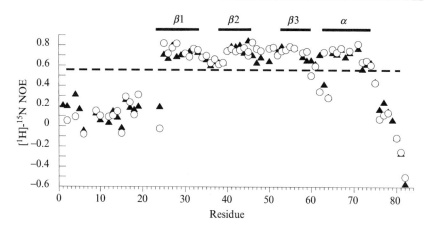

F<small>IG</small>. 2. Chromodomain backbone dynamics by [^{1}H]-^{15}N NOE measurements. Black triangles represent the free chromodomain while open circles represent the chromodomain-H3 peptide complex. The residues with [^{1}H]-^{15}N NOE value >0.55 are expected to be rigid and structured (above the dashed line).

and after addition of an H3 tail peptide (grey cross peaks) containing a dimethyllysine at residue 9. The large chemical shift perturbations are indicative of formation of a specific protein-peptide complex. In contrast, an unmethylated peptide or a peptide methylated at lysine 4 produces a few small chemical shift changes, indicating that these peptides do not specifically interact with the HP1 chromodomain.[8] NMR was also used to map the surface of the chromodomain which interacts with the H3 tail.[8] Figure 1B shows that a few key residues exhibit the greatest change in chemical shift, implicating them for direct peptide binding.

X-ray crystallography or NMR spectroscopy of a protein-peptide complex is greatly simplified if only minimally structured domains are studied. Backbone dynamics determined by NMR can establish which segments of the protein are ordered into a defined structure. Figure 2 shows a plot of HP1 chromodomain [^{1}H]-^{15}N NOE values for the free protein (filled triangles) and after complexation with a histone H3 peptide dimethylated at lysine 9 (open circles). Only residues 25–74 of the chromodomain are rigid in both the free and bound form as judged by the [^{1}H]-^{15}N NOE value >0.55. This information was then used to prepare a new compact construct, which dramatically improved growth of suitable crystals for high-resolution structure determination by X-ray crystallography.[11]

[11] S. A. Jacobs and S. Khorasanizadeh, *Science* **295,** 2080 (2002).

Protein Preparation

The coding regions of *Drosophila* HP1 residues 1–84 and 17–76 were sub-cloned into *Nde*I-*Bam*HI sites of pET16b vector (Novagen) and expressed in BL21(DE3) *Escherichia coli* (Stratagene) with an N-terminal His-tag. Isotopic labeling of the protein (^{15}N/^{13}C) was achieved by growing these cells in minimal media containing 7 g Na_2HPO_4, 3 g KH_2PO_4, 0.5 g NaCl, 0.1 g yeast extract, 4 g ^{13}C-glucose, 1 g ^{15}N-NH_4Cl, 50 mg thiamine-HCl, 0.5 g $MgSO_4$, and 20 mg $CaCl_2$ per liter of water. Cultures were grown at 37° to an optical density (at 600 nm) of 0.6, and protein expression was induced by adding isopropyl-β-D-thiogalactopyranoside (IPTG) to a concentration of 1 mM. Temperature was reduced and shaking continued for 4 h at 30° until cells were harvested by centrifugation, and the pellet was stored in −20° overnight. The frozen pellet (10 g) was resuspended in 50 ml of 50 mM potassium phosphate, pH 7.5, 10 mM imidazole, 500 mM NaCl, 5% glycerol, 1 mM benzamidine HCl, 1 mM phenylmethylsulfonylfluoride (PMSF), and the suspension was stirred in the cold room for 30 min. This suspension was then sonicated on ice with 30 pulses each 30 s using a Misonex sonicator XL. The crude lysate was then cleared by centrifugation at 35,000 g for 30 min and the supernatant loaded onto a column containing 3 ml of Ni^{+2}-agarose affinity resin (Qiagen). The column was pre-equilibrated with 30 ml of the same buffer. After all supernatant was loaded, the column was washed overnight with 1 L of 50 mM potassium phosphate, pH 7.5, 10 mM imidazole, 500 mM NaCl, and 5% glycerol. Bound material was eluted from the column in 1-ml fractions with 50 mM potassium phosphate, pH 7.5, 250 mM imidazole, 100 mM NaCl, and 5% glycerol, and fractions were analyzed for protein content and purity by SDS-PAGE and absorption spectroscopy at 280 nm. Protein fractions were pooled and concentrated using a Millipore ultrafree filtration device and further purified by gel filtration chromatography using a Superdex 75 column (Pharmacia) equilibrated with 50 mM potassium phosphate, pH 6.0, 25 mM NaCl, 1 mM sodium azide, and 0.1 mM DTT. Finally, column fractions containing HP1 were pooled and the sample concentrated to 1 mM for NMR studies. The HP1 chromodomain was prepared using this protocol for all of the assays described later.

Peptide Preparation

Histone tail peptides were prepared synthetically and purified by reverse-phase HPLC. A nonnative tyrosine residue is placed at the C-terminus of each histone peptide to allow determination of concentration by UV absorbance at 280 nm using the extinction coefficient $\varepsilon_{280} = 1280/$M/cm. All peptides were further purified in order to remove trace organic solvent by gel filtration chromatography. About 5 mg of the peptide was dissolved in 100 μl of distilled water and passed over a spin column containing 3 ml of Sephadex G10 resin (Pharmacia) equilibrated with distilled water. The pure peptides were then lyophilized for storage. This protocol for peptide preparation was used for all assays described later.

Protocol for Determination of the Specificity of the HP1 Chromodomain for K9 Methylated H3 Tail

All data were collected on a Bruker AVANCE (600 MHz) spectrometer with samples containing 0.2–1 mM HP1 1–84 in a buffer solution of 50 mM sodium phosphate, pH 6.0, 25 mM NaCl, 1 mM sodium azide, and 0.1 mM DTT at 25°. Spectra were processed with NMRPipe[12] and analyzed with NMRView.[13] First, the [^{1}H-^{15}N]-HSQC spectrum of the free chromodomain was collected. Next, a purified and dried peptide sample was resuspended in the ^{15}N-labeled chromodomain NMR sample in order to achieve molar ratio of 1:4 protein to peptide. The [^{1}H-^{15}N]-HSQC spectrum of this complex sample was collected, and compared with that of the free chromodomain in order to identify the sites with chemical shift changes associated with residues on the surface of peptide binding (see Fig. 1A).

Protocol for Mapping the H3 Tail Binding Surface of the Chromodomain

The first step in mapping the interaction surface is to assign the resonance of the protein in the free form as well as in complex with the peptide. To accomplish this, resonance assignments are carried out (using ^{15}N/^{13}C-labeled protein) by collecting and analyzing the three-dimensional NMR spectra HNCA, HNCO, HN(CO)CA, ^{15}N-edited NOESY, and ^{15}N-edited TOCSY for the free and complex samples.[14,15] Eq. (1) was used to tabulate the weighted average of the HN, N, C$^\alpha$, and C' chemical shift perturbations upon binding of the methylated H3 tail to HP1:[16]

$$\Delta_{\text{ave}} = 0.25 \left\{ (\Delta_{\text{HN}})^2 + \left(\frac{\Delta_{\text{N}}}{5}\right)^2 + \left(\frac{\Delta_{\text{c}^\alpha}}{2}\right)^2 + \left(\frac{\Delta_{\text{c}'}}{2}\right)^2 \right\}^{\frac{1}{2}}. \tag{1}$$

[12] F. Delaglio, S. Grzesiek, G. W. Vuister, G. Zhu, J. Pfeifer, and A. Bax, *J. Biomol. NMR* **6**, 277 (1995).
[13] B. A. Johnson and R. A. Blevins, *J. Biomol. NMR* **4**, 603 (1994).
[14] S. Grzesiek and A. Bax, *J. Magn. Reson.* **96**, 432 (1992).
[15] D. Marion, P. C. Driscoll, L. E. Kay, P. T. Wingfield, A. Bax, A. M. Gronenborn, and G. M. Clore, *Biochemistry* **28**, 6150 (1989).
[16] S. Grzesiek, S. J. Stahl, P. T. Wingfield, and A. Bax, *Biochemistry* **35**, 10256 (1996).

The residues which experience the greatest change ($\Delta_{ave} > 0.15$) were then mapped onto the chromodomain surface.

Protocol for Determination of Backbone Dynamics to Establish a Minimal Interacting Domain of HP1 with Methylated H3 Tail by NMR

The [^{1}H]-^{15}N steady-state heteronuclear NOE values for the backbone nuclei of the free and complexed chromodomain samples were measured by collecting a modified two-dimensional [^{1}H-^{15}N]-HSQC NMR spectrum with the water flip-back method.[17] Spectra were collected in an interleaved manner, one with a recovery delay of 6 s (reference spectrum) and the other with a recovery delay of 3 s of proton saturation (NOE spectrum). The [^{1}H]-^{15}N NOE value for each residue was calculated as the ratio of the cross-peak intensity NOE/reference, and error was estimated from the baseline noise in the two spectra.

Structure Determination of Protein-Histone Tail Complexes

Several NMR and X-ray crystal structures of protein modules interacting with modified histone tails have been solved.[10,11,18–21] These structures provide detailed views of the chemistry of acetyllysine and methyllysine recognition, as well as an explanation for the specificity of module/histone tail interactions. Figure 3 shows the structure of the HP1 chromodomain bound to a methylated H3 tail peptide at atomic resolution. The methyl ammonium group of the methyllysine is recognized by a three-member aromatic cage of the chromodomain. The binding is stabilized by π cation interaction between the π electrons of the aromatic rings, and the positive charge of the methyllysine.[11] Increasing methyl groups on the lysine side chain reduces the number of hydrogens available for hydrogen bonding. In the binding pocket, the dimethyllysine hydrogen bonds to one water molecule, whereas trimethyllysine is not capable of hydrogen bonding.

[17] S. Grzesiek and A. Bax, *J. Am. Chem. Soc.* **115,** 12593 (1993).
[18] C. Dhalluin, J. E. Carlson, L. Zeng, C. He, A. K. Aggarwal, and M. M. Zhou, *Nature* **399,** 491 (1999).
[19] B. P. Hudson, M. A. Martinez-Yamout, H. J. Dyson, and P. E. Wright, *J. Mol. Biol.* **304,** 355 (2000).
[20] R. H. Jacobson, A. G. Ladurner, D. S. King, and R. Tjian, *Science* **288,** 1422 (2000).
[21] D. J. Owen, P. Ornaghi, J. C. Yang, N. Lowe, P. R. Evans, P. Ballario, D. Neuhaus, P. Filetici, and A. A. Travers, *EMBO J.* **19,** 6141 (2000).

A   B

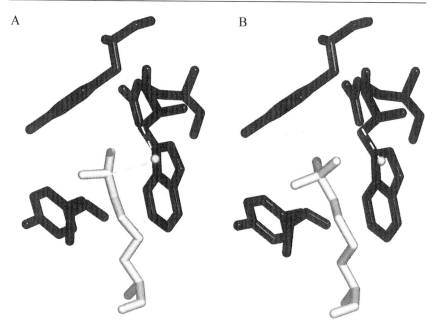

Fig. 3. The methyllysine binding aromatic cage (in black) of the HP1 chromodomain surrounding the H3 peptide methyllysine (in grey): (A) dimethyllysine and (B) trimethyllysine. Increasing methylation leads to a decrease in the potential to form hydrogen bonds with a nearby ordered water molecule (in grey).

Protocol for Determination of the Structure of the HP1 Chromodomain Complexed with a Methylated H3 Tail at Atomic Resolution

The purified HP1 fragment 17–76 was exchanged into a buffer of 10 mM potassium phosphate, pH 7.0, by gel filtration chromatography using Superdex 75 resin (Pharmacia) as a final purification step before concentrating to 1 mM. The purified H3 tail peptide residues 1–15 with dimethyllysine or trimethyllysine at residue 9 was resuspended in the protein solution to reach a molar ratio of 1:4 protein to peptide. As the HP1 chromodomain had not been previously crystallized, a broad initial screen of crystallization conditions was necessary. Several commercially available sparse matrix and grid screens (Hampton Research, Emerald Biostructures) were employed in initial screening using the hanging drop vapor diffusion method. In each case, 2 μl of the protein solution was combined with 2 μl of the reservoir solution and equilibrated against a 1-ml reservoir solution. Hexagonal crystals were obtained in various drops from an ammonium sulfate grid screen (Hampton Research), with optimal crystals growing when 3.2 M ammonium sulfate and 0.1 M MES buffer at pH 6.1 was used as the reservoir solution. Single crystals in space group C222$_1$ ($a = 33$ Å, $b = 76$ Å, $c = 75$ Å) were obtained. The crystals were cryo-protected in reservoir solution supplemented with 35% MPD, and flash-frozen in liquid nitrogen for data collection. Diffraction data were collected on a Rigaku RUH2R X-ray generator equipped with a Bruker SMART 6000 CCD detector. All data were processed and scaled with HKL2000.[22] The structure was solved by molecular replacement using CNS[23] and MOLREP[24] using the crystal structure of one monomer of the dimeric Swi6 chromo shadow domain[25] with all residues turned to alanines as the search model. Chain tracing and structure refinement was carried out using ARP/wARP,[26] CNS, and O.[27]

Fluorescence Spectroscopy

Macromolecular associations can be monitored by either the intrinsic tryptophan fluorescence of a protein or the fluorescence anisotropy of an extrinsic probe attached to one of the interacting species. The use of intrinsic fluorescence is somewhat limited since not all macromolecular associations produce a change in intrinsic fluorescence signal. However, intrinsic fluorescence does have an advantage in that there are no external probes introduced to the system. Changes in fluorescence intensity upon ligand titration can be fit to the following equation in order to attain dissociation constants:[28]

$$F_i = \frac{\left\{ F_s \left[([X_T] + [Y_T] + K_d) \pm \sqrt{([X_T] + [Y_T] + K_D)^2 - (4[X_T][Y_T])} \right] \right\}}{2[X_T]},$$

(2)

where F_i is the fluorescence change produced by fitting of increasing quantities of peptide, F_s the fluorescence change at saturation, $[X_T]$ the total

[22] Z. Otwinowski and W. Minor, *Methods Enzymol.* **276**, 307 (1997).

[23] A. T. Brunger, P. D. Adams, G. M. Clore, W. L. DeLano, P. Gros, R. W. Grosse-Kunstleve, J. S. Jiang, J. Kuszewski, M. Nilges, N. S. Pannu, R. J. Read, L. M. Rice, T. Simonson, and G. L. Warren, *Acta Crystallogr. D. Biol. Crystallogr.* **54**, 905 (1998).

[24] A. Vagin and A. Teplyakov, *J. Appl. Crystallogr.* **80**, 1022 (1997).

[25] N. P. Cowieson, J. F. Partridge, R. C. Allshire, and P. J. McLaughlin, *Curr. Biol.* **10**, 517 (2000).

[26] A. Perrakis, M. Harkiolaki, K. S. Wilson, and V. S. Lamzin, *Acta Crystallogr. D. Biol. Crystallogr.* **57**, 1445 (2001).

[27] T. A. Jones and M. Kjeldgaard, "O Version 5.9." 1994.

[28] K. Croce, S. J. Freedman, B. C. Furie, and B. Furie, *Biochemistry* **37**, 16472 (1998).

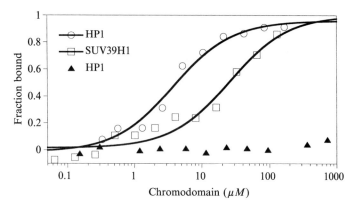

FIG. 4. Fluorescence anisotropy binding data. Binding curves for the interaction of the HP1 chromodomain (open circles) and SuVar(3-9) chromodomain (open squares) with fluoresceinated-H3 peptide residues 1–15 trimethylated at lysine 9. The HP1 binding assay with the unmodified H3 peptide is plotted with filled triangles. All binding measurements were performed at 15° in 50 mM potassium phosphate, pH 8.0, 25 mM NaCl, and 2 mM DTT.

concentration of protein, $[Y_T]$ the total concentration of titrated peptide, and K_D is the dissociation constant. This technique has been used successfully to quantitate the binding of the HP1 chromodomain to a methylated H3 tail peptide.[10]

Fluorescence anisotropy spectroscopy has become popular for studying protein-peptide, protein-protein, and protein-DNA interactions due to the use of relatively simple and inexpensive equipment, and the ability to provide rapid results. A series of fluorescence anisotropy experiments has demonstrated that the chromodomains of HP1,[8,11] Pdd1p and Pdd3p,[29] and SuVar(3–9) (shown in Fig. 4) proteins selectively bind to H3 peptide methylated at lysine 9. For mathematical and theoretical descriptions of this technique, the reader is directed to more specialized references.[30–32] Fluorescence anisotropy titrations are fit to the following equation to obtain dissociation constants:

$$A = A_f + (A_b - A_f)\left[\frac{[X]}{K_D + [X]}\right], \tag{3}$$

[29] S. D. Taverna, R. S. Coyne, and C. D. Allis, *Cell* **110,** 701 (2002).

[30] R. C. Cantor and P. R. Schimmel, *in* "Biophysical Chemistry. Part II: Techniques for the Study of Biological Structure and Function," p. 433. WH Freeman and Co., San Francisco, 1980.

[31] D. M. Jameson and W. H. Sawyer, *Methods Enzymol.* **246,** 283 (1995).

[32] T. Heyduk, Y. Ma, H. Tang, and R. H. Ebright, *Methods Enzymol.* **274,** 492 (1996).

where A is the measured anisotropy, A_f the anisotropy of the free state, A_b the anisotropy of the bound state, and $[X]$ is the protein concentration. This relationship is only true if the concentration of labeled species $\ll K_D$, and operates under the assumption that the peptide can exist in only two states, free and bound to the protein. A more complicated relationship is needed if the system to be studied necessitates that these assumptions are not met.

Figure 4 provides examples of interactions which have been successfully studied by fluorescence anisotropy. HP1 chromodomain interactions with two H3 peptides are shown. The HP1 chromodomain does not interact with the unmodified H3 peptide; instead there is strong binding to H3 peptide with a trimethyllysine at residue 9. This technique can effectively be used to rapidly examine predictions of binding properties based on sequence homology. The surface of the HP1 chromodomain which interacts with methylated H3 tail consists of residues that are conserved in SUV39H1 chromodomain, and thus we predicted that SUV39H1 also interacts with methylated H3 tail strongly.[11] Figure 4 shows that indeed the chromodomain of SUV39HI binds the H3 peptide with trimethylated lysine 9 with an affinity of 25 μM.

Practical Considerations

When using fluorescence anisotropy to study protein-protein interactions, the smaller of the two interacting proteins should carry the fluorescent probe. This experimental design results in the greatest difference in fluorescence anisotropy values between the free and bound states of the fluorophore, and thus maximal experimental sensitivity. Ideally the larger macromolecule is serially diluted to cover a concentration range of 100-fold below to 100-fold above the K_D of the interaction being studied. Most protein-histone tail interactions observed to date occur with an affinity of $\sim 10^{-6}$ M. Therefore, protein samples should range from approximately 0.1 μM to 1 mM for initial studies. As outlined earlier, it is very important to keep the concentration of the fluorescently labeled peptide at least 10-fold (preferably 100-fold) less than that of the K_D. Failure to do so could result in inaccurate affinity calculations. For protein-histone tail interactions, a concentration of ~ 100 nM has been found to be a good starting point. Concentrations of both the labeled and unlabeled proteins can then be fine tuned after initial experiments are conducted.

Instrumentation

We have used two different instruments (Beacon 2000 and Tecan SpectraFluor) for this type of measurement, and obtained comparable results. The Beacon 2000 fluorescence polarization system by PanVera has been used with much success in chromodomain/H3 tail interaction studies.[8,11,29] This system is extremely easy to use and relatively inexpensive. The Beacon 2000 system is capable of performing experiments at temperatures from 6–65°, making it ideal for monitoring the affinity of interaction over various temperatures. An alternative to the Beacon 2000 system is the Tecan SpectraFluor instrument which has the advantage that it can read samples in a 96-well plate format in a few minutes, therefore drastically reducing measurement time. However, the Tecan instrument is not capable of cooling below ambient temperature and is thus not suitable for evaluation of binding at low temperatures. Both instruments have minimal sample volumes of ~100 μl and comparable sensitivities of ~10^{-11} M fluorescein. In addition, both instruments offer numerous wavelength interference filters enabling many different fluorophores to be monitored.

Choice of Probes

Selection of fluorophores and coupling techniques are extensively described elsewhere.[33] Fluorescein is the most popular of the fluorescent probes used in fluorescence anisotropy assays due to its high quantum yield and long wavelengths of excitation and emission. The lifetime of the fluorescein excited-state is on the order of 4 ns,[30] making it possible to monitor protein complexes of molecular weight <100 kDa. Fluorophores with longer excited-state lifetimes should be selected in studies examining the formation of complexes larger than 100 kDa. A wide range of conjugation chemistries are available making it possible to conjugate fluorophores to amines, thiols, alcohols, aldehydes, ketones, hydrazines, and carboxylic acids.[33] The fluorescent label should be placed at a position on the peptide far enough away from the presumed interaction site to avoid interference with binding. In our experience, placing the fluorescein label at four residues away from binding segment of the peptide is sufficient to avoid interference in binding affinity.

Protocol for N-Terminal Fluorescein Succinimidyl Ester Labeling

Synthetic H3 tail peptides were N-terminally labeled with fluorescein using a kit from PanVera in 50-μl reaction volumes. Each reaction contained 100 μg of peptide, 2 mM fluorescein succinimidyl ester, and 100 mM potassium phosphate, pH 7.0. The reaction was carried out at 37° for 30 min. After 30 min, 10 μl of 1 M Tris-HCl, pH 8.0, was added to quench the reaction. The fluoresceinated peptide was then separated

---

[33] R. P. Haugland, "Handbook of Fluorescent Probes and Research Products." Molecular Probes Inc., Eugene, OR, 1996.

from the free fluorescein by gel filtration chromatography in a micro column containing 0.5–1.0 ml of sephadex G10 resin (Pharmacia) equilibrated with water. The labeled peptide (colored orange) elutes first from the column and is easily distinguished from the free fluorescein (yellow). In addition, the polarization of fractions from the gel filtration column can be used to distinguish these species. We have found that at 25°, a labeled peptide of 15–20 residues will have a polarization of 15 mP, while the free fluorescein molecule shows a polarization of −20 mP.

Protocol for the Determination of the Binding Affinities of the HP1 Chromodomain to Methylated H3 Tail Peptides by Fluorescence Anisotropy

The purified HP1 fragment 17–76 was dialyzed into 50 mM potassium phosphate, pH 8.0, 25 mM NaCl, and 2 mM DTT and concentrated to 500 μM to 1 mM. Serial twofold dilutions of this concentrated protein stock solution were then made to produce twelve 100-μl samples with the lowest concentration ∼0.1 μM. The fluoresceinated H3 peptide was added to the chromodomain samples to a final peptide concentration of 100 nM. A second set of chromodomain samples with no peptide added acts as a blank. Samples were centrifuged in order to avoid turbidity, which can make anisotropy measurements inaccurate. Following equilibration to 15° in a water bath, the anisotropy of the samples was read in a Beacon 2000 fluorescence polarization system.

The beacon system reports polarization values; these were converted to anisotropy using Eq. (4):

$$A = \frac{2 \cdot P}{3 - P}. \tag{4}$$

The data are then plotted as log chromodomain concentration versus anisotropy and fit to Eq. (3) by nonlinear least squares analysis using Kaleidagraph (Synergy Software). A_f, A_b, and K_D are fit as multiple, floating independent variables. For presentation purposes, anisotropy is converted to fraction bound (F_b) using the relationship of Eq. (5):

$$F_b = \frac{A - A_f}{A_b - A_f}. \tag{5}$$

Isothermal Titration Calorimetry

Isothermal titration calorimetry (ITC) is presently the only method capable of providing a complete thermodynamic profile of a protein-protein or protein-DNA interaction for interactions with binding affinities on the order of 10^{-8} to 10^{-3} M. This method has been exploited to examine the interaction of the HP1 chromodomain with the H3 tail methylated at lysine 9[8,11] and the TAF$_{\text{II}}$250 double bromodomain with acetylated histone H4 tail peptides.[20]

Small amounts of heat are either generated or absorbed by every intermolecular interaction. The change in enthalpy (ΔH), binding affinity (K_D), and stoichiometry (n) of a molecular interaction can be deduced from the measurement of such heats of interaction. These values can then be exploited to derive the changes in entropy (ΔS), free energy (ΔG), and heat capacity (ΔCp) of binding by the following relationships:

$$\Delta G = -nRT(\ln K_D), \tag{6}$$

$$\Delta G = \Delta H - T\Delta S, \tag{7}$$

$$\Delta C_p = \frac{\partial H}{\Delta T}. \tag{8}$$

The complete thermodynamic analysis of an interaction sets ITC apart from all other methods. A full theoretical description of ITC has been previously reviewed and should be sought out by interested readers.[34–36]

ITC analysis has been used to determine the full thermodynamic parameters of the binding of the HP1 chromodomain to H3 tail peptides containing either trimethyllysine or dimethyllysine at position 9.[11] Figure 5 shows an example of the data obtained from such an experiment where the peptide titrated into the chromodomain sample contained trimethyllysine at position 9. Table I summarizes the thermodynamics of chromodomain interaction with H3 peptides containing dimethyllysine or trimethyllysine at position 9 as measured by ITC.

The data clearly indicate that HP1 preferentially binds to the trimethyllysine containing peptide over the dimethyllysine peptide. The full

TABLE I
ITC ANALYSIS OF THE BINDING OF THE HP1 CHROMODOMAIN TO METHYLATED H3 PEPTIDES

H3 peptide	K_D (μM)	ΔH (kcal/mol)	N	ΔG (kcal/mol)	$T\Delta S$ (kcal/mol)
Me$_2$K9 H3	6.9 ± 0.2	-9.47 ± 0.04	1.02	-6.78 ± 0.01	-2.69 ± 0.05
Me$_3$K9 H3	2.5 ± 0.1	-9.83 ± 0.04	0.99	-7.36 ± 0.02	-2.46 ± 0.02

[34] H. F. Fisher and N. Singh, *Methods Enzymol.* **259**, 194 (1995).
[35] L. Indyk and H. F. Fisher, *Methods Enzymol.* **295**, 350 (1998).
[36] R. L. Biltonen and N. Langerman, *Methods Enzymol.* **61**, 287 (1979).

thermodynamic analysis by ITC aids in the analysis of this behavior. The interaction with trimethyllysine is more enthalpically ($\Delta\Delta H = -0.4$ kcal/mol) and entropically [$\Delta(T\Delta S) = 0.2$ kcal/mol] favorable than that with dimethyllysine. This is due to improved cation-π interactions and the van der Waals contacts.[11]

Binding experiments such as the one depicted in Fig. 5 were conducted at $16°$ and $25°$ in order to calculate the change in heat capacity upon complex formation. Large negative changes in heat capacity upon binding are assumed to be correlated with burial of nonpolar surfaces or large

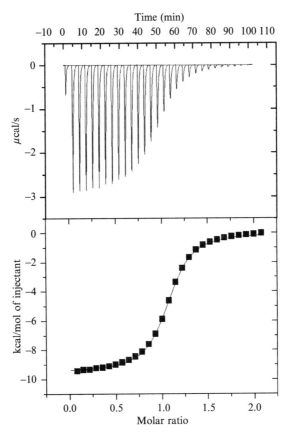

FIG. 5. Binding of the HP1 chromodomain to a trimethylated H3 tail peptide as measured by isothermal titration calorimetry. (A) Raw data showing the heats of injection. (B) Integrated heats of injection with the solid line representing the fit of the data to Eq. (10). Binding measurements were conducted at $25°$ in 50 mM potassium phosphate, pH 8.0, 25 mM NaCl, and 2 mM DTT.

conformational changes upon complexation.[37,38] The interaction of the chromodomain with the H3 tail shows a small ΔCp of -0.17 kcal/mol. This is consistent with lack of major conformational change in the chromodomomain upon peptide binding.[10,11]

Practical Considerations

ITC measurements are extremely sensitive to differences in buffer conditions; therefore, all protein and peptide solutions must be thoroughly dialyzed against a suitable buffer. The two interacting species must be dialyzed against the exact same buffer, and ideally in the same container. This is not always possible, particularly when one species is a small peptide. Alternatively, peptides can be exchanged into pure distilled water using a gel filtration or desalting column before being lyophilized. The lyophilized peptide can then be resuspended in the dialysis buffer of the protein sample. A buffer system should be chosen that ensures both the protein and peptide species are stable and soluble. A wide range of pH's, buffer, salt, and DTT concentrations are suitable for use in ITC experiments. Phosphate buffers are particularly useful in ITC due to low ionization enthalpy.[39] All solutions must be thoroughly degassed and devoid of any particulate matter.

The initial concentrations of the protein and peptide are critical to the success of the experiment. The proper concentrations of macromolecules are estimated using the following equation:

$$c = K_a \cdot M_{tot} \cdot n, \tag{9}$$

where c is a unit-less constant, K_a the binding association constant ($K_a = 1/K_D$), M_{tot} the total macromolecular concentration at the beginning of the experiment, and n is the stoichiometry of the interaction. The constant c can be used to estimate the shape of the binding isotherm produced by the experiment. Values of $c > 5000$ produce a rectangular curve with a height equal to exactly the ΔH of the interaction. It is not possible to accurately fit the K_D of an interaction when the c value is this high due to the sharp transition of the curve at the equivalence point. In the condition of $c < 1$, the curve becomes broad and featureless, making it impossible to extract the thermodynamics of the system. Therefore, to ideally measure the binding affinity and change in enthalpy, the c value should be above 1 and below 1000.[40] However, it is not possible to accurately estimate

[37] P. L. Privalov and G. I. Makhatadze, *J. Mol. Biol.* **213**, 385 (1990).

[38] G. I. Makhatadze and P. L. Privalov, *J. Mol. Biol.* **213**, 375 (1990).

[39] M. M. Pierce, C. S. Raman, and B. T. Nall, *Methods* **19**, 213 (1999).

[40] T. Wiseman, S. Williston, J. F. Brandts, and L. N. Lin, *Anal. Biochem.* **179**, 131 (1989).

the c value for initial experiments unless the binding affinity and stoichiometry are known fairly accurately. This may not be practical and several preliminary experiments may be needed to obtain general estimates of the stoichiometry and affinity.

Several control experiments should be conducted.[39,41] First, the buffer solution should be titrated into the cell containing the protein and the heat observed. Additionally, the reverse experiment, where the peptide is titrated into the buffer solution containing no protein should also be performed if sufficient amounts of peptide are available. These control experiments determine the heats of dilution of the system and the results are substracted from the experimental data to obtain accurate heats.

Protocol for Analyzing the Thermodynamics of the Interactions of the HP1 Chromodomain with H3 Tail Peptides Containing Dimethyllysine 9 or Trimethyllysine 9 by ITC

All calorimetry measurements were performed using a VP-ITC instrument from MicroCal (NorthHampton, MA).[40] Purified HP1 17–76 was concentrated to a volume of 3 ml yielding a concentration of 70 μM. This concentration was used because it produces a c value of $\sim$15 [Eq. (9)] assuming a stoichiometry of 1:1 and a K_a of 2^5 L/mol, where the latter was easily estimated by fluorescence anisotropy binding studies (see earlier). The sample was dialyzed against 3×4 L of a buffer solution of 50 mM sodium phosphate, pH 8.0, 25 mM NaCl, and 2 mM DTT. Synthetic H3 peptides were resuspended in the HP1 dialysis buffer to a concentration of 700 μM. The reference cell was filled with dialysis buffer and the sample cell with the 70 μM HP1 17–76 solution. The H3 peptide solution was then titrated into the sample cell with 25 injections of 10 μl each. Injections were spaced at 3-min intervals. The integrated heats of injection were fit to the quadratic equation:

$$Q = (n[cd]_t \Delta H V_0)/2 \cdot \left\{ 1 + [H3]_t/(n[cd]_t) + 1/(nK_a[cd]_t) \right.$$

$$\left. - \left\{ (1 + [H3]_t/(n[cd]_t) + 1/(nK_a[cd]_t))^2 - 4[H3]_t/(n[cd]_t) \right\}^{\frac{1}{2}} \right\}, \quad (10)$$

where Q is the heat evolved, n the stoichiometry of the interaction, ΔH the enthalpy change, $[cd]_t$ the total concentration of the chromodomain, $[H3]_t$ the total concentration of the H3 peptide, K_a the association constant, and V_0 is the initial volume. ΔH, n, and K_a are fit as floating multiple

[41] R. O'Brien, B. Z. Chowdhry, and J. E. Ladbury, *in* "Protein-Ligand Interactions: Hydrodynamics and Calorimetry," p. 263. Oxford University Press, London, 2001.

independent variables using the analysis software Origin which is used to normalize the heats of binding and perform a volume correction due to dilution before integration. A control curve representing the heat of dilution of the buffer titrated into the chromodomain sample was subtracted from the data before fitting. Eqs. (6) and (7) were then used to obtain ΔG and ΔS for the interaction.

Acknowledgments

W.F. is a Robert Black fellow of the Damon Runyon Cancer Research Foundation. This work was supported in part by NIH grant GM116635 to SK.

[10] Expression, Purification, and Biophysical Studies of Chromodomain Proteins

By Peter R. Nielsen, Juliana Callaghan, Alexey G. Murzin, Natalia V. Murzina, and Ernest D. Laue

Chromodomains[1,2] are one of a number of similar domains that are involved in chromatin structure.[3] They can be divided into two subfamilies: the chromodomains and the chromo shadow domains.[4] The chromodomain is found in a diverse selection of proteins, but the chromo shadow domain is only associated with the HP1-family proteins, where it is always found C-terminal to a chromodomain. For some of these domains it has been possible to identify an interacting partner. In these cases it is possible to use biophysical techniques to understand how these proteins work. Alternatively, one can use biophysical methods to study interactions with (or identify) less well characterized interacting partners that might bind to other chromodomains.

In the first section we will describe the structural features of chromodomains and how we can use this information to identify chromodomains from amino acid sequences. We have employed various biophysical techniques to study the chromodomain and chromo shadow domain of HP1β. Here we describe how we express and purify the individual domains as well as the intact protein. We also describe how we characterize the purified

[1] J. C. Eissenberg, *Gene* **275**, 19 (2001).
[2] R. Paro and D. Hogness, *Proc. Natl. Acad. Sci.* **88**, 263 (1991).
[3] S. Maurer-Stroh, N. J. Dickens, L. Hughes-Davies, T. Kouzarides, F. Eisenhaber, and C. P. Ponting, *Trends Biochem. Sci.* **28**, 69 (2003).
[4] R. Aasland and A. F. Stewart, *Nucl. Acids Res.* **23**, 3168 (1995).

proteins using steady-state fluorescence spectroscopy, analytical ultracentrifugation (AUC), and nuclear magnetic resonance (NMR) spectroscopy. There will not be room for a description of how we use NMR to determine 3D structures, but the procedures we use in our laboratory are the subject of a recent review.[5]

Sequence Analysis and Recognition of Chromodomains

The chromodomain (chromo box) was first defined as a conserved sequence motif found in the *Drosophila* HP1 and Pc proteins.[2] Since then several chromodomain structures have been determined,[6–9] all of which show that the domain extends well beyond the most conserved residues that defined the original chromo box used in sequence searches. The availability of these 3D structures has now allowed us to define the sequence requirements that are needed to form the chromodomain fold.

The chromodomain structure comprises an N-terminal three-stranded β-sheet that packs against a C-terminal α-helix.[6–9] The conserved chromo box sequence corresponds to the second/third strands, which are connected by a loop of variable length (Fig. 1). The structural core and packing between the β-sheet and the α-helix is well conserved—it is determined by the pattern of hydrophobic residues and by the presence of conserved bulges in the first strand (Fig. 1B). These bulges are crucial for defining the geometry of the interaction between the sheet and the helix, and thus the domain's overall 3D structure (Fig. 1A). The structural core of the domain also includes part of the linker between the β-sheet and the α-helix. Consequently, variations in the length of the linker are limited to either one or two turns of helix, as exemplified by the chromodomain and chromo shadow domain of HP1β (aka MOD1 or M31), respectively (Fig. 1). Neither longer nor shorter linkers connecting the sheet to the helix can be incorporated without disrupting the structure.

[5] D. Nietlispach, H. R. Mott, K. M. Stott, P. R. Nielsen, A. Thiru, and E. D. Laue, *in* "Methods in Molecular Biology, Protein NMR Techniques. 2nd Edition: Structure Determination of Protein Complexes by NMR" (K. A. Downing, ed.), Vol. 278, Humana Press Inc., Totowa, NJ, (in press) (2004).

[6] L. J. Ball, N. V. Murzina, R. W. Broadhurst, A. R. C. Raine, S. J. Archer, F. J. Stott, A. G. Murzin, P. B. Singh, P. J. Domaille, and E. D. Laue, *EMBO J.* **16,** 2473 (1997).

[7] S. V. Brasher, B. O. Smith, R. H. Fogh, D. Nietlispach, A. Thiru, P. R. Nielsen, R. W. Broadhurst, L. J. Ball, N. V. Murzina, and E. D. Laue, *EMBO J.* **19,** 1587 (2000).

[8] N. P. Cowieson, J. F. Partridge, R. C. Allshire, and P. J. McLaughlin, *Curr. Biol.* **10,** 517 (2000).

[9] D. A. Horita, A. V. Ivanova, A. S. Altieri, A. J. S. Klar, and R. A. Byrd, *J. Mol. Biol.* **307,** 861 (2001).

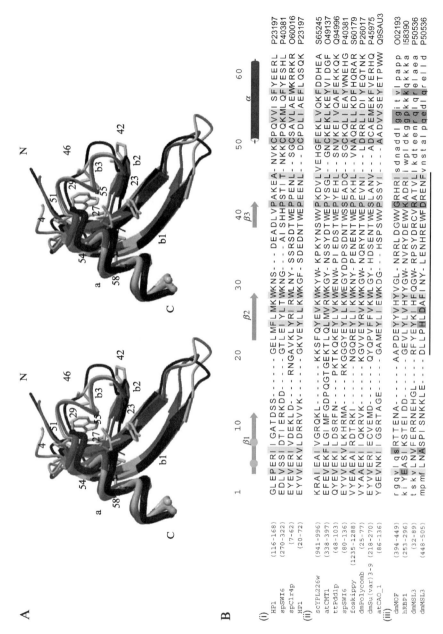

Fig. 1. (*continued*)

The structural studies, together with a large number of biochemical experiments, originally suggested that chromodomains might act as protein interacting modules, which function by targeting different protein complexes to particular sites on chromatin.[6] It is now clear that a complete chromodomain structure is essential for these functions. First, there are functionally impaired HP1 and Pc mutants with point substitutions in the hydrophobic core which appear to destabilize the chromodomain structure.[6,10,11] Second, in the HP1β and Swi6 chromo shadow domains, the α-helix provides both the interface for dimerization as well as the binding site for interaction with other proteins.[7,8] Finally, both the HP1 and Swi6 chromodomains have recently been shown to specifically recognize and bind to Lys-9 methylated histone H3.[12–14] The hydrophobic groove formed between the β-sheet and the α-helix provides the binding-site for this interaction,[15,16] that is, the formation of this binding site requires an intact chromodomain.

[10] S. Messmer, A. Franke, and R. Paro, *Genes Dev.* **6**, 1241 (1992).

[11] J. S. Platero, T. Hartnett, and J. C. Eissenberg, *EMBO J.* **14**, 3977 (1995).

[12] M. Lachner, D. O'Carrol, S. Rea, K. Mechtler, and T. Jenuwein, *Nature* **410**, 116 (2001).

[13] A. J. Bannister, P. Zegerman, J. F. Partridge, E. A. Miska, J. O. Thomas, R. C. Allshire, and T. Kouzarides, *Nature* **410**, 120 (2001).

[14] J.-I. Nakayama, J. C. Rice, B. D. Strahl, C. D. Allis, and S. I. Grewal, *Science* **292**, 110 (2001).

[15] P. R. Nielsen, D. Nietlispach, H. R. Mott, J. Callaghan, A. Bannister, T. Kouzarides, A. G. Murzin, N. V. Murzina, and E. D. Laue, *Nature* **416**, 103 (2002).

[16] S. A. Jacobs and S. Khorasanizadeh, *Science* **295**, 2080 (2002).

FIG. 1. Sequence alignment and structures of chromodomains. (A) Superposition of the structures of the HP1β chromodomain (black, PDB code 1apo), the HP1β chromo shadow domain (grey, PDB code 1dz1), and the Swi6 chromo shadow domain (white, PDB code 1e0b). (B) Comparison of sequences containing the chromo box consensus motif (i) chromodomains with known structures, (ii) selected chromodomain sequences that clearly fit the requirements of the structure, and (iii) the putative chromodomains from MSL3, MOF1, and RBP1. In (B) the positions of secondary structure elements are indicated by green arrows (*extended* strands) and dark blue cylinders (α-helices). The blue dots indicate the positions of the conserved bulges in the first strand. Conserved hydrophobic core residues that define the fold are shown in yellow, while those determining the borders of the turn between the β-strand and the helix are shown in green. In MSL3, MOF1, and RBP1, the grey residues indicate absence of conserved residues important for the formation of the 3D chromodomain fold. The chromo box used in previous sequence analysis is indicated by a black line. For each sequence the residue and SWISS-PROT/TrEMBL access numbers are given. Residues are color coded similarly in (A) and (B)—the residue numbers in (A) correspond to the positions of the amino acids in the sequence alignment. Panels (A) and (B) were produced using MOLSCRIPT [P. J. Kraulis, *J. Appl. Crystallogr.* **24**, 946 (1991).] and ALSCRIPT [G. J. Barton, *Protein Eng.* **6**, 37 (1993).], respectively. (See color insert.)

Based on this analysis we can now construct a structure-based sequence profile (the highlighted residues in Fig. 1B), which allows us to readily detect intact, and therefore functional, chromodomains in proteins in which it has been predicted. As an example, consider the MSL-3 protein, and the histone acetyl-transferase MOF, which have been recently found to contain protein-RNA interaction modules.[17] Sequence analysis originally suggested that the region of these proteins involved in RNA binding contains a chromo box motif.[4,18,19] However, on examination of the structure-based sequence profile it is clear that there are no suitable conserved sequences flanking the chromo box in the MSL3, MOF, and RBP1 families (Fig. 1B). In fact, the only part of the sequence that is conserved in these families maps on to the second/third strands and their connecting loop. This loop varies between the known chromodomains and does not determine the fold. We conclude that the RBP1, MOF, and MSL-3 proteins cannot contain canonical chromodomains. These results suggest that the MOF and MSL-3 proteins bind RNA sequences via a novel RNA binding domain that contains a conserved sequence motif found also in chromodomains.

The ability to recognize domains from amino acid sequences is clearly a useful tool when characterizing proteins with unknown function. However, as we demonstrate here, sequence motifs have to be used with care and should be validated against structural and functional data when it is possible.

In the following sections, we describe the procedures that we use to express and purify chromodomains, as well as the biophysical methods we employ to study these domains and their interactions with other peptides and proteins.

Expression and Purification Procedures

Expression of HP1 Proteins and their Domains

Ideally for structural studies, for example, for making NMR samples, one needs 5–10 mg of protein that is at least 90% pure and stable at room temperature for about 2 weeks. This puts exacting requirements on the methods employed for protein production. Optimizing the level of protein expression is the first step. With both the chromodomain and chromo shadow domain we had unexplained protein expression before induction that could not be prevented, for example, by adding more glucose to the growth media, by changing the *Escherichia coli* host strain, or by

[17] A. Akhtar, D. Zink, and P. B. Becker, *Nature* **407,** 405 (2000).
[18] E. V. Koonin, S. Zhou, and J. C. Lucchesi, *Nucleic Acids Res.* **23,** 4229 (1995).
[19] A. Hilfiker, D. Hilfiker-Kleiner, A. Pannuti, and J. C. Lucchesi, *EMBO J.* **16,** 2054 (1997).

overexpression of the Lac repressor. This "leaky" expression resulted in eventual plasmid loss and a low yield of protein—presumably due to toxicity of the proteins. We found that we could not obtain the desired level of protein expression using overnight or non-stationary liquid starter cultures to inoculate growth media used for protein expression. Instead, we chose to inoculate the growth media with a large number of colonies directly from plates. Freshly transformed cells were plated out so as to get large numbers of isolated colonies—typically we used 4 plates per 0.5 L of growth media. All the isolated colonies, but not the cell lawn, were used to inoculate the growth flasks with an approximately equal number of cells—$OD_{600} \sim 0.1$–0.3. In addition, we added glucose to 1% w/v. Protein expression was induced with 1 mM IPTG at $OD_{600} \sim 0.7$ and the cells were grown for another 3 h before collection.

Purification of HP1 Proteins and Their Domains

The cells were then resuspended in ice cold lysis buffer (0.5 M NaCl, 20 mM Tris of pH 7.9, and 5% (v/v) Sigma General Protease Inhibitor Cocktail) and broken in a French Press (or using an equivalent method). NP-10 was added to 0.5% (w/v), β-mercaptoethanol (BME) to 2 mM, imidazole to 4 mM and the cells spun at 15,000 rpm for 30 min in a precooled rotor to remove insoluble material. (Positively charged chromodomains also require suitable treatment with DNAse I before centrifugation.) The supernatant was then loaded onto a Ni-column—typically we use a Ni-NTA resin (Qiagen, Novagen, Pierce). To make a new column, 4 ml of 50% agarose beads are placed in a plastic 20-ml column (Bio-Rad) which already contains about 5 ml of water. Once the agarose has settled a hydrophobic disc is placed on the top of the gel. The column is then washed with 20 ml of water, 20 ml of buffer A (300 mM NaCl in 20 mM Tris of pH 7.9), 20 ml of 1 M imidazole in buffer A, and then 20 ml of buffer A again. This washing procedure will remove excess Ni from the column. It is optional for new agarose, but obligatory after recharging with Ni (see later). The columns can be reused many times for the same protein purification, but should be recharged with Ni using 20 ml of 0.5 M NiSO$_4$ following use. Sometimes it is advisable to additionally repack the column.

The supernatant is loaded to the top of a column equilibrated in buffer A. After loading, the column is washed with 50 ml of buffer A containing 2 mM BME, 4 mM imidazole and 100 μl of Sigma General Protease Inhibitors. It is then washed with 50 ml of 20 mM imidazole in buffer A (without the BME and protease inhibitors) and then with 25–30 ml of 80 mM imidazole in buffer A. The later washes should be collected separately in 5 and 25 ml aliquots and analyzed by SDS gel electrophoresis.

Most of the unwanted proteins will typically elute in the first 5 ml, but the 25-ml aliquot may contain some protein. Next, the protein is eluted with 25 ml of 0.5 M imidazole in buffer A. Analysis of the various fractions by SDS gels is then needed to determine which should be pooled.

The eluted solutions contain Ni leached from the column, which may cause precipitation during protein concentration. To avoid this we usually pass the solutions through a Chelex resin column (Bio-Rad). The protein is then concentrated to $\sim$5 ml in an Amicon 60-ml cell (Millipore) using a Diaflo YM3 membrane (Millipore) and buffer exchanged on a PD-10 column (Pharmacia) into Factor Xa cutting buffer (100 mM NaCl, 50 mM Tris of pH 8.0). At this stage the OD$_{280}$ is measured to estimate the yield. ($A_{280} = 1.0$ is $\sim$0.5 mg/ml for the chromodomain, 0.66 mg/ml for the chromo shadow domain and 0.75 mg/ml for full length HP1β.) Factor Xa (Roche) is dissolved in the Factor Xa buffer and added to the protein at between 1:50 and 1:100 (w/w). After addition of CaCl$_2$ to 2 mM final, the sample is left overnight at 25°. (The cleaved His-tag can then be removed by passing the sample through the Ni-column again if required.) At this stage the protein is usually sufficiently pure for most biochemical applications. However, these proteins are not sufficiently stable for structural studies and after a few weeks, storage, degradation bands appear on SDS gels. We have found that the chromodomains can be safely kept at $-20°$, whereas the shadow domain and full-length proteins tend to aggregate after being frozen (even with flash freezing or in the presence of glycerol). To further purify the protein, an ion exchange chromatography step (MonoQ or MonoS (Pharmacia) columns) helps to improve the protein stability. After Factor Xa digestion we concentrate the protein solution to 5 ml using an Amicon 60-ml cell, prior to buffer exchange on the PD-10 column into buffer B—10 mM Na phosphate (pH 7.5–8.0 for MonoQ and pH 5.5–6.5 for MonoS). Alternatively, one can dialyze the protein in buffer B either overnight or for 2–3 h to reduce the salt concentration. A 1-ml MonoQ or MonoS column is washed/equilibrated with 5 ml buffer B (see earlier), 10 ml buffer C (1 M NaCl in buffer B) and 5 ml buffer B. The sample is loaded onto the column in buffer B and run on an FPLC system at 0.5 ml/min, collecting 1-ml fractions. The column is then washed with 10 ml of buffer A and a gradient applied from 0% to 60% C over 60 ml and from 60% to 100% C over 5 ml. The resulting fractions are run on an SDS gel and pooled as appropriate. The shadow domain elutes from the column in lower-salt concentration than the chromodomain and full-length HP1β protein, which have higher surface charge. Mass spectrometry of the samples should then be used to identify whether any of the protein sample has become degraded or clipped at either the N- or C-termini.

Sample Preparation

Protein aggregation will interfere with most protein-protein interaction assays and will decrease the quality of NMR spectra or the likelihood of crystallization. The full-length HP1 protein and the shadow domain usually require gel filtration to remove any aggregated protein, whereas the chromodomains samples do not. After prolonged storage at 4°, samples may need to be re-run on the gel filtration column to remove further aggregates that form. For gel filtration of NMR samples, we use a Superdex S75 column (120-ml bed volume from Pharmacia) in 100 mM NaCl, 20 mM phosphate of pH 5.5–8.0 at 1 ml/min over 150 ml, collecting 1-ml fractions between 35 and 125 ml. The protein fractions are pooled, concentrated, and buffer exchanged if required. To check for any remaining aggregation, and to carry out binding assays, we run an analogous 2.4-ml bed volume gel filtration column using a Smart system (Pharmacia). For NMR sample preparation, the protein-containing S75 fractions were concentrated to 0.4 ml in a 10-ml Amicon cell and buffer exchanged into NMR buffer (10 mM Na phosphate with pH as required by the experiment). The A_{280} of the solution is measured using a quartz capillary (Shimadzu) with 0.5-mm diameter (sample volume of 7–10 μl) to estimate the protein concentration and 50 μl D_2O, ^{2}H-DTT to 10 mM, ^{2}H-EDTA to 1 mM, and 0.01% NaN_3 are added to the protein sample. Finally, the sample is made up to 0.55 ml with NMR buffer.

Making Mixed Dimer Chromo Shadow Domain Samples

The shadow domain is a symmetrical dimer and equivalent residues in each monomer give rise to the same set of signals in the NMR spectra. Accordingly, it is not possible to distinguish between signals that arise from NOEs between intra- and inter-molecular proton interactions at the dimer interface. To overcome this problem we made a protein sample where one monomer is unlabeled and the other ^{13}C/^{15}N-labeled. With such a sample one can employ NMR experiments that detect interactions between two protons, one which is attached to a ^{13}C-nucleus and the other to a ^{12}C-nucleus. This allows direct and unambiguous determination of inter-molecular interactions at the dimer interface. Normally one can make such a sample by mixing equal amounts of the differently labeled proteins and leave them until they reach equilibrium. In the case of the shadow domain this failed. The two forms of the shadow domain were mixed to final concentrations of either 0.1–1 mM or 100 nM to 1 μM. The mixed samples were left to equilibrate for 1 week. No exchange of monomers was detected by NMR experiments (for the high-concentration range) or by Coomassie stained pull-down assays (for the low-concentration range). In addition, an AUC experiment showed that all of the shadow domain exists as a dimer without detectable trace of monomers—the K_d for dimer dissociation being below the detection range. Taking all these results together we concluded that the shadow domains form very stable dimers with a low exchange rate and an estimated K_d of below 150 nM. These findings also suggest that, *in vivo*, exchange of subunits between the different isoforms of HP1 is unlikely.

To prepare a mixed dimer for NMR experiments we denatured a mixture of equal amounts of the ^{12}C- and ^{13}C-labeled proteins in 6 M guanidium HCl (GuHCl) and then refolded the protein by subsequent dilution to reduce the denaturant concentration. To determine the GuHCl concentration where unfolding/refolding is occurring, we measured UV absorption spectra of the shadow domain at different GuHCl concentrations and found that there was a large intensity change between 4 and 5 M GuHCl. For denaturation, 0.5 ml of the protein solution was added to 10 ml of 6 M GuHCl in 10 mM Na phosphate of pH 8.0, 1 mM EDTA and 1 mM DTT, and left for 10 min at room temperature. For subsequent refolding, GuHCl solutions of decreasing concentration in the same buffer were very slowly added (drop-by-drop—each solution taking about 10 min to add) as follows: 10 ml of 4.5 M GuHCl, 10 ml of 4 M GuHCl, 20 ml of 3 M GuHCl, and finally 30 ml of buffer alone. The refolded protein was then concentrated in an Amicon cell and purified by gel filtration (as mentioned earlier).

Expression and Purification of the Histone H3 N-Terminal Peptides

Histone H3 N-terminal peptides (amino acids 1–14 or 1–17) were expressed as GST fusions in BL21·pREP4 (pREP4 is Qiagen plasmid with an extra copy of the *LacI* gene) at 24°. The GST-fusion peptide was purified by conventional procedures except that the cell lysate was treated with DNAse I and the glutathione agarose beads were extensively washed after binding the protein with 1 M NaCl and 0.5% NP-40 in 20 mM Tris of pH 7.5 to remove DNA contamination. After washing, the beads were transferred into thrombin cutting buffer (as recommended by the manufacturer, Novagen) and the peptide was cleaved by overnight incubation of the bead suspension at 4°. The peptide was washed from the beads with 100 mM NaCl in 10 mM Tris of pH 8.0, loaded directly onto an octadecyl (C18) column, and eluted with a 0–100% water/acetonitrile gradient containing 0.1% trifluoroacetic acid. The eluted peptide was freeze dried and redissolved in Na phosphate of pH 8.0.

Reductive Methylation and Production of ^{13}C-Methyl–Labeled Histone H3 N-Terminal Peptides

It is possible to produce peptides that are dimethylated on all the amino groups in an unmodified substrate by reductive alkylation.[20] At high pH, and in the presence of sodium borohydride as a reducing agent, the N-terminal α-NH$_2$ and the ε-NH$_2$ groups of lysine side chains will receive first one and then another methyl group from formaldehyde. We used this method to produce peptides that were specifically ^{13}C-labeled in all the methyl positions of the dimethylated amino groups. This enabled us to unambiguously identify signals from the Lys-9 methyl groups in NMR spectra of histone H3. One hundred microliters of 3.5 mM purified peptide were mixed on ice with 900 μl of freshly made 1 mg/ml NaBH4 in 200 mM NaHBO3 of pH 9.0 and 5 μl of 20% ^{13}C-labeled formaldehyde. The mixture was incubated on ice for 6 min and then transferred to a new tube containing 1 mg of NaBH4 powder. A further 5 μl of the formaldehyde solution was added and the cycle was repeated eight times. The final product was checked by mass spectrometry for uniform methylation and the absence of cross-linking.

Chromodomain–Peptide Complex NMR Samples

Samples of the complexes between chromodomains and peptides were prepared in the following way. For the chromodomain–histone H3 peptide complex, the protein obtained after gel filtration was buffer exchanged as for chromodomain samples and concentrated in an Amicon cell but this time to 0.25 ml. The histone H3 peptide was then added from a 2-mM stock solution to the required level. For studies of the shadow domain–CAF-1 complex, the 25-mer CAFp150 peptide is not very soluble in aqueous buffer. This meant that the complex had to be made at low concentration. To avoid aggregation the peptide was added to a protein solution diluted ~20 times compared to the NMR sample concentration before concentrating the complex using an Amicon cell equipped with a YM1 membrane. To both types of sample 55 μl D$_2$O was added and the sample volume adjusted to 550 μl with the appropriate phosphate buffer. Usually the samples also contained 10 mM [2]H-DTT, 1 mM [2]H-EDTA, and 0.01% NaN$_3$.

Fluorescence Spectroscopy to Study Ligand Binding

Fluorescence spectroscopy is an ideal method by which to study the interaction of the HP1β chromodomain with histone H3.[21] The interaction is mediated via the methyl groups of lysine 9 of histone H3, which fit into a hydrophobic pocket on the surface of the chromodomain formed by the

aromatic residues Tyr 21, Phe 45, and Trp 42. Protein fluorescence origi-
nates from the aromatic residues, Phe, Tyr, and Trp. However, the fluorescence of proteins containing all three aromatic amino acids is usually
dominated by tryptophan. Upon methyl lysine binding to the chromodomain, the microenvironment of the tryptophan changes and thus the intensity of the fluorescence emission alters. This provides a mechanism,
intrinsic to the protein, by which the interaction can be studied. Steady-
state fluorescence not only identifies the occurrence of an interaction, but
can also be used to determine the affinity of the binding.

Sample Preparation

In order to perform steady-state fluorescence analysis of the chromodomain–histone H3 complex the protein must be present in a suitable buffer.
The buffer must maintain the protein in the correctly folded state and must
not affect the intensity of the light emitted from the tryptophan residues.
For this reason it is important to remove any imidazole left in the sample
after Ni^{2+} column purification. (However, it is worth noting that the presence of glutathione does not have this effect.) Thus, during the final stages
of chromodomain purification the protein was buffer exchanged into
10 mM Na phosphate of pH 7.4 and 150 mM NaCl. The solution of the
HP1β chromodomain was made up to a concentration of 3 μM, as confirmed by amino acid analysis. The solution of the histone H3 peptide, to
which binding of the chromodomain was to be investigated, was made up
to a concentration of 1 mM within a solution of the 3 μM chromodomain.
This eliminates the need to account for any dilution effect upon addition of
the peptide to the chromodomain.

Experiments

In order to perform the titrations the chromodomain solution was
placed in a 1-cm × 1-cm quartz cuvette in the spectrofluorimeter (Perkin
Elmer, LS55 Luminescence Spectrometer) equipped with a thermostated
cell holder. All solutions used were previously incubated in a water bath
and all experiments were performed at 23°. After the protein solution
was transferred to the cuvette and had been inserted into the spectrofluorimeter, the sample was left for ~10 min in order to ensure that the temperature had equilibrated. (Fluctuations in temperature are to be avoided since
they affect the thermodynamics and kinetics of the interaction, and thus the
fluorescence intensity emitted on excitation.) The sample was excited at

[20] G. E. Means and R. E. Feeney, *Biochemistry* **7**, 2192 (1968).
[21] M. R. Eftink, *Methods Enzymol.* **278**, 221 (1997).

280 nm and emission spectra were recorded between 300 and 400 nm—excitation and emission slits of 5 nm, and a scan speed of 200 s were used. Following the initial measurement of the chromodomain alone, small aliquots of 0.5–2.5 μl of the 1 mM histone H3 peptide solution were added sequentially (increasing the peptide concentration by 0.5–2.5 μM at a time) to the solution of the chromodomain within the cuvette. Upon peptide addition the solution was mixed using a plastic sterile loop and allowed to settle prior to recording the next emission spectrum. The peak fluorescence intensity is detected at ~335 nm confirming that the chromodomain is folded (the value for tryptophan in aqueous solution is for comparison, 348 nm). During the titration the tryptophan fluorescence intensity increased on addition of the histone H3 peptide. Additions of the peptide were then continued until no change in the fluorescence intensity was detected and saturation of the chromodomain H3 binding site had been achieved. Figure 2 shows a typical titration series. In this way it was

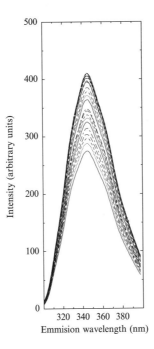

FIG. 2. Tryptophan fluorescence emission spectra of a HP1β (Y21W) chromodomain on addition of histone H3 peptide. Binding of the H3 peptide to the chromodomain gives rise to an increase in the intensity of the fluorescence. The fluorescence intensity ceases to increase on addition of further peptide once saturation of the binding site has been achieved.

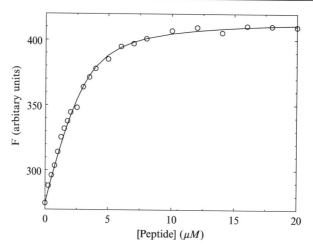

FIG. 3. Binding curve for the interaction of histone H3 tri-methylated at Lys-9 with the chromodomain of HP1β. The K_d for the interaction is 0.65 μM and has been determined using Eq. (1) (see text).

possible to compare the interaction between the chromodomain and a number of histone H3 peptides with different modifications. We studied H3 peptides, each methylated by different amounts at different lysine residues. One was tri-methylated at both Lys-4 and Lys-9, whereas in the other both residues were di-methylated. A third peptide was tri-methylated solely at Lys-9 and a fourth tri-methylated only at Lys-27.

Data Analysis

The change in fluorescence intensity upon addition of an interacting species provides a measure of the degree to which the environment of the Trp residues of the protein under investigation become altered due to the interaction. The resultant fluorescent intensities can be plotted as a function of the concentration of the interacting species (Fig. 3), and the K_d can then be determined for the interaction by least squares fitting using the equation[21]:

$$F_{obs} = F_P + (F_{PL} - F_P)/(2[P]_{tot})([P]_{tot} + [L]_{tot} + K_d$$
$$-\text{sqrt}\left(([P]_{tot} + [L]_{tot} + K_d)^2 - 4[P]_{tot}[L]_{tot}\right), \qquad (1)$$

where F_{obs} represents the fluorescence observed, F_p represents the fluorescence of the free protein, F_{PL} represents the fluorescence of the bound species, and $[P]_{tot}$ and $[L]_{tot}$ represent the total protein and ligand concentrations, respectively.

Results

The studies of the interaction of the chromodomain with bis-methylated H3 peptides showed that the binding affinity of the chromodomain to the H3 peptide di-methylated at both Lys-4 and Lys-9 was very similar to that of the analogous tri-methylated peptide. The K_d s for the chromodomain binding to the two peptides were found to be 2.1 and 1.9 μM, respectively. The K_d for the interaction of the chromodomain with the H3 peptide tri-methylated at only Lys-9 was found to be only slightly lower than those determined for the interaction of the chromodomain with the bis-methylated H3 peptides, with a K_d of 0.7 μM. This implies that the methylation of Lys-4 neither contributes to nor greatly interferes with the recognition of Lys-9 methylated histone H3 by the HP1β chromodomain. The binding studies of the chromodomain with the H3 peptide methylated at Lys-27 showed that this peptide binds much more weakly ($K_d \sim 12$ μM). Despite the sequence similarity to the Lys-9 position, it is clear that the sequence surrounding Lys-27 does not provide a suitable interaction partner for the chromodomain of HP1β, as was predicted from the structure.[15,16]

Characterizing Protein Samples by AUC

Analytical ultracentrifugation is a well-established technique for measuring the hydrodynamic properties of proteins (or complexes) in solution. The equations describing the results obtained using the ultracentrifuge rest on a solid thermodynamic foundation that has been well described.[22] Experiments performed in the ultracentrifuge give information on the size and shape of solutes in well-defined solvents. It is possible to measure accurate molecular weights, characterize interactions, or to investigate complex mixtures. In our laboratory we mainly use AUC for two purposes: to investigate the oligomeric state of purified protein samples, and to confirm complex formation and the stoichiometry of interactions. It follows that we only analyze highly purified proteins and we normally do not carry out sedimentation velocity experiments since we already know what the components in the solution are. Instead, we investigate the behavior of proteins by sedimentation equilibrium analysis.

In sedimentation equilibrium experiments the sample is spun at a velocity where it is possible to achieve an equilibrium between sedimentation in the gravitational field and diffusion in the opposing direction. When equilibrium is attained the concentration of the solute is measured as a

[22] T. Svedberg and K. O. Pedersen, *in* "The Ultracentrifuge" (R. H. Fowler and P. Kapitza eds.), Clarendon Press, Oxford, 1940.

function of the radial position. For a single ideal solute the radial dependence of the concentration can be described by[22,23]:

$$c(r) = c(r_0) \exp\left[M(1 - v_{bar}\rho)\omega^2\left(r^2 - r_0^2\right)/(2RT)\right] + K, \qquad (2)$$

where v_{bar} is the partial specific volume of the solute, ρ is the density of the solvent (both of which can be calculated from the amino acid and buffer composition, respectively, using Sednterp, Tom Laue and David B. Hayes, University of New Hampshire),[24] ω is the angular velocity, $c(r)$ is the concentration at the radial position r in centimeters, M is the molecular weight of the solute, R is the gas constant, and T is the temperature in degrees Kelvin. K is a constant added to take account of imperfections in the optical system. The first point included in the calculations is at r_0, with $c(r_0)$ being the concentration at this point and r is the distance from the center of the rotor in centimetres. With a knowledge of v_{bar} and ρ it is possible to derive M by least squares fitting. This expression can be expanded to describe mixtures of molecules, the behavior of a system in equilibrium between several species and also non-ideal behavior.[22,25]

Experimental Design

Normally we design our experiments such that it would be possible to detect multimerization in a reliable manner. This essentially means performing nine separate experiments to obtain enough data to allow the determination of dissociation constants. This is done by preparing three samples, containing different concentrations of the protein in the same buffer, and performing equilibrium experiments at three different rotor speeds. The average molecular mass of the samples should change with starting concentration and with angular velocity for interacting systems, but not if it is a single species.

When choosing the optimal concentration for the experiments there are three main issues to be considered. The first is to make sure that it is possible to detect the protein in a concentration dependent manner. Second, when looking at interactions it is necessary to work at concentrations where all the species in the equilibrium are present in significant amounts. Finally, it has to be taken into consideration that at high concentrations proteins will exhibit non-ideal behavior.

The Beckman XL-I instrument that we use employs two different types of detection: absorption (180–800 nm) and Rayleigh interference optics. It

[23] J. W. Williams, K. E. Van Holde, R. L. Baldwin, and H. Fujita, *Chem. Rev.* **58,** 715 (1958).

[24] All software mentioned in this section can be obtained from the RASMB web site: http://www.bbri.org/RASMB/

[25] D. K. McRorie and P. J. Voelker, "Self-Associating Systems in the Analytical Ultracentrifuge." Beckman Instruments Inc., Palo Alto, California.

is recommended that the maximal adsorption does not exceed 1.2—although in our experience the detector appears to be linear up to an absorption of 2.0. We normally choose starting concentrations that give an absorption at 280 nm of between 0.2 and 0.5 in a 1-cm cell. (In our hands the signal-to-noise ratio gets too low, if we use a sample with a lower absorption than this.) For the interference optics the maximal concentrations that can be employed are around 5 mg/ml.[26]

The next thing to decide is what rotor speeds to apply. The goal is to choose the speed such that there is a good curvature on the measured concentration profile (large concentration difference between the meniscus and the bottom of the cell) while ensuring that the sample does not pellet. This is achieved by selecting rotor speeds that will give a reduced apparent molecular weight [the term in the exponent of Eq. (2)] of between 1.5 and 5.0. When studying homo dimerization the speeds are chosen based on the molecular weight of the monomer—making sure not to pellet the dimer.

Sample Preparation

It is essential to use samples of high purity and quality if accurate information about size or affinity is to be obtained. Any impurities or non-specific aggregates can contribute to the signal measured in the experiments compromising the interpretation of the data. The best results are obtained, if the sample is subjected to gel filtration chromatography shortly before performing the AUC experiments to remove any non-specific higher-order aggregates.

It is important that the buffer used as the reference in each cell corresponds exactly to the solvent that holds the protein of interest. This can be achieved in a number of ways. Typically, the fractions containing the protein after gel filtration chromatography will need to be concentrated. If this is done by filter retention, the buffer that comes through the filter can be used for any dilutions needed as well as for the reference buffer. Alternatively, the protein sample can be dialyzed extensively against the required final buffer and this can be used as the reference. Another method, that is recommended by Laue,[27] is to use spin columns for buffer exchange.[28]

The choice of buffer will also affect the result of the experiment. It is important not to introduce substances that interfere with the detection method used to follow the concentration of the protein. Thus, DTT cannot be used in the buffer if protein concentrations are monitored by absorption

[26] G. Ralston, "Introduction to Analytical Ultracentrifugation." Beckman Instruments, Palo Alto, California, 1993.

[27] T. Laue, "Short Column Sedimentation Equilibrium Analysis for Rapid Characterization of Macromolecules in Solution." *Technical Information DS-835.* Beckman Coulter, Palo Alto, California, 1992.

[28] R. I. Christopherson, M. E. Jones, and L. R. Finch, *Anal. Biochem.* **100,** 184 (1979).

at 280 nm. Breakdown products of DTT absorb light at this wavelength and even very careful calibration of sample and reference solutions cannot remove all of the contribution to the final signal that comes from these substances (i.e., interpretation of the absorbance in terms of protein concentration will not be possible). Since experiments are performed over several days, some samples require the presence of antioxidants. In these cases we use TCEP—Tri(2-carboxyethyl)-phosphine[29] (obtained from Pierce). In order to minimize the effects of non-ideality on the results, buffers should also contain more than 150 mM NaCl.[26]

AUC Studies of a Typical Chromodomain

To investigate the oligomeric state of a chromodomain, three samples were prepared by dilution to concentrations corresponding to an A_{280} value of 0.4, 0.6, and 0.8. For each sample the left chamber in a two-sector centerpiece of an assembled cell is loaded with 100 μl of the sample and the right chamber with 110 μl of the reference buffer. The samples are brought to equilibrium at three speeds of 25,000, 30,000, and 34,000 rpm at 4°. The radial concentration gradient of protein in the sample chamber is measured both as the absorbance at 280 nm and by the number of fringes from bottom to top of the cell. For each speed the samples are left for 16 h prior to the first measurement. Another set of data is then collected after 20 h and the two are compared to verify that there is equilibrium between sedimentation and diffusion in all of the samples. Ideally the samples should, after the data sets for the highest rotor speed have been collected, be spun at a velocity that sediments the protein, and a further data set is collected. This is done in order to measure the contribution to the signal from imperfections in the optical system in the centrifuge. In practice this can be difficult to achieve when working with small proteins since the material of the standard centerpieces (charcoal filled Epon, Beckman) cannot tolerate the high speeds required. It can, however, be overcome by using centerpieces made from aluminum or titanium.

The collected data can then be analyzed by least squares fitting to the model described by Eq. (2). The values of $c(r_0)$ and K are fitted locally for each data set, but the apparent molecular weight is determined globally. We use the program WinNonlin.[30] Models of increasing complexity are used to describe the data until one is found where there is good agreement between measured and calculated data. This means that we first assume that there is only one ideal component in the sample. Based on the value obtained compared to the calculated molecular weight of the protein under investigation the

[29] J. A. Burns, J. C. Butler, J. Moran, and G. M. Whiteside, *J. Org. Chem.* **56**, 2648 (1991).
[30] M. L. Johnson, J. J. Correira, D. A. Yphantis, and H. R. Halvorson, *Biophys. J.* **36**, 575 (1981).

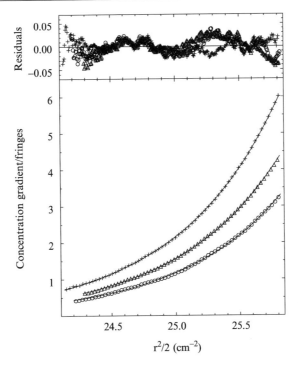

Fig. 4. AUC data for a chromodomain. Three data sets recorded using the Rayleigh interference optics at equilibrium at 30,000 rpm are shown. The starting concentrations for the three samples were: $(\circ)A_{280} = 0.4$, $(\triangle)A_{280} = 0.6$, and $(+)A_{280} = 0.8$. The solid lines are calculated using the parameters obtained by fitting all nine data sets simultaneously to Eq. (2) as described in the text. For clarity only every fifth point is plotted. The residual values left when subtracting the calculated values of the concentration from the measured values are shown in the top half of the graph. An even distribution around zero indicates that the data are described well by the model. The remaining data sets were recorded at 25,000 and 34,000 rpm.

next step can be decided (Fig. 4). The shape of the residuals obtained from the difference between the measured and calculated data can also be used as an indication of the behavior of the system under study.[25] In this case the molecular weight calculated for the chromodomain is 10,600 Da compared to 10,149 Da as calculated based on the amino acid sequence.

Characterization of Chromodomains by NMR Spectroscopy

A detailed characterization of the binding site for ligands of the chromodomains can be carried out using NMR spectroscopy.

In NMR spectroscopy it is possible to take advantage of the fact that the nuclei of atoms with spin quantum number equal to $1/2$, distribute between two energy levels when placed in a magnetic field.[31] The energy difference between these two states can be measured and is characteristic of the strength of the magnetic field of the spectrometer used for the experiment and the type of nucleus. The local electronic environment changes the size of the magnetic field experienced by a nucleus. This means that each atom in a protein will have a characteristic and different value. The value of this local contribution is called the chemical shift. Because of this it is in principle possible to identify signals in the NMR spectrum corresponding to all the different spin 1/2 nuclei in the protein (or complex) under investigation. For proteins the nuclei commonly used are the proton (^{1}H) and particular isotopes of nitrogen (^{15}N) and carbon (^{13}C), respectively. Whereas 99.98% of hydrogen is ^{1}H only 0.37% is ^{15}N, and 1.11% is ^{13}C. Thus, it is necessary to enrich the fraction of these isotopes to nearly 100% by expressing the protein in a host that grows on media containing either one or both of these isotopes as the sole nitrogen and carbon source, respectively.[32]

Characterizing Interactions

A ligand will alter the environment of the atoms in the amino acids that it interacts with. This means that the chemical shift of these atoms will change when the ligand binds and this binding can be followed by NMR spectroscopy. From the theory of NMR it can be shown that these changes can be categorized into one of three types—which one occurs depends on the difference in the chemical shift between the bound and the free form compared to the exchange rate of the ligand. If the exchange is slower than the size of the change in chemical shift between the free and bound form, then signals corresponding to both the free and the bound forms can be observed and the relative populations can be estimated by measuring the intensity of the two signals. In this case the system is said to be in the slow exchange limit. This will be the case for strong binding. In the fast exchange limit where the exchange is fast compared to the chemical shift difference, there will only be one observable signal for a given nucleus with a chemical shift that is the weighted average of the two forms. The intermediary exchange regime gives rise to highly broadened signals that are difficult to analyze.

This means that, unless the system is in intermediate exchange, it is possible to obtain dissociation constants for an interaction by adding the ligand to the protein and following the change in chemical shift (or intensity) of a well-resolved signal during the titration. One-dimensional proton spectra can in principle be used. If it is possible to obtain isotope-labeled samples, however, it is advantageous to record 2D heteronuclear experiments such as a 2D ^{1}H-^{15}N HSQC.[33–35] This spectrum will contain one signal

[31] J. Cavanagh, W. J. Fairbrother, A. G. Palmer, III, and N. J. Skelton, "Protein NMR Spectroscopy. Principles and Practice." Academic Press, San Diego, 1996.
[32] L. P. Mcintosh and F. W. Dahlquist, *Q. Rev. Biopys.* **23,** 1 (1990).

for each backbone amide proton and its attached nitrogen nucleus in the protein. The limited number of signals, combined with the improved resolution gained from distributing the signals into two dimensions, make these spectra some of the most convenient to analyze. The spectrometer time needed for each experiment can be as short as 10 min, provided the sample is sufficiently concentrated, making it feasible to perform a detailed titration.

The value of the K_d can be measured without doing any assignment of signals in the NMR spectrum to particular atoms in the protein. If an assignment is available, however, it is possible to identify the amino acids in the protein that are likely to be involved in the interaction. This is simply done by looking to see if a signal corresponding to a given residue changes its chemical shifts on addition of the ligand. The chemical shift is, however, a very sensitive marker and even small structural rearrangements upon binding can in some cases give rise to significant changes for residues far from the actual binding site. A good example of this is the chromo shadow domain where the addition of the PXVXL-containing peptide ligand breaks the symmetry and alters the structure of the dimer causing two thirds of the backbone amide signals to shift in the NMR spectra.[7] In these cases the binding site can normally still be identified by ignoring small changes, and, if the structure is known, by only considering candidates that are known to be surface exposed.[5]

Histone H3 Binding to the HP1β Chromodomain

As part of the initial characterization of histone H3 peptide binding to HP1β we mapped the binding site using the approach described earlier. A 1-mM sample of uniformly ^{15}N-labeled HP1β chromodomain was prepared in 20 mM Na phosphate of pH 7.0, with 150 mM NaCl and 10% D_2O. A 2D ^{1}H–^{15}N HSQC spectrum was recorded on a Bruker DRX600 spectrometer (Bruker, Karlsruhe) at 298 K, collecting 4 scans for each of the 150 time points in the indirect dimension. The spectral width was set to 10,000 Hz in the proton dimension and 2000 Hz in the indirect nitrogen dimension. Titration experiments are performed by gradually adding aliquots of concentrated solutions of ligand to the NMR sample. In cases where the solubility of the ligand makes it impossible to achieve very high concentrations the effect of dilution of the NMR sample can be alleviated by having the chromodomain present in the ligand stock solution at the NMR sample concentration. Additions are performed either using a spatula that will fit in the NMR tube or by transferring the solution to an Eppendorf tube.

[33] G. Bodenhausen and D. J. Ruben, *Chem. Phys. Lett.* **69,** 185 (1980).

[34] M. Piotto, V. Saudek, and V. Sklenar, *J. Biomol. NMR* **2,** 661 (1992).

[35] V. Sklenar, M. Piotto, R. Leppik, and V. Saudek, *J. Magn. Reson. Series A* **102,** 241 (1993).

For each addition the sample is transferred to an Eppendorf tube and mixed with peptide solution before being transferred back into the NMR sample tube. An experiment identical to the one described earlier was then recorded. Most signals were in the fast exchange regime. This means that the titration can be followed by changes in chemical shift, and the K_d of the interaction can be obtained by fitting those chemical shift changes to the same function (1) as is used for steady state fluorescence data. Peptide solutions were prepared from lyophilized material and the concentrations were determined by amino acid analysis. (*Note:* We find that it is possible to get a reasonable estimate of the amount of the peptide by weighing it prior to dissolving it in the buffer and then dividing the mass obtained by two to account for the residual water.)

Identifying the Binding Site

Figure 5 shows the first and last point of such a titration. The spectrum had been assigned for the free chromodomain[6] and, since the exchange regime was at the fast exchange limit, the signals for the complex could

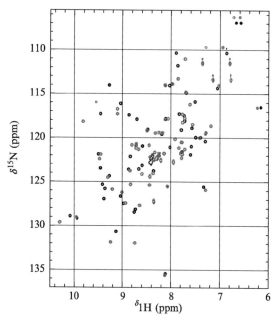

Fig. 5. 2D ^{1}H-^{15}N HSQC spectra identifying the peptide binding site. Signals from the ^{15}N-labeled HP1β chromodomain (10–80) without (black) and with (grey) a 2:1 excess of a bis-(Lys-4,9) di-methylated histone H3 (1–18) peptide: (NH$_2$-ARTK(Me)$_2$QTARK (Me)$_2$STGGKAPGG-COOH). Spectra recorded on samples where the added peptides have bis-tri-methylated (Lys-4,9) or tri-methylated Lys-9 give similar results.

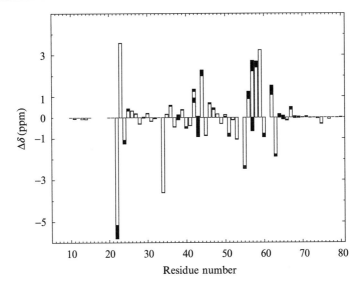

Fig. 6. Chemical shift differences for the ^{15}N (open bars) and ^{1}H$_N$ (filled bars) atoms of the HP1β chromodomain, in the presence of histone H3, were determined from the spectra in Fig. 5.

be assigned by following them as they moved when the ligand concentration was gradually increased:

$$\delta_{obs} = p_{free} * \delta_{free} + p_{complex} * \delta_{complex},$$

where p_{free} and $p_{complex}$ are the molar fractions of free protein and protein in the complex, respectively, δ_{obs} is the measured chemical shift, δ_{free} is the chemical shift of the free form, and $\delta_{complex}$ is the chemical shift in the complex. These assignments were later confirmed by sequential assignment.[15] This meant that it was possible to assign a value for the change in the chemical shift values ($\Delta\delta$) for most amino acid residues in the chromodomain (Fig. 6). As can be seen from Fig. 7 this enabled us to confirm that the peptide binds in the region originally proposed based on the structural similarities between the chromodomain and Sac7d.[6]

A similar experiment was performed using the unmodified histone H3 peptide (data not shown). From the changes in the spectrum it can be seen that the unmodified peptide interacts with the same binding site as the methylated species. It is clear, however, that the interaction is very weak. Even with a 2.5-fold excess of the peptide the average chemical shift is still only approximately halfway between the free form and the shift for

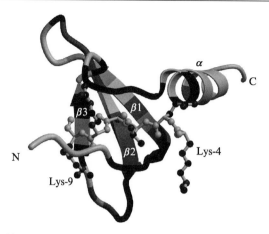

FIG. 7. The histone H3 binding site mapped onto the structure of the HP1β chromodomain using the chemical shift differences plotted in Fig. 6. The chromodomain is shown as a ribbon cartoon with the peptide represented as a ball-and-stick model. The peptide backbone is shown in white, whereas the side chain atoms are darker. Black regions of the chromodomain ribbon show changes of more than 1 ppm. Lighter regions show smaller changes. The figure was prepared using Molscript [P. J. Kraulis, *J. Appl. Crystallogr.* **24,** 946 (1991)].

the complex with Lys-9 methylated peptides. Since the experiments were conducted at 0.1 mM this indicates that the K_d for the interaction is around 0.2 mM, explaining why it was not possible to detect this interaction in pull-down assays.[12,13]

It was possible to characterize the binding site further. The same 2D ^{1}H-^{15}N HSQC experiment was performed on a complex between the chromodomain and a peptide that was tri-methylated at Lys-9. In this case the chemical shifts were compared to the values from the sample that was di-methylated at Lys-9. Only a subset of the residues was further affected by the presence of the extra methyl group. These were Tyr-21, Trp-42, and residues 47–51. This suggested that the methyl lysine binding site was situated primarily in the loop between the second and third β-strand. When the structure was solved we were able to confirm that residues 21, 42, and 51 are indeed some of the main residues that interact with the methyl lysine side chain.

[11] Quantitative Assays for Characterization of the Sir2 Family of NAD+-Dependent Deacetylases

By MARGIE T. BORRA and JOHN M. DENU

Introduction

Within nucleosomes, histones are targets of enzyme-catalyzed post-translational modifications such as acetylation, phosphorylation, methylation, and ubiquitination.[1–4] Dynamic modification of nucleosomes regulates processes such as DNA synthesis, DNA repair, and transcription. Among these modifications, reversible acetylation has emerged as a major mechanism in controlling these processes. Hyperacetylation has been correlated with transcriptionally active genes, while hypoacetylation has been correlated with silenced loci. Acetylation is carried out by histone acetyltransferases, which catalyzes the transfer of the acetyl group from acetyl-CoA to the ε-amino group of lysine residues.[5,6] Histone/protein deacetylases catalyze the removal of the acetyl group from the lysine residues.[7,8]

Histone/protein deacetylases have been classified based on their similarity to the yeast enzymes: the Rpd3-like (class I), the HDA-like (class II), and the Silent Information Regulator 2 (Sir2)-like (class III) deacetylases.[7–9] Both class I and class II, which are conserved in many eukaryotes and are commonly referred to as HDACs, likely catalyze the removal of the acetate group from the acetylated lysine residue by activation of a water molecule for direct hydrolysis,[10] generating free acetate and the deacetylated protein (Fig. 1). Both class I and class II have been found in large multiprotein complexes that are recruited by DNA-binding proteins to specific-chromatin domains. Class I and class II enzymes are sensitive to the inhibitor Trichostatin A (TSA).[7,8]

---

[1] J. Wu and M. Grunstein, *Trends Biochem. Sci.* **25,** 619 (2000).

[2] K. Luger and T. J. Richmond, *Curr. Opin. Genet. Dev.* **8,** 140 (1998).

[3] A. P. Wolffe and J. J. Hayes, *Nucleic Acids Res.* **27,** 711 (1999).

[4] P. Cheung, C. D. Allis, and P. Sassone-Corsi, *Cell* **103,** 263 (2000).

[5] M. H. Kuo and C. D. Allis, *Bioessays* **20,** 615 (1998).

[6] S. Y. Roth, J. M. Denu, and C. D. Allis, *Annu. Rev. Biochem.* **70,** 81 (2001).

[7] S. G. Gray and T. J. Ekstrom, *Exp. Cell Res.* **262,** 75 (2001).

[8] C. M. Grozinger and S. L. Schreiber, *Chem. Biol.* **9,** 3 (2002).

[9] S. Khochbin, A. Verdel, C. Lemercier, and D. Seigneurin-Berny, *Curr. Opin. Genet. Dev.* **11,** 162 (2001).

[10] M. S. Finnin, J. R. Donigian, A. Cohen, V. M. Richnon, R. A. Rifkind, P. A. Marks, R. Breslow, and N. P. Pavletich, *Nature* **401,** 188 (1999).

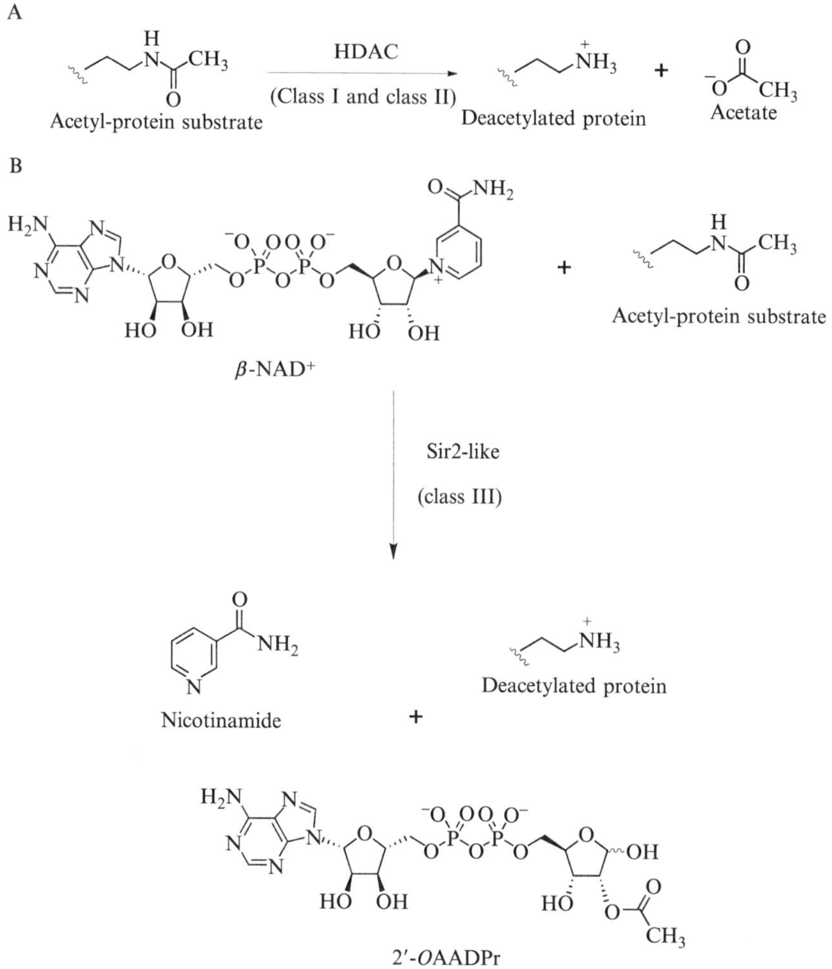

A

Acetyl-protein substrate   Deacetylated protein   Acetate

B

β-NAD⁺   Acetyl-protein substrate

Sir2-like

(class III)

Nicotinamide   Deacetylated protein

2′-OAADPr

FIG. 1. Comparison of the HDAC and Sir2-like reactions. (A) Class I and class II HDACs likely activate a water molecule for direct hydrolysis of the acetyl group from the acetylated substrate to form deacetylated protein and free acetate as products. (B) The Sir2 family of deacetylases catalyzes the tight coupling of NAD⁺ cleavage and substrate deacetylation to produce nicotinamide, deacetylated product, and O-acetyl-ADP-ribose (OAADPr).

Among the deacetylases, the Sir2 family (class III) of enzymes, also referred to as sirtuins,[11,12] are unique, as they are TSA-insensitive and NAD⁺-dependent.[8,13–17] These enzymes are conserved in many organisms ranging from some bacteria to humans.[12] The founding member of this family, yeast Sir2, has been found in multiprotein complexes and is required for gene silencing in yeast.[18] Sir2-like enzymes tightly couple the cleavage of NAD^+ and the deacetylation of a substrate to produce nicotinamide, deacetylated product, and the novel compound, O-acetyl-ADP-ribose (OAADPr)[16,19–22] (Fig. 1). Stoichiometric analysis of the enzymatic reaction showed that for every mole of NAD^+ cleaved and substrate deacetylated, 1 mol of nicotinamide, deacetylated product, and OAADPr are produced.[16,19,23,24] The enzymatically derived OAADPr has been reported to be $2'$-O-acetyl-ADPribose ($2'O$AADPr),[20–22] which through an intramolecular transesterification reaction, equilibrates with $3'O$AADPr in solution.[20,21] Although recent investigations demonstrated OAADPr bioactivity and OAADPr metabolism by specific nuclear and cytoplasmic enzymes, the physiological functions for this metabolite remain to be established.[24,25]

Although histones are considered targets for so-called histone deacetylases, enzymes within these deacetylase families also target nonhistone proteins. Recently, various acetylated proteins such as p53,[26,27] the transcription factor

[11] R. A. Frye, *Biochem. Biophys. Res. Commun.* **260**, 273 (1999).

[12] R. A. Frye, *Biochem. Biophys. Res. Commun.* **273**, 793 (2000).

[13] S. Imai, C. M. Armstrong, M. Kaeberlein, and L. Guarente, *Nature* **403**, 795 (2000).

[14] J. S. Smith, C. B. Brachmann, I. Celic, M. A. Kenna, S. Muhammad, V. J. Starai, J. L. Avalos, J. C. Escalante-Semerena, C. Grubmeyer, C. Wolberger, and J. D. Boeke, *Proc. Natl. Acad. Sci. USA* **97**, 6658 (2000).

[15] J. Landry, A. Sutton, S. T. Tafrov, R. C. Heller, J. Stebbins, L. Pillus, and R. Sternglanz, *Proc. Natl. Acad. Sci. USA* **97**, 5807 (2000).

[16] K. G. Tanner, J. Landry, R. Sternglanz, and J. M. Denu, *Proc. Natl. Acad. Sci. USA* **97**, 14178 (2000).

[17] J. M. Denu, *Trends Biochem. Sci.* **28**, 41 (2003).

[18] D. Moazed, *Mol. Cell* **8**, 489 (2001).

[19] J. C. Tanny and D. Moazed, *Proc. Natl. Acad. Sci. USA* **98**, 415 (2001).

[20] A. A. Sauve, I. Celic, J. Avalos, H. Deng, J. D. Boeke, and V. L. Schramm, *Biochemistry* **40**, 15456 (2001).

[21] M. D. Jackson and J. M. Denu, *J. Biol. Chem.* **277**, 18535 (2002).

[22] J. H. Chang, H. C. Kim, K. Y. Hwang, J. W. Lee, S. P. Jackson, S. D. Bell, and Y. Cho, *J. Biol. Chem.* **277**, 34489 (2002).

[23] J. Landry, J. T. Slama, and R. Sternglanz, *Biochem. Biophys. Res. Commun.* **278**, 685 (2000).

[24] M. T. Borra, F. J. O'Neill, M. D. Jackson, B. Marshall, E. Verdin, K. R. Foltz, and J. M. Denu, *J. Biol. Chem.* **277**, 12632 (2002).

[25] L. A. Rafty, M. T. Schmidt, A. L. Perraud, A. M. Scharenberg, and J. M. Denu, *J. Biol. Chem.* **277**, 47114 (2002).

[26] J. Luo, A. Y. Nikolaev, S. Imai, D. Chen, F. Su, A. Shiloh, L. Guarente, and W. Gu, *Cell* **107**, 137 (2001).

[27] H. Vaziri, S. K. Dessain, E. Ng Eaton, S. I. Imai, R. A. Frye, T. K. Pandita, L. Guarente, and R. A. Weinberg, *Cell* **107**, 149 (2001).

TAF$_1$68,[28] acetyl-CoA synthetase,[29] and tubulin[30] have been shown to be substrates for some Sir2-like enzymes *in vitro* and/or implicated *in vivo*.

Differences in the chemistry of Sir2- and HDAC-catalyzed reactions necessitate the utilization of distinct assays, as approaches commonly used for evaluation of HDAC activity may not be applicable for characterization of Sir2-like enzymes. In this chapter, we will briefly overview general HDAC assays, then detail methodologies for quantitative analysis of the Sir2-catalyzed reaction. These quantitative assays permit steady-state kinetic analysis, which can be used for comparison of different Sir2 homologues, determination of the enzyme's kinetic mechanism, examination of substrate specificity, and analysis of enzyme inhibitors. High-performance liquid chromatography (HPLC)-based, charcoal-binding and TLC-based assays are used to monitor and quantify the substrates and/or products of the reaction. Enzyme-catalyzed nicotinamide-NAD$^+$ exchange reaction, in which nicotinamide condenses with an enzyme-ADP-ribose–like intermediate to regenerate the substrate NAD$^+$ and acetylated protein, allows for assessment of the catalytic step preceding the formation of the deacetylated product and *O*AADPr. Given that nicotinamide is a strong product inhibitor of the Sir2-catalyzed reaction, the outlined nicotinamide inhibition analyses can be used to distinguish the Sir2 family of deacetylases.

General Approaches for Monitoring HDAC Activity

Detection of HDAC-Produced Acetate

Detection of the liberated acetate using ethyl acetate extraction is a common method of monitoring HDAC activity.[31–37] The reactions typically involve the incubation of the enzyme with radiolabeled-acetylated

[28] V. Muth, S. Nadaud, I. Grummt, and R. Voit, *EMBO J.* **20**, 1353 (2001).

[29] V. J. Starai, I. Celic, R. N. Cole, J. D. Boeke, and J. C. Escalante-Semerena, *Science* **298**, 2390 (2002).

[30] B. J. North, B. L. Marshall, M. T. Borra, J. M. Denu, and E. Verdin, *Mol. Cell* **11**, 437 (2003).

[31] D. Kolle, G. Brosch, T. Lechner, A. Lusser, and P. Loidl, *Methods* **15**, 323 (1998).

[32] S. Imiliani, W. Fischle, C. Van Lint, Y. Al-Abed, and E. Verdin, *Proc. Natl. Acad. Sci. USA* **95**, 2795 (1998).

[33] S. J. Darkin-Rattray, A. M. Gurnett, R. W. Myers, P. M. Dulski, T. M. Crumley, J. J. Allocco, C. Cannova, P. T. Meinke, S. L. Colletti, M. A. Bednarek, S. B. Singh, M. A. Goetz, A. W. Dombrowski, J. D. Polishook, and D. M. Schmatz, *Proc. Natl. Acad. Sci. USA* **93**, 13143 (1996).

[34] M. J. Hendzel, G. P. Delcuve, and J. R. Davie, *J. Biol. Chem.* **266**, 21936 (1991).

[35] R. Sendra, I. Rodrigo, M. L. Salvador, and L. Franco, *Plant Mol. Biol.* **11**, 857 (1988).

[36] A. Kervabon, J. Mery, and J. Parello, *FEBS Lett.* **106**, 93 (1979).

[37] J. Taunton, C. A. Hassig, and S. L. Schreiber, *Science* **272**, 408 (1996).

histones or peptides. The reaction is quenched by addition of acid (HCl/ acetic acid) to denature the enzyme, and the radiolabeled acetate is extracted using the organic solvent ethyl acetate. Radioactivity of the organic phase is determined by scintillation counting. Although ethyl acetate extraction has been successfully used to determine general HDAC activity, its applicability is limited to class I and class II HDACs, in which acetate is a direct product of the reaction (Fig. 1A). For the Sir2-like–catalyzed reaction, the acetyl group is transferred to the ADP-ribose portion of NAD$^+$ to generate OAADPr (Fig. 1B), which cannot be extracted in ethyl acetate (Borra and Denu, unpublished data).

Detection of Acetylated and Deacetylated Substrates

Besides ethyl acetate extraction, general histone/protein deacetylase activity has been analyzed by various approaches such as gel electrophoresis, western blot analysis, and filter-binding assays. The western blot analysis utilizes antibodies that recognize the acetylated forms of histones or other substrates.[38,39] Reaction mixtures are loaded onto SDS-PAGE gels and subjected to western blotting using antibodies against the acetylated forms of proteins. Using radiolabeled, acetylated protein/histone, SDS-PAGE gels can also be dried and exposed to photographic film, after which the loss of radioactivity on the substrate can be visualized.[15,40] Alternatively, the reaction mixture can be spotted onto a Whatman P81 cation exchange paper, which traps the radiolabeled, acetylated protein substrate.[15,40] After extensive washing, the paper is subjected to scintillation counting, and a decrease in the counts reflects the level of deacetylation.[15,40] Another alternative is to quench and precipitate the proteins by adding 20% trichloroacetic acid (TCA). Samples can be filtered through a glass filter on a vacuum manifold. The precipitated proteins bind to the filter, and radioactivity in the filter is determined by scintillation counting.[40]

Other less utilized approaches that rely on detection of the deacetylated protein product have also been developed and include the scintillation proximity assay, and nonisotopic, fluorescence-based assays. The scintillation proximity assay does not rely on the separation of substrates and products and is therefore suitable for higher throughput screening.[41] For

[38] Y. Zhang, G. LeRoy, H. P. Seelig, W. S. Lane, and D. Reinberg, *Cell* **95**, 279 (1998).

[39] Y. Zhang, N. Li, C. Caron, G. Matthias, D. Hess, S. Khochbin, and P. Matthias, *EMBO J.* **22**, 1168 (2003).

[40] J. Landry and R. Sternglanz, *Methods* **31**, 33 (2003).

[41] B. Nare, J. J. Allocco, R. Kuningas, S. Galuska, R. W. Myers, M. A. Bednarek, and D. M. Schmatz, *Anal. Biochem.* **267**, 390 (1999).

this assay, the substrate used is a biotinylated [³H]acetyl histone H4 peptide, which was shown to bind to streptavidin-coated SPA beads impregnated with scintillant. Binding of the [³H]acetyl peptide stimulates the light emission via the scintillant, which can be detected using a scintillation counter. Decrease in the radioactive signal indicates deacetylation of the substrate. The fluorescence-based deacetylase assay relies on the release of a fluorophore on deacetylation of the substrate. One fluorogenic assay relies on the use of N-(4-methyl-7-coumarinyl)-N-α-(*tert*-butyloxycarbonyl)-N-Ω-acetyllysinamide (MAL) as a substrate.[42] Upon substrate deacetylation, the fluorogenic product is separated from the reaction mixture and quantitated using a reversed-phase HPLC. A more recently developed fluorogenic assay is based on ability of trypsin to cleave deacetylated substrates.[43] For this assay, the peptide substrates have the acetylated lysine residue attached to a 4-methylcoumarin-7-amide (MCA) moiety at the carboxy terminus. Upon substrate deacetylation, trypsin cleaves the peptides proximal to the lysine residues, releasing the MCA moiety, whose fluorescence can be monitored at $\lambda_{em} = 460$ nm ($\lambda_{ex} = 390$ nm).

Although the scintillation proximity and the nonisotopic, fluorescence-based assays allow for high-throughput analysis, the substrates (biotinylated [³H]acetyl histone H4 peptide, MAL, and protein-MCA complex) used for these assays are not commercially available and may therefore limit their general use.

Approaches for Characterization of the Sir2 Family of Deacetylases

In this section, we will describe various quantitative approaches for specifically monitoring the enzymatic activity of the Sir2 family of NAD^+-dependent histone/protein deacetylases. First, a method for producing and quantifying a [³H]acetylated substrate will be described. Then quantitative methods for monitoring deacetylase activity of Sir2-like enzymes will be discussed. Using these methods, products formed during the Sir2-catalyzed reaction can be quantified and the initial velocity of the reaction can be determined. These approaches can, therefore, be utilized for steady-state kinetic analysis. Methods that are applicable to the class I and class II deacetylases will be indicated where appropriate. Nicotinamide-utilizing assays that are specific for the Sir2 family of deacetylases will also be discussed.

[42] K. Hoffmann, G. Brosch, P. Loidl, and M. Jung, *Nucleic Acids Res.* **27**, 2057 (1999).

[43] D. Wegener, F. Wirsching, D. Riester, and A. Schwienhorst, *Chem. Biol.* **10**, 61 (2003).

Production and Quantitation of [³H]acetylated H3 Peptide

The acetylated histone H3 peptide is a good substrate for many Sir2 homologues.[13,15,16,21,23,24] This section provides a detailed protocol for the p300/CBP-associating factor (P/CAF)-catalyzed lysine-14 [³H]acetylation of the H3 peptide, corresponding to the N-terminal residues of histone H3.[24] The reaction is carried out in the presence of 10–15 μM [2-³H₃]acetyl-CoA (NEN Life Science Products), 250 μM H3 peptide, 6–8 μM P/CAF in 100 mM Tris, pH 7.5, with 1 mM DTT at 37° for 1 h. The reaction is quenched by the addition of trifluoroacetic acid (TFA) to a final concentration of 1%. The quenched reaction is injected onto a reversed-phase HPLC column (e.g., Vydac C18 Small Pore column, 10 × 250 mm, 201SP1010) to resolve the acetylated and unacetylated H3 peptide, which can be monitored by measuring the absorbance at 214 nm. The HPLC is run with 100% solvent A (0.05% TFA/H₂O) for 1 min followed by a linear gradient of 0–40% solvent B (0.02% TFA in acetonitrile) over 40 min. Fractions are collected, and radioactivity is determined by scintillation counting. Fractions containing the [³H]acetylated H3 peptide ([³H]AcH3) are pooled and lyophilized to dryness. The lyophilized [³H]AcH3 can be resuspended in water or buffer of choice.

To quantify the [³H]AcH3 produced, the radioactivity of 3–5 μl aliquots of the resuspended [³H]AcH3 is determined by scintillation counting. Because there is a 1:1 stoichiometry between the moles of [³H]acetyl-CoA consumed and the moles of [³H]AcH3 produced, the total moles of [³H]AcH3 in solution can be determined by dividing the CPM of the peptide aliquots by the specific activity (CPM/mol) of the [³H]acetyl-CoA stock solution and taking into account the dilution factor.

If nonradiolabeled peptides are desired, the amount of the synthetic acetylated peptide in the stock can be determined using several methods. To determine the amount of peptide present in the sample, the peptide substrate can be subjected to amino acid analysis, which is usually performed after peptide synthesis when the peptide is purchased commercially. If not, the amino acid analysis can also be performed by most university protein core facilities. Alternatively, Sir2 enzymes, which catalyze a 1:1 stoichiometry between NAD⁺ cleaved and substrate deacetylated, can also be employed for quantification of stock peptides. A limited amount of peptide and a known amount of NAD⁺ can be reacted with Sir2-like enzymes until complete substrate deacetylation is achieved. The amount of acetyl substrate present can be calculated from determining the amount of NAD⁺ consumed after complete deacetylation of a sub-stoichiometric amount of acetyl substrate. If [³²P]NAD⁺ is utilized in the assay, remaining NAD⁺ can be quantified by measuring the radioactivity. Alternatively, remaining

NAD^+ can be quantified using commercially available NAD^+-dependent dehydrogenases such as pyruvate dehydrogenase[44] or α-ketoglutarate dehydrogenase.[45] These NAD^+-dependent dehydrogenases convert NAD^+ to NADH, which can be spectrophotometrically monitored at 340 nm. Subtracting the concentration of remaining NAD^+ from the initial concentration yields the amount of NAD^+ consumed and, thus, the initial concentration of the acetylated peptide substrate. A time course of deacetylation may be necessary to ensure complete conversion of peptide substrate.

Deacetylase Assays

HPLC-based Deacetylase Assay. This assay relies on the separation of substrates and products of the deacetylase reaction by a reversed-phase HPLC. Quenched reaction mixtures are injected onto a C18 column and, using a gradient of increased levels of organic solvent, substrates, products, and enzyme can be resolved. This assay is applicable to all three classes of deacetylases.

For characterization of most Sir2-like enzymes, monoacetylated H3 and H4 peptides, corresponding to the 20 N-terminal residues of histone H3 and H4, can be used. The 110-μl reactions are carried out at 37° in 50 mM Tris (or phosphate), pH 7.5, with 1 mM DTT. NAD^+ and acetylated peptide are mixed and pre-incubated in a 37° water bath for 5 min. Typical concentrations of NAD^+ and acetylated peptide have ranged from 0.25 μM to 1 mM; however, the range should be determined empirically, as the K_m values for these substrates may differ by orders of magnitude depending on the Sir2 homologue (the kinetic parameter K_m is discussed in the *Kinetic Analysis of the Sir2-Mediated Reaction* section). The reaction is initiated by the addition of enzyme. Although typical enzyme concentrations ranging from 0.25 to 0.5 μM and reaction times ranging from 1 to 15 min have been used, these parameters should be empirically determined. The concentration of enzyme and reaction time should be adjusted such that less than 10–15% of the substrate is converted to products to maintain steady-state conditions. The reaction is quenched by the addition of TFA to a final concentration of 1%. Quenched samples are kept on ice or stored in −20° if not immediately injected onto the HPLC column. Samples are injected onto a reversed-phase HPLC column (e.g., a Vydac C18 column, 4.6 × 250 mm, 201SP104) to resolve substrates and products. A 100-μl loop is typically used for the injections and the flow rate is set to 1 ml/min. After injection, the system is run isocratically with solvent A

[44] Y. Kim, K. G. Tanner, and J. M. Denu, *Anal. Biochem.* **280,** 308 (2000).
[45] G. A. Hunter and G. C. Ferreira, *Anal. Biochem.* **226,** 221 (1995).

(0.05% TFA/H$_2$O) for 1 min followed by increasing levels of solvent B (0.02% TFA in acetonitrile). The gradient used for each assay may vary depending on the type of peptide substrate used. For efficient separation of reactions containing the 20-mer AcH3 N-terminal peptides, a gradient of 0–20% B over 20 min is used. For efficient resolution of reactions containing the 20-mer H4 peptides, the following gradient is used: 0–10% solvent B for 4 min followed by 10–25% B for 25 min. Following each run, the column is washed with 100% B for 3–5 column volumes followed by re-equilibration with solvent A for 3–5 column volumes. Elution of substrates and products is monitored by measuring the absorbance at 214 nm (to monitor all substrates and products) or at 260 nm (to specifically monitor nicotinamide, NAD$^+$, and *O*AADPr). Using the above gradients, a good resolution between the monoacetylated and deacetylated peptides can be achieved. Deacetylated H3 peptide typically elutes at 16 min, acetylated H3 at 18 min, deacetylated H4 at 15 min, and acetylated H4 at 17 min. Substrates and products elute at the approximate percentages of solvent B: nicotinamide at 5%, deacetylated H3 at 16%, deacetylated H4 at 13%, acetylated H3 at 18%, and acetylated H4 at 14%, NAD$^+$ at 12%, and *O*AADPr at 8%. The areas of the peaks are integrated for quantification. To calculate the percent deacetylation, the area of the deacetylated peptide peak is compared to the combined areas of the acetylated and deacetylated peptide peaks. Because a known amount of acetylated peptide is used, the percentage of the deacetylation is then used to determine the amount of deacetylated product formed over the particular time of the assay, to obtain an initial rate.

This HPLC-based assay is also applicable for separating radiolabeled substrates and products. For example, [^{3}H]acetylated histone peptides or [^{14}C]- or [^{32}P]-labeled NAD$^+$ can be utilized. [^{14}C]NAD$^+$ can be synthesized using the Sir2-catalyzed nicotinamide-NAD$^+$ exchange procedure (described later) or purchased commerically, and [^{32}P]NAD$^+$ can be obtained from NEN Life Science Products (800 Ci/mmol). The amounts of the radiolabeled substrates and products can be monitored and quantified by collecting the fractions eluted from the HPLC, adding a constant volume of each fraction into scintillation vials and determining the radioactivity of each fraction by scintillation counting. To calculate the amounts of product formed, the radioactivity of the product can be divided by the total radioactivity of all the fractions collected, which corresponds to the radioactivity of the substrate prior to the reaction. The percent product is then multiplied by the concentration of radiolabeled substrate to obtain the concentration of product formed. Alternatively, the total radioactivity of the product can be divided by the specific activity (CPM/mol) of the radiolabeled substrate to obtain the moles of product formed.

One of the limitations of this HPLC-based assay is its relatively time-consuming nature, where each HPLC run can take approximately 1 h. Only a limited number of injections can be performed over a typical working day. Use of an autoinjector can greatly facilitate the analysis of a large number of samples. Another limitation of this assay is the detector's inability to detect low concentrations of substrates when absorbance is used to quantitate levels of deacetylation. With low peptide concentrations (below 5 μM) larger injection volumes (e.g., 2 ml) are required in order to have sufficient substrate and products to accurately detect. Moreover, because the retention times may change over column useage, it is necessary to check the elution of standards routinely.

Charcoal-binding Deacetylase Assay. The charcoal-binding assay is based on the idea of differential binding of substrates and products to activated charcoal; acetylated and deacetylated peptides, NAD$^+$, nicotinamide, *O*AADPr, and enzyme will bind to activated charcoal. The charcoal-binding assay is performed using [^{3}H]acetylated substrate. This assay takes advantage of the fact that under high pH (≥ 9.5) and heat, the [^{3}H]acetyl group from [^{3}H]*O*AADPr is hydrolyzed. The [^{3}H]acetate does not bind to charcoal and can, therefore, be separated from the rest of the charcoal-bound substrates and products. This charcoal-binding assay allows for hydrolysis of the pH labile ester of *O*AADPr and is similar in basic principle to the modified ethyl acetate extraction method developed by Landry and Sternglanz,[40] where *O*AADPr hydrolysis by NaOH precedes ethyl acetate extraction in acid.

The charcoal-binding assay is performed by reacting [^{3}H]acetylated peptide/protein, NAD$^+$, and enzyme in 50 mM Tris (or phosphate) pH 7.5, with 1 mM DTT in an 80-μl reaction at 37$°$. A control solution where no enzyme is added should be included. To quench the reaction, 70 μl of the reaction mixture is added into an aliquot (typically 70–100 μl) of charcoal slurry (1/3 w/v of charcoal (Sigma, C-5260) and 2/3 of 2 M glycine, pH 9.5). The mixture is then heated at 95$°$ for 1 h to hydrolyze the [^{3}H]acetate from [^{3}H]*O*AADPR. The samples are centrifuged and a fix volume (typically 100 μl) of the supernatant, which contains the [^{3}H]-acetate, is transferred into another aliquot of charcoal slurry (determination of the amount/volume needed is discussed later). The addition of the supernatant to the second charcoal slurry is performed to remove the residual peptides that may not have bound to the first charcoal slurry. The samples are mixed, centrifuged, and the supernatant containing the [^{3}H]-acetate is transferred into a clean tube. The samples are then centrifuged to remove residual charcoal from the solution, and a constant volume of each supernatant is transferred into a scintillation vial containing scintillation cocktail. The radioactivity from the supernatant corresponds to the [^{3}H]*O*AADPr produced. In order to determine the specific activity (CPM/mol) of the acetylated peptide in the reaction, 3–5 μl of the reaction mixture is placed into a scintillation vial and 2 M glycine, pH 9.5, buffer is added up to the same volume as the volume of supernatant. This is done to minimize differences in the quenching during scintillation counting. The CPM obtained from the reaction mixture is divided by the moles of acetylated peptide to obtain the specific activity (CPM/mol). Given the 1:1 stoichiometry between the amount of substrate deacetylated and products formed, dividing the CPM of the supernatant by the specific activity of the peptide and taking into account the dilution would yield the amount of [^{3}H]OAADPr produced. For this assay, it is important to note the volumes of the charcoal slurry and the volume of the supernatant transferred after the first spin to account for the dilution.

Prior to detailed analyses, the volume of charcoal slurry needed should be determined. The amount of charcoal to be used in the assay depends on the concentrations of substrates employed. To determine the volume or amount of charcoal needed for a given concentration of substrate, 80-μl control solutions are prepared with the desired concentrations of [^{3}H]acetylated substrate and NAD$^+$ in 50 mM Tris (or phosphate), pH 7.5, with 1 mM DTT. In these control solutions where no enzyme is added, no [^{3}H]OAADPR is produced. A 70-μl aliquot from each control solution is added into varying volumes or amounts of charcoal slurry. Samples are treated as described earlier. Because no [^{3}H]OAADPR is expected to form, only background counts should be present in the supernatant. The volume or amount of charcoal slurry that produces no significant radioactivity in the supernatant is desirable for this method. Significant counts in the supernatant could be indicative of unbound [^{3}H]acetylated substrate; the concentrations of substrates are higher than the capacity of the charcoal. Also, the [^{3}H]acetylated peptide will not normally liberate free acetate under the conditions of this assay, but the presence of free [^{3}H]acetate in the stock solution may be observed, in which case, the radioactivity from the control solutions can be subtracted from the reaction samples.

Comparison of the charcoal-binding assay and the HPLC-based assay showed that the two approaches are in excellent agreement. Charcoal-binding and HPLC-based assays were utilized in an NAD$^+$ saturation experiment, in which 5 μM of the 20-mer AcH3 peptide and increasing NAD$^+$ concentrations (0.25–20 μM) were reacted with in the presence of 25 nM HST2. The amount of products formed was quantified as described earlier, and initial velocity data were fitted into the Michaelis-Menten equation to obtain the kinetic parameters K_m, V_{max}, and V_{max}/K_m (these kinetic parameters will be discussed later). Using the charcoal-binding assay, the K_m and V_{max} values are 2.5 $\pm$ 0.1 μM and 0.18 $\pm$ 0.003 s^{-1}, respectively. Using the HPLC-based assay, the K_m and V_{max} values are $2.8 \pm 0.4\ \mu M$ and $0.24 \pm 0.01\ s^{-1}$, respectively. A typical saturation curve using the charcoal-binding assay is illustrated in Fig. 2.

TLC-based Deacetylase Assay. This assay is based on the ability to resolve small coenzyme metabolites using thin-layer chromatography. Analyses of Sir2 activity using TLC have been previously described.[15,19] Sir2 reactions are performed in the presence of radiolabeled NAD^+; selection of the label's location depends on the desired molecule to be monitored. To monitor the formation of nicotinamide specifically, a $[^{14}C]NAD^+$ in which the $[^{14}C]$-label is located in the nicotinamide moiety should be used. To monitor the formation of OAADPr or ADP-ribose, a $[^{32}P]NAD^+$, in which the label is located in one of the phosphate groups, should be used. To perform this assay, the radiolabeled NAD^+ and the acetylated substrate are mixed and pre-incubated in a 37° water bath for 5 min. The reaction is initiated by addition of the enzyme and is quenched with TFA to a final concentration of 1%. An aliquot of 3–5 μl of the reaction mixture, containing approximately 1000–10,000 CPM, is spotted onto a TLC plate (Whatman aluminum-backed silica gel plates, 250-μm layer), approximately 2 cm from the bottom. Samples are spotted with at least 1 cm distance from each other. To increase the radioactivity, it may be necessary to apply a small volume at a time and dry the plate prior to spotting additional volumes. The TLC plate is dried prior to placement in the chamber with

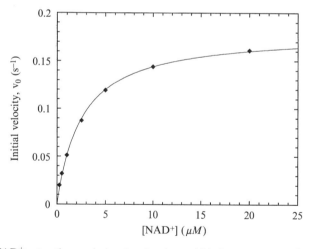

Fig. 2. NAD^+ saturation analysis using the charcoal-binding assay. NAD^+ saturation of the Sir2-like reaction rate was performed at 37° in the presence of varying NAD^+ concentrations (0.25–20 μM), 25 nM HST2, a yeast Sir2 homologue, 5 μM AcH3(K14), and 1 mM DTT in 50 mM Tris, pH 7.5 (at 37°). The reaction mixture was quenched by addition of the charcoal slurry. The product formed was quantified, and the initial velocities of the reaction at various NAD^+ concentrations were plotted and fitted using the Michaelis-Menten equation.

solvent containing certain percentages of ethanol and ammonium acetate (or any desired solvent). Typical solvents used in this assay are 80:20 or 70:30 mixture of ethanol:2.5 M ammonium acetate.[15,19] After placement of the plate into the chamber, the chamber is sealed, and the chromatography is run until the solvent front is approximately an inch from the top of the plate to achieve a good resolution between NAD^+, nicotinamide, and OAADPr. Once the run is completed, the TLC plate is removed from the chamber and allowed to air dry. The plate is exposed to a phosphoimaging screen or photographic film, for at least 16 h. Using a densitometer or imager, the level of radioactivity of the substrate and product spots is determined, and the amounts of substrates converted and products formed are quantified. Using the 80:20 solvent, the following R_f values were obtained: 0.75 for nicotinamide, 0.29 for NAD^+, and 0.58 for OAADPR. Using the 70:30 solvent, the following R_f values were obtained: 0.84 for nicotinamide, 0.49 for NAD^+, 0.74 for OAADPR, and 0.69 for ADPR.

This TLC-based assay is fast and convenient to use and multiple samples can be applied per plate. Another advantage of using this approach is that the reaction can be performed using various acetylated substrates. Using this TLC-based assay, we have looked at the ability of various Sir2-like enzymes to deacetylate 5-mer, 7-mer, and 9-mer versions of an H3 peptide, containing acetylated lysine-14. This approach is also applicable for high-throughput assays, as multiple samples can be applied onto a single TLC plate. Because this assay relies on the detection of radioactive NAD^+, nicotinamide, and OAADPr, this assay is not applicable for the class I and class II HDACs.

Nicotinamide-Utilizing Assays

Nicotinamide-NAD$^+$ Catalyzed Exchange Assay. This assay was first described by Landry *et al.*[15,23] and is specific for the Sir2 family of deacetylases. The assay is performed in the presence of NAD^+, acetylated substrate, enzyme, and [carbonyl-^{14}C]nicotinamide. Sir2-like enzymes follow a sequential mechanism in which both NAD^+ and acetylated substrate have to bind prior to any catalytic step (Borra *et al.*, manuscript in preparation). Upon binding of both substrates, NAD^+ is cleaved, nicotinamide is released, and an ADP-ribose–like intermediate is formed. The exogenous [^{14}C]nicotinamide condenses with the ADP-ribose–like intermediate and drives the reverse reaction to form [^{14}C]NAD^+.[46,47] This assay can be used for generating [^{14}C]NAD^+ (with the radioisotope on the nicotinamide moiety).

[46] A. A. Sauve and V. L. Schramm, *Biochemistry* **42**, 9249 (2003).

[47] M. D. Jackson, M. T. Schmidt, N. J. Oppenheimer, and J. M. Denu, *J. Biol. Chem.* (in press) (2003).

To perform the [^{14}C]nicotinamide-NAD$^+$ exchange reaction, typical concentrations of substrates and enzyme are as follows: 0.2 μM enzyme, 300 μM acetylated substrate (concentration should be close to saturating, meaning that the concentration should be at least 5–10× the K_m for the substrate), 500 μM NAD$^+$, 30 μM to 1 mM [^{14}C]nicotinamide (approximately 10–100 CPM/pmol). The reaction is carried out at 37° in 50 mM Tris (or phosphate) pH 7.5, with 1 mM DTT. After quenching the reaction by the addition of TFA to a final concentration of 1%, [^{14}C]NAD$^+$ and [^{14}C]nicotinamide can be resolved and quantified by spotting the samples onto a TLC plate as described earlier. The TLC plate can then be exposed to a phosphoimaging screen or photographic film. The amounts of [^{14}C]NAD$^+$ and [^{14}C]nicotinamide can be quantified by using a densitometer. Alternatively, the quenched samples can be injected and resolved by reversed-phase HPLC. Using the typical gradients described in the *HPLC-based Deacetylase Assay* section, nicotinamide and NAD$^+$ elute at approximately 5% and 12% solvent B, respectively. Radioactivity of collected fractions can be determined by scintillation counting (Fig. 3). As illustrated in Fig. 3, the radioactivity on the nicotinamide fraction decreases while

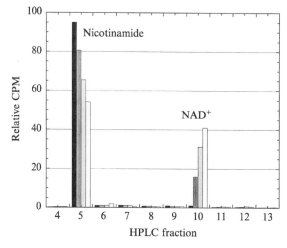

FIG. 3. [^{14}C]Nicotinamide-NAD$^+$ exchange reaction. The reaction was performed at 37° in the presence of 500 μM NAD$^+$, 300 μM AcH3(K14), 0.2 μM HST2, and 50 μM [^{14}C]nicotinamide in 50 mM Tris, pH 7.5, with 1 mM DTT for the following time points: 0 min (black bar), 1 min (dark gray bar), 2 min (light gray bar), and 3 min (white bar). The reaction was quenched with TFA to a final concentration of 1% and injected into the reversed-phase HPLC. Radioactivity of the fractions collected from the HPLC was determined by scintillation counting.

radioactivity on the NAD^+ fraction increases over time. Fractions containing $[^{14}C]NAD^+$ and $[^{14}C]$nicotinamide can be pooled separately and lyophilized for further use.

Nicotinamide Inhibition Assays. Among the different deacetylase families, only the Sir2 family produces nicotinamide as one of the products. Therefore, Sir2-like enzymes, but not the class I and class II HDACs, are susceptible to product inhibition by nicotinamide. In reactions involving Sir2 homologues, nicotinamide has been shown to be a noncompetitive inhibitor against NAD^+.[23,48] We have also demonstrated that nicotinamide exhibits noncompetitive inhibition with respect to the acetylated substrate (Borra *et al.*, manuscript in preparation). To determine nicotinamide inhibition against the acetylated substrate, the deacetylase assay is performed in the presence of saturating concentrations (i.e., at least 5- to 10-fold higher than the K_m) of NAD^+, increasing acetylated substrate concentrations (ranging 5- to 10-fold lower and higher than the K_m) and various but fixed concentrations of nicotinamide. To determine nicotinamide inhibition against NAD^+, the deacetylase assay is performed in the presence of saturating concentrations of the acetylated substrate, increasing concentrations of NAD^+, and various but fixed concentrations of nicotinamide. The assay is performed in 50 mM Tris (or phosphate), pH 7.5, with 1 mM DTT at 37°. Because the nicotinamide inhibition assay requires many samples, the charcoal-binding and TLC-based approaches should be utilized as described earlier. The assays are carried out under steady-state conditions, in which less than 10% of substrates are converted to products. The amounts of products formed are quantified as described in the charcoal-binding or TLC-based methods. The initial velocity of the reaction is determined by dividing the concentration of product formed by the reaction time and by the concentration (or total protein amount) of the enzyme used (initial velocity calculation is discussed in the *Kinetic Analysis of the Sir2-Mediated Reaction* section later). A double reciprocal plot of 1/velocity versus 1/substrate concentration is constructed to determine the type of inhibition. Because nicotinamide is a noncompetitive inhibitor of the Sir2 family of deacetylases, the double reciprocal plot is expected to produce a series of lines that intersect to the left of the $1/v$ axis (Fig. 4). Fitting the inhibition data using the KinetAsyst software (IntelliKinetics, State College, PA) or any equivalent software package can provide the inhibition constants K_{ii} and K_{is}, which reflect changes in the V_{max} and V_{max}/K_m steady-state parameters, respectively.

[48] K. J. Bitterman, R. M. Anderson, H. Y. Cohen, M. Latorre-Esteves, and D. A. Sinclair, *J. Biol. Chem.* **277**, 45099 (2002).

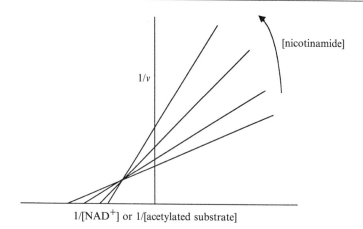

[nicotinamide]

$1/v$

$1/[NAD^+]$ or $1/[acetylated\ substrate]$

Fig. 4. Schematic representation of a nicotinamide inhibition reaction. The reaction is performed in the presence of varying concentrations of either NAD^+ or acetylated substrate, saturating concentration of the other substrate, and various fixed levels of nicotinamide, typically ranging from 25 to 100 μM. The initial velocities of the reactions are determined, and a double reciprocal plot of 1/velocity versus $1/[NAD^+]$ or 1/[acetylated substrate] should yield a series of lines that intersect to the left of the 1/velocity axis, indicating noncompetitive inhibition.

Kinetic Analysis of the Sir2-Mediated Reaction

The quantitative approaches described earlier allow for accurate determination of the initial velocity, and thus, are suitable for performing steady-state kinetic analyses. To calculate the initial velocity of each reaction, the concentration of product formed is divided by the reaction time and the amount of enzyme used. A graph of initial velocity versus substrate concentrations, which typically straddles the K_m by 5-fold in either direction, can be generated and the data fitted into the Michaelis-Menten equation [Eq. (i)], using Kaleidagraph (Synergy software, Reading, PA) or an equivalent software, to determine the kinetic parameters V_{max}, K_m, and V_{max}/K_m

$$V_0 = (V_{max} * S)/K_m + S. \tag{1}$$

The V_{max} is the maximal velocity achieved when both NAD^+ and acetylated substrate are at saturating conditions (i.e., the concentrations are at least 5- to 10-fold higher than K_m), and the K_m value is the substrate concentration at half-maximal velocity. The V_{max}/K_m is a second-order rate constant that reflects the catalytic efficiency of the enzyme at low substrate concentration and is the best parameter for examination of substrate specificity, as this term includes both substrate binding and catalysis. For the

Sir2-like reaction, the V_{max}/K_m parameter includes all catalytic steps from substrate binding up to and including the nicotinamide release step. The kinetic parameter k_{cat}, which reflects the maximal turnover rate of the enzyme, can be directly obtained by dividing V_{max} by the concentration of the pure enzyme. For several Sir2-like enzymes that we have examined, the V_{max}, K_m, and V_{max}/K_m values varied depending on the enzyme and acetylated substrate used. We have determined V_{max} values ranging from 0.2 to 0.7 s^{-1}, and depending on the enzyme-substrate pair, K_m values have ranged from <2 to 400 μM. The catalytic efficiency of the enzyme also varied depending on the acetylated substrate used, with V_{max}/K_m value of $\sim 1 \times 10^5$ M^{-1} s^{-1} for good substrates and 100-fold lower for poor substrates.

Concluding Remarks

Steady-state kinetic analyses require the accurate measurement of the substrates consumed or products formed during the reaction. The products of the Sir2-like reaction can be identified and quantified using the HPLC-based, charcoal-binding, and TLC-based assays described here. These quantitative approaches can be used to compare the reactivity of various Sir2 homologues, to determine substrate specificity, and to explore inhibitor susceptibility for Sir2-like enzymes. Though beyond the scope of this chapter, these assays can be utilized to perform more detailed kinetic analyses that may include examination of various features of the Sir2-like chemical mechanism, identification of the critical ionizations, mutational analysis, establishing the basic kinetic mechanism, and elucidating the order of substrate binding and product release.[49–52]

Acknowledgment

We would like to acknowledge Dr. Michael D. Jackson and Manning Schmidt for helpful discussion.

[49] I. H. Segel, "Enzyme Kinetics. Behavior and Analysis of Rapid Equilibrium and Steady-State Enzyme Systems." Wiley-Interscience, New York, 1993.

[50] R. A. Copeland, "Enzymes. A Practical Introduction to Structure, Mechanism, and Data Analysis," 2nd Ed., Wiley-VCH, New York, 2000.

[51] A. Cornish-Bowden, "Fundamentals of Enzyme Kinetics." Portland Press, London, 1995.

[52] A. Fersht, "Structure and Mechanism in Protein Science. A Guide to Enzyme Catalysis and Protein Folding." Freeman, New York, 1999.

[12] Selective HAT Inhibitors as Mechanistic Tools for Protein Acetylation

By Yujun Zheng, Paul R. Thompson, Marek Cebrat, Ling Wang, Meghann K. Devlin, Rhoda M. Alani, and Philip A. Cole

Histone acetyltransferases (HATs) regulate gene expression by the targeted acetylation of histones and other proteins. It has been a difficult challenge to identify the functional effects of protein acetylation in specific pathways. We have introduced the use of selective HAT inhibitors as mechanistic tools to probe the catalytic features of HATs and their roles in various cellular pathways. We describe the design, synthesis, and applications of these inhibitors in different biological contexts. Lys-CoA, a selective p300/CBP inhibitor, has been particularly useful in elucidation of the role of p300/CBP HAT activity in gene regulation. It is hoped that HAT inhibitors might ultimately serve useful clinical roles in the treatment of cancer and other diseases.

Histone acetyltransferases (HATs), which catalyze the transfer of acetyl groups from acetyl-CoA to the ε-amino groups of lysine residues in proteins (Fig. 1), play a major role in the regulation of transcriptional machinery and the modulation of gene expression. While HATs were initially thought of as enzymes that catalyze the acetylation of histones on lysine residues, it is increasingly recognized that they act on a wide variety of protein substrates.[1] These acetylation events may in some cases affect protein-protein interactions. Protein modules such as bromodomains have been suggested to specifically bind to acetyl-lysine–containing sequences.[2] Furthermore, protein acetylation is a reversible modification. At least two types of histone deacetylases (the HDACs and sirtuins) have been discovered that have led to an understanding of "chromatin signaling" as a dynamic process. In many ways, parallels can be drawn with reversible protein phosphorylation in cellular signaling.[2] In addition to their importance in fundamental biology, the linkage between protein acetylation and disease, especially cancer, is being intensively studied and it may be that HATs will represent potential therapeutic targets.

The use of small molecules to modulate cellular and *in vivo* systems, sometimes called "chemical genetics," has received renewed interest in the biology community in part because of continuing improvements in

[1] R. Marmorstein, *Nat. Rev. Mol. Cell. Biol.* **2**, 422 (2001).
[2] S. L. Schreiber and B. E. Bernstein, *Cell* **111**, 771 (2002).

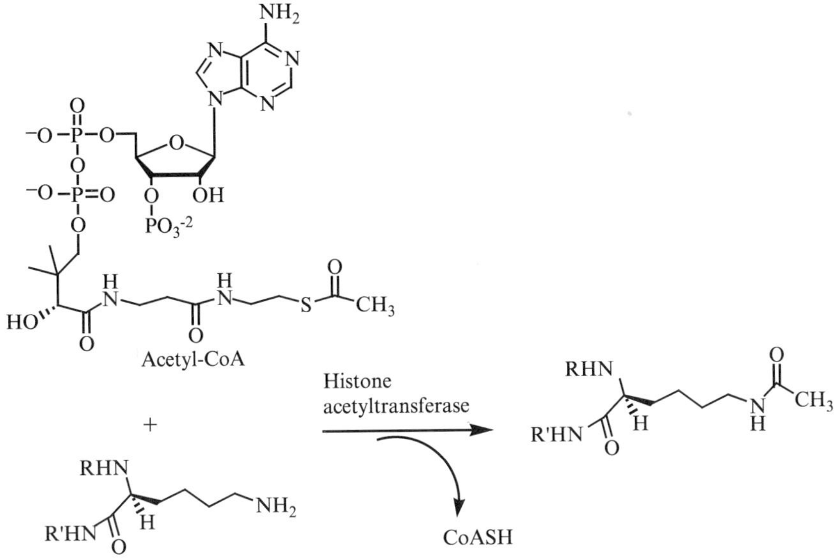

Fig. 1. Histone acetyltransferase catalyzed reaction.

synthetic chemistry as well as molecular design.[3] Histone deacetylase inhibitors such as trapoxin, trichostatin, and butyrate have played a major role in assessing the general role of HDACs and acetylation in a variety of transcriptional systems.[2] Indeed, the use of HDAC inhibitors preceded the isolation of the first histone deacetylases and was critical in the molecular identification of these proteins.[4,5] The current reported compounds have not proven especially useful in distinguishing among the nearly a dozen HDACs (Class I and Class II) and thus have shown limited value in unravelling specific pathways.

Unlike HDACs, the molecular identification of HATs was carried out without the benefit of selective inhibitors.[6] The first selective HAT inhibitory compounds were reported 3 years ago,[7] 4 years after the first reports of nuclear HATs. The widely divergent structural and mechanistic features

[3] R. L. Strausberg and S. L. Schreiber, *Science* **300,** 294 (2003).

[4] M. Yoshida, M. Kijima, M. Akita, and T. Beppu, *J. Biol. Chem.* **265,** 17174 (1990).

[5] J. Taunton, C. A. Hassig, and S. L. Schreiber, *Science* **272,** 408 (1996).

[6] J. E. Brownell, J. Zhou, T. Ranalli, R. Kobayashi, D. G. Edmondson, S. Y. Roth, and C. D. Allis, *Cell* **84,** 843 (1996).

[7] O. D. Lau, T. K. Kundu, R. E. Soccio, S. Ait-Si-Ali, E. M. Khalil, A. Vassilev, A. P. Wolffe, Y. Nakatani, R. G. Roeder, and P. A. Cole, *Mol. Cell* **5,** 589 (2000).

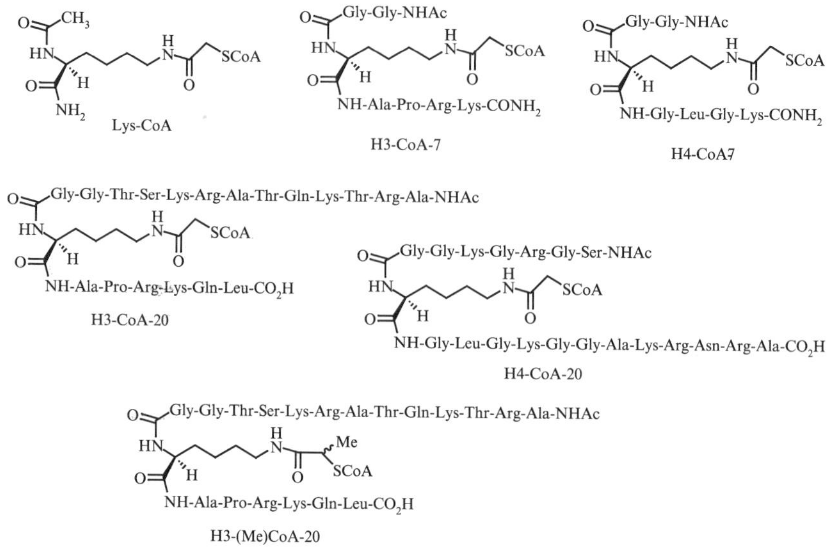

FIG. 2. Designed peptide-CoA conjugates for HAT inhibition

of the different HAT enzymes have allowed for a high degree of specificity to be achieved in designing HAT inhibitors without intensive efforts. The HAT inhibitory compounds reported to date fall under the category of bisubstrate analogs. While simply designed, these compounds have proven quite useful in the analysis of HAT function in a wide array of *in vitro* and biological systems.

Design and Synthesis

The bisubstrate analogs targeted in the first generation of HAT inhibitors[7] involved the covalent linkage of the peptide substrate and acetyl coenzyme A (acetyl CoA) substrate (Fig. 2). This is an approach that has proven quite powerful in a number of different acetyltransferases.[8] Note that the linkage is somewhat unnatural in that the attacked carbonyl of acetyl-CoA is spaced one methylene unit from the CoA sulfur atom. The primary motivation for this functionality comes from its track record with other acetyltransferases and the synthetic simplicity and stability of the target compound. Interestingly, this spacing may in fact be beneficial for potency. For example, in a related system the loss of the methylene group

[8] W. Zheng and P. A. Cole, *Curr. Med. Chem.* **9,** 1187 (2002).

appears to reduce potency.[9] Since the nature of the interaction between the lysine-containing peptide substrate and the HAT enzymes was not known at the outset of these efforts, a range of peptide moieties were examined. The amino terminal tails of histones H3 and H4 were known to be likely physiologic acetylation targets of the HATs PCAF and GCN5 (close homologs) and p300 and CBP (close homologs) so these sequences were selected for the inhibitors. The length of the recognition motifs were not known so a range of up to 20 amino acids was tested. The original synthetic approach exploited the synthesis of the peptides on solid phase with differential protection of the lysine earmarked for CoA modification.[7] In these earlier efforts the lysine was coupled to bromoacetic acid after cleavage of the peptide from the solid support. In more recent efforts (see Fig. 3), bromoacetylation is done while the peptide is still immobilized on the resin and the bromoacetamide intermediate is then cleaved from the solid phase.[10,11] For Lys-CoA, Rink amide resin is utilized, whereas for the more complex peptide-CoA conjugates, Wang resin is used resulting in a C-terminal carboxylic acid. Solution-phase coupling between the bromoacetamide and CoASH is generally a clean reaction and the final product purified by reversed-phase HPLC on a C-18 column and characterized by MS. Multimilligrams of the desired compounds have been generated in this fashion.

Kinetic and Structural Studies

Initial determination of the potency and selectivity of these compounds was carried out with enzyme assays involving full-length recombinant p300 and PCAF as enzymes, mixed histone substrates, and ^{14}C-acetyl-CoA.[7] Separation of the acetylated products was carried out using SDS-PAGE and quantitation was done employing phosphorimage analysis referenced to an internal ^{14}C-BSA standard. The results of these inhibition experiments are shown in Table I. Lys-CoA was shown to be a rather potent p300 inhibitor (IC$_{50}$ 0.4 μM) and selective (>100-fold) versus PCAF. In more recent assays that utilize the amino-terminal peptides rather than the full-length histone substrates, it has been shown that Lys-CoA shows good selectivity versus recombinant yeast GCN5 and EsaI, with IC$_{50}$s greater than 40 μM for these enzymes (Table I). Peptide-CoA conjugates

[9] E. M. Khalil, J. De Angelis, M. Ishii, and P. A. Cole, *Proc. Natl. Acad. Sci. USA* **96**, 12418 (1999).

[10] A. N. Poux, M. Cebrat, C. M. Kim, P. A. Cole, and R. Marmorstein, *Proc. Natl. Acad. Sci. USA* **99**, 14065 (2002).

[11] M. Cebrat, C. M. Kim, P. R. Thompson, M. Daugherty, and P. A. Cole, "Synthesis and Analysis of Potential Prodrugs of Coenzyme A Analogs for the Inhibition of Histone Acetyltransferase p300." *Bioorg. Med. Chem.* **11**, 3307–3313 (2003).

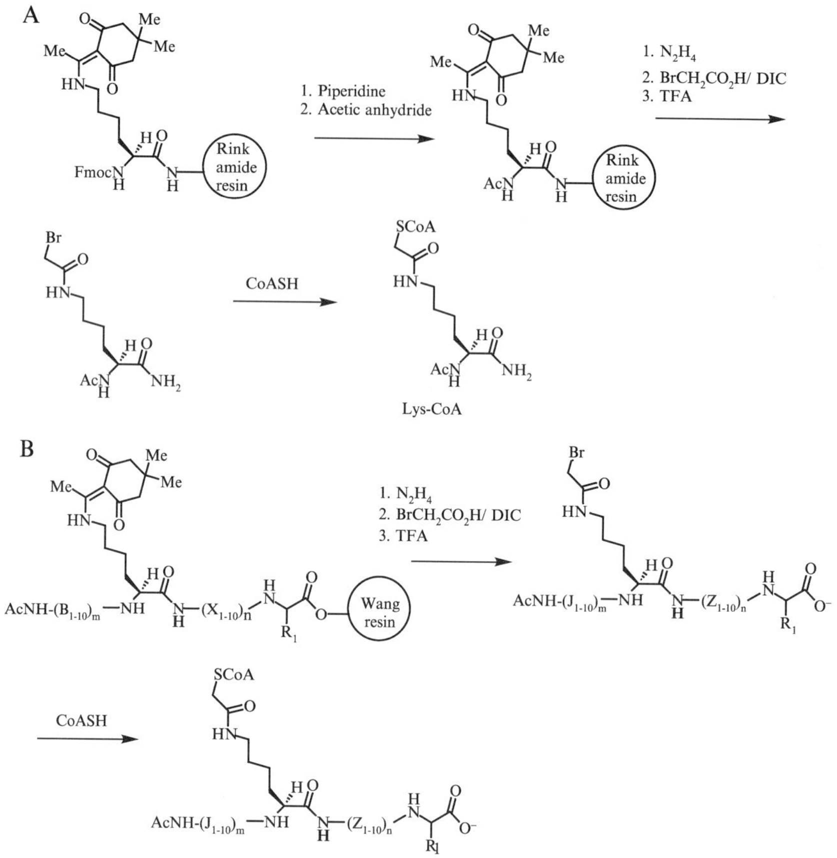

FIG. 3. Synthetic schemes for (A) Lys-CoA and (B) assorted peptide-CoA conjugates (B, Z, J represent variable amino acids that lead to the structures in Fig. 2).

have also proved to be weak inhibitors of the EsaI homolog Mof (unpublished data), consistent with the nonsequential catalytic mechanisms of this HAT family. As expected, Lys-CoA shows high potency against the p300 homolog CBP.[12] In more extensive kinetic analysis, Lys-CoA was shown to exhibit slow, tight-binding behavior with a half-life of interaction with p300 of 4.5 min and a K_i^* of 19 nM.[13] Since product release appears to be rate-limiting for p300 HAT activity, this long half-life could be rationalized as due to a slow conformational change.

[12] A. Polesskaya, I. Naguibneva, A. Duquet, S. Ait-Si-Ali, P. Robin, A. Vervish, P. Cole, and A. Harel-Bellan, *EMBO J.* **20,** 6816 (2001).

[13] P. R. Thompson, H. Kurooka, Y. Nakatani, and P. A. Cole, *J. Biol. Chem.* **276,** 33721 (2001).

TABLE I

IC_{50} VALUES FOR SYNTHETIC COMPOUNDS AND CoASH[a]

Compound	p300 IC_{50} (μM)	PCAF IC_{50} (μM)	GCN5 IC_{50} (μM)	EsaI IC_{50} (μM)
CoASH	200	>20	NA	NA
H3-20	NA	>20	NA	NA
H3-CoA-7	>30	>20	NA	NA
H3-CoA-20	200	0.75^b	0.5^b	200^c
H4-CoA-20	>10	>10	NA	$>100^c$
Lys-CoA	0.5	200	200	$>100^c$
H3-(Me)-CoA-20	60	0.36^a	0.3^b	NA

[a] These measurements were made using mixed histones (33 μg/ml) and acetyl-CoA (10 μM) as substrates, and full-length HAT enzymes as described in ref.[7] unless otherwise noted. NA: not assayed.

[b] This measurement was made using recombinant PCAF and GCN5 catalytic domains with 50 μM H3-20 peptide substrate and 10 μM acetyl-CoA as described.[10]

[c] These measurements were made using recombinant EsaI HAT domain [Y. Yan, N. A. Barlev, R. H. Haley, S. L. Berger, and R. Marmorstein, *Mol. Cell* **6**, 1195 (2000)] (a kind gift of R. Marmorstein) and mixed histones (33 μg/ml) and acetyl-CoA (10 μM) as substrates as described in ref.[7]

The more complex peptide-CoA conjugates such as H4-CoA-7 were considerably weaker inhibitors of p300 than the simple Lys-CoA structure. Since these longer peptides are better p300 substrates, bisubstrate analogs with these longer peptide moieties might have been expected to be preferred. In later work, it was discovered that p300 most likely follows a ping-pong kinetic mechanism.[13] In ping-pong kinetic mechanisms for two-substrate enzymes, binding of both substrates to the enzyme simultaneously should be unfavorable, suggesting why the more complex peptide-CoA conjugates are weak p300 inhibitors. The structural basis for the high-affinity of lysyl moiety of Lys-CoA is still unknown, however.

In contrast to p300, PCAF is potently (IC_{50} 0.5 μM) inhibited by the complex H3-CoA-20 derivative but not shorter analogs.[7] H3-CoA-20 thus represents a selective inhibitory agent, and is also a weak inhibitor versus EsaI. However, it is about equipotent at blocking the PCAF homolog GCN5 (Table I).[10] Detailed kinetic studies revealed that H3-CoA-20 shows a K_i of 28 nM and is a linear competitive inhibitor versus acetyl-CoA and a linear noncompetitive inhibitor versus peptide substrate.[14] This was expected since PCAF follows an ordered BiBi ternary complex kinetic

[14] O. D. Lau, A. D. Courtney, A. Vassilev, L. A. Marzilli, R. J. Cotter, Y. Nakatani, and P. A. Cole, *J. Biol. Chem.* **275,** 21953 (2000).

mechanism.[14,15] Its relatively short half-life (<1 min) for binding to PCAF was different from the behavior of Lys-CoA and p300 and was consistent with the fact that product release does not appear to be rate-determining for PCAF.[14,15] One interesting finding was that shorter peptide-CoA conjugates such as H3-CoA-7 were rather weak inhibitors of PCAF and GCN5. This result nicely correlates with studies showing that histone H3–derived peptides considerably shorter than 20 amino acids are weak PCAF substrates even though X-ray structures of the ternary complex of GCN5 reveal that most of the key interactions are mediated by the proximal residues near Lys-14.[16,17]

Based on previous work on the GNAT superfamily member serotonin N-acetyltransferase,[9] a bisubstrate analog (H3-(Me)CoA-20) containing an isopropionyl linker was synthesized[10] as an epimeric mixture and shown to be 4-fold more potent in blocking PCAF and GCN5 compared to the original compound (assuming only one stereoisomer to be potent). An X-ray structure of the H3-(Me)CoA-20/GCN5 complex revealed that it was the S-isomer (within the isopropionyl linker) of H3-(Me)CoA-20 that was bound.[10] More interestingly, the structure showed diminished interaction between the peptide moiety of the bisubstrate analog and the enzyme compared to the ternary complex.[10] Relatively few residues of the peptide (4–5) made contact with GCN5 and this structure seemed paradoxical in light of the increased potency of longer peptide-CoA conjugates. However, site-directed mutagenesis studies supported the notion that this complex corresponds to a late catalytic intermediate in which the peptide substrate may be dislodged in preparation for product release.[10] Thus, it is likely that the initial encounter complex between the bisubstrate analog and GCN5, not captured by the X-ray structure, involves more extensive peptide-enzyme interactions.

Applications of HAT Inhibitors in Gene Expression Studies

The ultimate value of the development of selective HAT inhibitors is to address biologically meaningful questions about HAT functions. While genetic approaches involving embryonic knockouts, RNAi, and "dominant negatives" have given insights into the roles of HAT activity in a variety of transcriptional pathways, they have limitations. For example, knockouts and RNAi would delete the whole of p300 and CBP protein, including their

[15] K. G. Tanner, M. R. Langer, and J. M. Denu, *Biochemistry* **39**, 11961 (2000).

[16] R. C. Trievel, F. Y. Li, and R. Marmorstein, *Anal. Biochem.* **287**, 319 (2000).

[17] J. R. Rojas, R. C. Trievel, J. Zhou, Y. Mo, X. Li, S. L. Berger, C. D. Allis, and R. Marmorstein, *Nature* **401**, 93 (1999).

many recruiting domains, and would not distinguish physiologic effects due specifically to the HAT domains. Furthermore, since p300 and CBP (and PCAF and GCN5) may show overlapping function, the more technically demanding double knockouts may be required. Dominant negative mutants can always be difficult to interpret because of the unwanted side effect of protein overexpression. Moreover, none of these methods provide the temporal control of dynamic processes or reversibility attainable with small molecule-protein inhibitors. Thus, synthetic compounds can be quite useful in assessing the role of HAT function. The lack of cell permeability seen in many cases[18] provides a significant limitation of the synthetic inhibitors described in this review. Nevertheless, both Lys-CoA and H3-CoA-20 have been used in a series of transcriptional studies as highlighted later. Their delivery in biological experiments can be divided into three major approaches:

1. *In vitro* transcription studies with simple addition to cell-free systems
2. Microinjection into cells
3. Concomitant addition of a lipid permeabilizing agent

HAT Inhibitors in *In Vitro* Transcription Systems

The applications of HAT inhibitors in the *in vitro* setting are straight-forward since the need for cell permeability does not come into play and because of the precise possibility of temporal control. These studies have typically involved assessing the role of p300/CBP in a chromatin template. In studies by Roeder and co-workers, the compound Lys-CoA has been most useful in establishing the role of targeted histone acetylation in transcriptional coactivation of p300, distinguishing the contributions of exogenous versus nuclear extracts containing p300 and CBP, and revealing the importance of histone acetylation in initiation rather than elongation of transcription.[19,20] In these studies, it was found that Lys-CoA (10 μM) could block transcription from a chromatin template but not a naked DNA template and that the PCAF/GCN5 selective inhibitors had only a small effect.

In related studies, Montminy and co-workers showed that CREB-dependent activation of chromatin transcription was blocked by Lys-CoA

[18] K. Subbaramaiah, P. A. Cole, and A. J. Dannenberg, *Cancer Res.* **62,** 2522 (2002).
[19] T. K. Kundu, V. B. Palhan, Z. Y. Wang, W. An, P. A. Cole, and R. G. Roeder, *Mol. Cell* **6,** 551 (2000).
[20] W. An, V. B. Palhan, M. A. Karymov, S. H. Leuba, and R. G. Roeder, *Mol. Cell* **9,** 811 (2002).

(10 μM), also supporting the role for p300 HAT activity in this process.[21] Recently, Brady's group and Nyborg and co-workers each demonstrated the role of p300/CBP HAT activity as critical in transcriptional activation by the HTLV-I Tax protein using Lys-CoA in a dosage-dependent fashion.[22,23] Nyborg's study was especially noteworthy because it demonstrated the role of HAT activity in transcriptional activation even within a "tailless" chromatin template.[23] Numerous potential p300/CBP substrates were found in nuclear extracts and their acetylation was blocked by Lys-CoA. Somewhat contrasting the findings of Roeder and co-workers,[20] Nyborg's results hint at the potential for multiple acetylation targets in transcriptional activation, at least in certain contexts.[23]

Kraus and co-workers have examined the role of p300/CBP in transcriptional activation by thyroid hormone and retinoic acid receptors in an *in vitro* transcriptional system.[24] These studies showed a clear role for the p300/CBP HAT activity which was inhibited by Lys-CoA. Interestingly, the potential HAT activity of steroid receptor coactivator (SRC) proteins was not important in transcriptional activations, although the SRCs play an important role in recruiting p300/CBP to the hormone-regulated promoters. The selective PCAF/GCN5 inhibitor H3-CoA-20 was also able, albeit to a lesser extent, to block transcriptional activation in this system, possibly suggesting a role for the PCAF/GCN-5 HAT activity in this process.[24]

HAT Inhibitors and Microinjection Delivery

While cellular microinjection is a relatively specialized technique in the cell biology community, it has played a significant part in mechanistic studies on transcriptional activation. It, thus, has been embraced by a number of labs examining the role of p300/CBP HAT activity in various processes. In general, concentrated ($\sim$1–10 mM) solutions of Lys-CoA are needed since the compound is typically diluted by 100-fold upon microinjection into the cells. By microinjection into the *C. elegans* gonadal arm, Shi and co-workers have used the relative effects of the CBP inhibitor Lys-CoA to establish a role for CBP HAT activity in *C. elegans* development.[25] In

[21] H. Asahara, B. Santoso, K. Du, P. A. Cole, I. Davidson, and M. Montminy, *Mol. Cell Biol.* **21**, 7892 (2001).

[22] H. Lu, C. A. Pise-Masison, T. M. Fletcher, R. L. Schiltz, A. Nagaich, M. Radonovich, G. Hager, P. A. Cole, and J. N. Brady, *Mol. Cell Biol.* **22**, 4450 (2002).

[23] S. A. Georges, H. A. Giebler, P. A. Cole, K. Luger, P. J. Laybourn, and J. K. Nyborg, *Mol. Cell Biol.* **23**, 3392 (2003).

[24] J. Li, K. C. Lee, P. A. Cole, J. Wong, and W. L. Kraus, *Mol. Endocrinol.* **17**, 908 (2003).

[25] M. Victor, Y. Bei, F. Gay, D., C. Mello, and Y. Shi, *EMBO Rep.* **3**, 50 (2002).

contrast, H3-CoA-20 had no effect in this system. These findings were corroborated by a "knock-in" experiment where CBP lacking a functional HAT domain showed an identical phenotype.[25]

The role of p73 acetylation by p300/CBP was also probed by microinjection studies.[26] It had been proposed that p300 (but not CBP) was responsible for acetylation of p73 in the context of DNA damage induced by doxorubicin through a c-abl–dependent pathway. The site-specific acetylation of p73 was hypothesized to be responsible for apoptosis in this setting. Consistent with these proposals, microinjected Lys-CoA was able to inhibit p73-induced apoptosis in p53−/− mouse embryonic fibroblasts in response to doxorubicin.[26]

Ott and co-workers[27] have used Lys-CoA to investigate the role of Tat acetylation by p300 in transcriptional activation. Microinjection of only 8 μM Lys-CoA into HeLa cells caused a reduction in Tat-mediated transcriptional activation of a luciferase reporter. In parallel, it was shown that siRNA specific for p300 could abolish Tat-mediated transcriptional activation, whereas siRNA specific for CBP had no effect. Taken together, these results argue for a specific role for p300 HAT activity in Tat-mediated transcriptional activation. Connected to these findings, Ott's group also showed that Tat acetylation modulated its interaction with CyclinT1 and proposed that this could allow for Tat to more readily be transferred to the elongating RNA polymerase II.

Recent studies from Wong's group highlight the value of Lys-CoA in a frog oocyte transcription system investigating the mechanisms of androgen receptor and thyroid hormone receptor.[28] By microinjection of Lys-CoA, Wong demonstrated that histone acetylation was critical in transcriptional activation and cofactor recruitment but was not key in altering DNA supercoiling mediated by SWI/SNF.[28] These studies support a sequential mechanism for chromatin remodeling.

Intracellular Delivery of Lys-CoA with Sphingosylphosphoryl Choline

While the charges on Lys-CoA reduce its bioavailability, Harel-Bellan and co-workers discovered that the reagent sphingosylphosphoryl choline (SPC, previously marketed by Invitrogen as "Transport") could be used to enhance its cellular uptake.[12] Presumably by poking small holes in

[26] A. Costanzo, P. Merlo, N. Pediconi, M. Fulco, V. Sartorelli, P. A. Cole, G. Fontemaggi, M. Fanciulli, L. Schiltz, G. Blandino, C. Balsano, and M. Levrero, *Mol. Cell* **9**, 175 (2002).

[27] K. Kaehlcke, A. Dorr, C. Hetzer-Egger, V. Kiermer, P. Henklein, M. Schnoelzer, E. Loret, P. A. Cole, E. Verdin, and M. Ott, "Acetylation of Tat defines a Cyclin T1-independent step in HIV transactivation." *Mol. Cell* **12**, 167–176 (2003).

[28] Z.-Q. Huang, J. Li, L. M. Sachs, P. A. Cole, and J. Wong, *EMBO J.* **22**, 2146 (2003).

cell membranes, SPC was found to allow for intracellular p300/CBP HAT activity to be blocked. Surprisingly, the enzyme remains blocked even after immunoprecipitation and *in vitro* HAT assay, suggesting a very slow dissociation of inhibitor. As expected, PCAF was not inhibited by Lys-CoA under these conditions.

Using this approach, Harrel-Bellan and co-workers showed that p300/CBP HAT activity was essential for terminal differentiation of muscle cells.[12] This contradicted a previous study arguing against a role for p300/CBP HAT activity in muscle cell differentiation.[29] The Lys-CoA inhibitor was also valuable in providing temporal information related to muscle differentiation. Thus, Lys-CoA blocked cell fusion and the expression of late muscle specific markers like myosin heavy chain and myosin creatine kinase.

Following Harel-Bellan's approach, Medrano and co-workers successfully employed SPC to introduce Lys-CoA into melanocytes.[30] These studies revealed that p300/CBP HAT activity was critical in preventing senescence in melanocytes and may play a vital role in the immortalization of these cells associated with malignant conversion. Thus, p300/CBP inhibitors may have therapeutic value in pigmented cell neoplasia.

Summary

Selective and potent bisubstrate analog inhibitors are now available for the PCAF/GCN5 and p300/CBP enzymes. The development of these inhibitors has provided for new insights into the mechanisms of histone acetyltransferases and the role of protein acetylation in gene regulation. They have been particularly powerful in dissecting the role of p300/CBP HAT activity in the context of its overall contributions to gene regulation. Despite their utility, several challenges still exist. No potent HAT inhibitors have been reported for the EsaI family of HATs and no compounds can distinguish between the close homologs p300 and CBP or PCAF and GCN5. Moreover, increasing the cell permeability properties of existing compounds is an important future direction. Ultimately, it will be of great importance to know whether small molecule HAT inhibitors can make an impact on the treatment of human disease.

---

[29] T. A. McKinsey, C. L. Zhang, and E. N. Olson, *Curr. Opin. Cell Biol.* **14,** 763 (2002).

[30] D. Bandyopadhyay, N. A. Okan, E. Bales, L. Nascimento, P. A. Cole, and E. E. Medrano, *Cancer Res.* **62,** 6231 (2002).

Acknowledgments

We are grateful to our many collaborators whose names are mentioned in the references. We thank Dr. R. Marmorstein for a gift of EsaI. We thank K. Miller for technical assistance. This work was supported in part by the NIH and Ellison Medical Foundation. P.R.T. was supported in part by a Canadian Institutes for Health Research post-doctoral fellowship.

[13] Histone Deacetylase Inhibitors: Assays to Assess Effectiveness *In Vitro* and *In Vivo*

By Victoria M. Richon, Xianbo Zhou, J. Paul
Secrist, Carlos Cordon-Cardo, W. Kevin Kelly, Marija
Drobnjak, and Paul A. Marks

While it has been almost four decades since the discovery that core nucleosomal histones are post-translationally modified by acetylation and methylation,[1] the publications on histone modification have increased exponentially following the cloning of histone deacetylase 1 (HDAC1), and histone acetyltransferase A (HAT A) in 1996.[2,3] It is well established that post-translational modifications, for example, acetylation, methylation, and ubiquitination of lysine residues, phosphorylation of serine and threonine residues, and methylation of arginine residues of the core histone tails, play pivotal roles in regulating cellular functions involving chromatin, such as transcription, replication, and DNA repair. Moreover, the interplay between multiple histone tail modifications has led to the histone code hypothesis[4,5] which states "that distinct histone modifications, on one or more tails, act sequentially or in combination to form a "histone code," that is, read by other proteins to bring about distinct downstream events."[4]

The ability to efficiently monitor changes in histone acetylation has become increasingly important as the roles this complex event plays in several diseases, including cancer,[6] neurodegenerative diseases,[7] and autoimmune diseases,[8] have become better understood. Additionally,

---

[1] V. G. Allfrey, R. Faulkner, and A. E. Mirsky. *Proc. Natl. Acad. Sci. USA* **51**, 786 (1964).

[2] J. Taunton, C. A. Hassig, and S. L. Schreiber, *Science* **272**, 408 (1996).

[3] J. E. Brownell, J. Zhou, T. Ranalli, R. Kobayashi, D. G. Edmondson, S. Y. Roth, and C. D. Allis, *Cell* **84**, 843 (1996).

[4] B. D. Strahl and C. D. Allis, *Nature* **403**, 41 (2000).

[5] B. M. Turner, *Bioessays* **22**, 836 (2000).

[6] P. Marks, R. A. Rifkind, V. M. Richon, R. Breslow, T. Miller, and W. K. Kelly, *Nat. Rev. Cancer* **1**, 194 (2001).

[7] J. P. Taylor and K. H. Fischbeck, *Trends Mol. Med.* **8**, 195 (2002).

[8] N. Mishra, D. R. Brown, I. M. Olorenshaw, and G. M. Kammer, *Proc. Natl. Acad. Sci. USA* **98**, 2628 (2001).

inhibitors of histone deacetylases have shown efficacy in several animal models of human disease[9–11] and are currently undergoing clinical evaluation as anti-cancer agents.[6] These histone deacetylase inhibitors fall into four structural classes that include the short chain fatty acids (e.g., butyrate, valproate), hydroxamic acids (e.g., suberoylanilide hydroxamic acid or SAHA, Trichostatin A), cyclic tetrapeptides (e.g., FK-228), and benzamides (e.g., CI-994 and MS-275). The agents currently under evaluation in clinical trials span all four classes of inhibitors. Several in-depth reviews on histone deacetylase inhibitors have recently been published.[6,12,13]

Studies investigating the effects of histone deacetylase inhibitors normally involve measuring changes in the histone acetylation levels induced by their treatment. For example, samples have been obtained from cells cultured *in vitro* with HDAC inhibitors, from various tissues, including spleen, brain, liver, and tumors isolated from animals treated *in vivo* with HDAC inhibitors, and from peripheral mononuclear cells, bone marrow cells, and biopsy samples from patients participating in clinical trials with HDAC inhibitors. Protocols for the isolation of histones and the analysis of the cellular and tissue effects of histone deacetylase inhibitors are described in this chapter.

Isolation of Histones

A common method of monitoring histone acetylation is through the purification and evaluation of cellular histone extracts. The method mentioned later details the isolation of histones from cell suspensions. When isolated tissues are used, the samples must first be homogenized to cell suspensions. When using peripheral blood mononuclear cells or bone marrow aspirates, the cells must first be isolated from the matrix by density gradient centrifugation. Since the acetylation of histones is a reversible protein modification, samples should be either processed immediately or flash frozen at $-80°$ for later processing. Although cell type–dependent, the typical protein yield is approximately 10–100 μg of histone extract per 10^7 cells.

[9] J. S. Steffan, L. Bodai, J. Pallos, M. Poelman, A. McCampbell, B. L. Apostol, A. Kazantsev, E. Schmidt, Y. Z. Zhu, M. Greenwald, R. Kurokawa, D. E. Housman, G. R. Jackson, J. L. Marsh, and L. M. Thompson, *Nature* **413,** 739 (2001).

[10] E. Hockly, V. M. Richon, B. Woodman, D. L. Smith, X. Zhou, E. Rosa, K. Sathasivam, S. Ghazi-Noori, A. Mahal, P. A. Lowden, J. S. Steffan, J. L. Marsh, L. M. Thompson, C. M. Lewis, P. A. Marks, and G. P. Bates, *Proc. Natl. Acad. Sci. USA* **100,** 2041 (2003).

[11] N. Mishra, C. M. Reilly, D. R. Brown, P. Ruiz, and G. S. Gilkeson, *J. Clin. Invest.* **111,** 539 (2003).

[12] W. K. Kelly, O. A. O'Connor, and P. A. Marks, *Expert Opin. Investig. Drugs* **11,** 1695 (2002).

[13] R. W. Johnstone, *Nat. Rev. Drug Discov.* **1,** 287 (2002).

Histone Extraction Protocol

Collect cells ($2.5–10 \times 10^6$), wash with phosphate-buffered saline (PBS), and centrifuge at $600g$ for 5 min. Discard the supernatant, resuspend the cell pellet in ice cold lysis buffer (1 ml/10^7 cells) [10 mM MgCl$_2$, 10 mM Tris-HCl, 25 mM KCl, 1% Triton X-100, 8.6% sucrose, protease inhibitors, pH 6.5], and incubate on ice for 5 min. Centrifuge at $600g$ for 5 min at 4°, remove the supernatant, wash the nuclear pellet in TE buffer ($\sim$1 ml/10^7 cells) [10 mM Tris-HCl, pH 7.4, 13 mM EDTA], centrifuge at $600g$ for 5 min at 4°, and discard the supernatant. Resuspend the pellet in ice cold water containing 0.4 N H$_2$SO$_4$ (100 μl/10^7 cells), vigorously vortex for 5 s, and incubate on ice for 1 h—vortexing intermittently during the incubation. Centrifuge at $10,000g$ for 10 min at 4°, collect the supernatant, add ice cold acetone (1 ml/100 μl of supernatant), and incubate at $-20°$ for 1 h. Centrifuge at $10,000g$ for 10 min at 4° and carefully discard the supernatant without disturbing the histone pellet. After the pellet has been air-dried, resuspend in water (100 μl/10^7 cells) and measure the protein concentration. Store isolated histones in $-80°$ freezer.

Analysis of Histone Acetylation Levels

Histone acetylation can be evaluated in a variety of ways depending on the sample to analyze and on the information desired. Described later are protocols for three popular methods of detecting histone acetylation: differential migration via acid-urea-triton (AUT) gel electrophoresis, western blotting, and immunohistochemistry.

Differential Migration

Histone acetylation results in the loss of a positive charge on the acetylated lysine residue. This loss of charge can be visualized as a reduction in the migration rate of the histone subtypes (H2A, H2B, H3, and H4) during electrophoresis through an AUT slab gel. Because each histone can be acetylated at several potential lysine residues, AUT gels display a ladder pattern detailing the extent of acetylation for each histone subtype. AUT gel electrophoresis is the preferred method of analysis when evaluation of the stoichiometry of histone acetylation is desired. The protocol described later is modified from the procedure described by Yoshida et al.[14] Due to the position and resolution of the various histone subtypes, this AUT gel electrophoresis protocol is especially useful for visualizing the acetylated forms of histone H4.

[14] M. Yoshida, M. Kijima, M. Akita, and T. Beppu, *J. Biol. Chem.* **265**, 17174 (1990).

AUT Protocol

The resolving gel [1 M acetic acid, 8 M urea, 0.5% Triton X-100, 45 mM NH$_4$OH, 16% acrylamide] is prepared as follows:

> 1.7 ml glacial acetic acid
> 14.4 g urea
> 150 μl 100% Triton X-100
> 0.9 ml saturated NH$_4$OH
> 4.64 g acrylamide
> 0.16 g bis-acrylamide

> Add H$_2$O to 30 ml (low heat may facilitate dissolution)
> 200 μl 10% ammonium persulfate
> 20 μl TEMED
> Layer exposed gel edge with H$_2$O, let polymerize at 4° for ≥16 h

The stacking gel [1 M acetic acid, 8 M urea, 0.5% Triton X-100, 45 mM NH$_4$OH, 8.2% acrylamide] is prepared as follows:

> 1.7 ml glacial acetic acid
> 14.4 g urea
> 150 μl 100% Triton X-100
> 0.9 ml saturated NH$_4$OH
> 2.3 g acrylamide
> 0.16 g bis-acrylamide

> Add H$_2$O to 30 ml (low heat may facilitate dissolution)
> 200 μl 10% APS
> 20 μl TEMED
> Let polymerize at room temperature ~4 h

The gel should be pre-run at 170 V for 4 h at 4° in running buffer [0.2 M glycine, 1 M acetic acid]. The terminals should be switched from conventional electrophoresis so that the positive charged histones migrate toward the cathode.

Normally, 1–20 μg of histone extract should be used per lane depending on the lane size and the method of protein staining to be used. The histone samples are diluted 1:1 with AUT sample buffer [7.4 M urea, 1.4 M NH$_4$OH, 10 mM dithiothreitol], incubated at room temperature for 5 min, and then 2.4 μl of 1% pyronine Y dye in glacial acetic acid is added per 10 μl of sample. The samples are electrophoresed at 170 V for 20–30 h at 4°. The protein bands are then visualized via Coomassie Blue or silver staining.

An example of the differential migration of histone H4 through an AUT gel due to acetylation induced by SAHA is shown in Fig. 1.

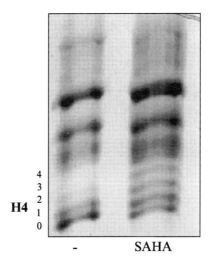

Fɪɢ. 1. SAHA induces histone acetylation as visualized by Coomassie staining after AUT gel electrophoresis. Murine erythroleukemia cells were treated with 2.5 μM SAHA for 6 h. Histone extracts were prepared and evaluated via AUT gel electrophoresis as described in this chapter. The numbers denote the bands generated by laddering of histone H4 due to acetylation at multiple lysine residues.

Immunological Detection

Immunological detection has become the method of choice to determine histone acetylation due to the expanding commercial availability of antibodies that recognize key histone modifications. A plethora of antibodies, both monoclonal and polyclonal, has been raised against both native histones as well as peptides corresponding to individual acetylation sites. The availability of these antibodies makes it possible to evaluate global histone acetylation as well as the acetylation of specific residues on specific histone subtypes. Detailed later are two immunological detection methods, western blotting and immunohistochemistry, used to evaluate histone acetylation changes.

Western Blotting Protocol

Standard western blotting procedures can be used to evaluate acetylation levels in histone extracts prepared as described earlier. Typically, 1–5 μg of histone extract are subjected to standard SDS-polyacrylamide gel electrophoresis using 15% mini-gels. After transferring to nitrocellulose or PVDF, the membranes are blocked with 4% non-fat milk in PBS for 15 min, and then incubated with anti-acetylated histone primary

[SAHA], μM

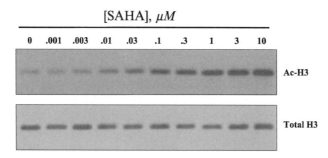

Fig. 2. SAHA induces histone H3 acetylation in cultured T24 cells. T24 human bladder carcinoma cells were treated with the indicated concentration of SAHA for 4 h. The cells were then collected and histones extracted according to the described protocol. Western blotting was performed using a polyclonal anti-acetylated histone H3 antibody from Upstate (Ac-H3) or a polyclonal histone H3 antibody from Abcam (Total H3).

antibody overnight at 4°. The membrane is then washed and incubated with an appropriate biotinylated secondary antibody at room temperature for 1 h. After extensive washing, the samples are developed using an avidin-biotin-peroxidase system (Vector Laboratories, Burlingame, CA).

Examples of histone H3 acetylation detected by western blotting analysis from histone samples isolated from SAHA treated cells are shown in Fig. 2.

Immunohistochemistry Protocol

Immunohistochemistry can be performed on cell or tissue preparations and is useful for evaluating histone acetylation in the context of the whole cells and tissues. For example, for samples made up of several cell types, such as tissues, the only feasible way to discern relative histone acetylation levels among the various cell types present is by immunohistochemistry. For tissues and cells, samples can be fixed in buffered formalin and paraffin embedded. For isolated cells, a less laborious method is to cytospin the cells and fix to microscope slides.

Paraffin-embedded samples are cut into $5\text{-}\mu M$ sections, deparaffinized by conventional methods, quenched in 0.1% H_2O_2 to block endogenous peroxidases, and subjected to antigen retrieval by boiling in 0.01 M citric acid (pH 6.0) for 15 min. Isolated cell samples are prepared by fixation with methanol/acetone prior to quenching in 0.1% H_2O_2. Both types of samples are then blocked in 10% normal goat serum, followed by incubation with rabbit-derived, anti-acetylated histone antibodies of interest. After washing, the samples are incubated with a 1:1000 dilution of a biotinylated goat anti-rabbit IgG secondary antibody. After extensive washing, the samples are developed using an avidin-biotin-peroxidase system

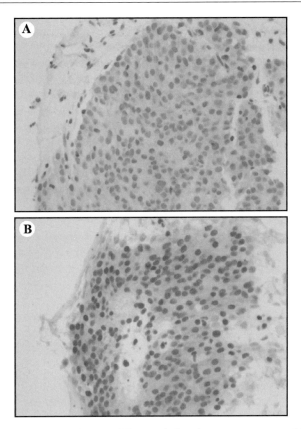

FIG. 3. SAHA induces histone H3 acetylation in prostate tumor biopsies. 40× magnification of prostate tumor biopsies from a patient before (A) and after (B) treatment with 900 mg/m²/day for 3 days with the HDAC inhibitor, SAHA. Paraffin-embedded biopsy tissue was sectioned and stained using a 1:2500 dilution of an anti-acetylated histone H3 antibody (Upstate). Positive staining was visualized with diaminobenzidine (brown color). Cell nuclei are counterstained with hematoxylin (blue/gray color). (See color insert.)

(Vector Laboratories, Burlingame, CA). Diaminobenzidine is used as the final chromogen and hematoxylin is used as the nuclear counterstain.

An example of immunohistochemical staining of acetylated histone H3 from a prostate tumor biopsy of a patient treated with SAHA is shown in Fig. 3.

Acknowledgments

We wish to thank Albert Cupo and Eddie Rosa for technical assistance.

Section II

Immunochemical Assays of Chromatin Functions

[14] Immunochemical Analysis of Chromatin

By Michael Bustin, Richard C. Robinson, and Fred K. Friedman

Introduction

Ever since antibodies were first used to study the organization of histones in chromatin,[1,2] immunochemical approaches have repeatedly proved to be important tools for elucidating the structure and function of chromatin. Accordingly, their use is constantly increasing, concomitant with the growing repertoire of antibodies specific to nucleosomal components, to modified histones, nucleic acids, nonhistone proteins, chromatin-modifying enzyme complexes, and other components that modulate the structure and activity of chromatin. Immunochemical reagents are especially versatile because antigen–antibody reactions occur under a relatively wide range of conditions that do not markedly change the structure of the chromatin fiber. These considerations are important in view of the dynamic nature of the chromatin fiber, the transient binding of nonhistone and regulatory factors to chromatin, and the reversible posttranslational modification of core histones.

Antibodies have a wide range of applications for the study of chromatin structure and function, exemplified by the list in Table I. A plethora of immunochemical protocols used to study histones, nucleosomes, chromatin, and chromosomes has already been described[3–7] and additional protocols are presented in detail in the present volume of Methods in Enzymology. Ultimately, all these procedures involve antigen–antibody reactions in the context of chromatin. This chapter focuses on the major factors that govern these reactions and that need to be considered when employing immunochemical approaches to gain insights into questions related to chromatin.

Antigen–Antibody Reactions in Chromatin

The basis of all immunochemical reactions in chromatin analysis is the binding of an IgG molecule to the nucleosome. Figure 1 presents a computer graphics model of two antibody molecules with their "active

[1] M. Bustin, Nat. New Biol. 245, 207 (1973).
[2] M. Bustin, D. Goldblatt, and R. Sperling, Cell 7, 297 (1976).
[3] M. Bustin, Methods Enzymol. 170, 214 (1989).
[4] S. Muller and M. H. Van Regenmortel, Methods Enzymol. 170, 251 (1989).
[5] V. Orlando, Trends Biochem. Sci. 25, 99 (2000).
[6] D. A. White, N. D. Belyaev, and B. M. Turner, Methods 19, 417 (1999).
[7] N. Suka, Y. Suka, A. A. Carmen, J. Wu, and M. Grunstein, Mol. Cell 8, 473 (2001).

METHODS IN ENZYMOLOGY, VOL. 376   0076-6879/04 $35.00

TABLE I

IMMUNOCHEMICAL METHODS FOR STUDIES ON CHROMATIN AND
CHROMATIN-INTERACTING COMPONENTS

Method	Description
Purification	Immunoaffinity purification of chromatin binding proteins, chromatin-modifying multiprotein complexes, nucleosomes containing unique proteins, or modification in histones or DNA
Detection	Western analysis Enzyme-linked immunosorbent assays (ELISAs) Antibody-based histology Immunofluorescence (regular and confocal)
Demonstration of complex formation	Coimmunoprecipitation Supershift assays with specific antibodies Confocal immunfluorescence microscopy
Intranuclear localization	Immunofluorescence (regular and confocal) Immune electron microscopy
Detection of binding to nucleosomes or chromatin	Supershift assays Chromatin immunoprecipitation (ChIP) assays Immune sedimentation Immune adsorption
Analysis of posttranslational modification in histones	ChIP assays Immunofluorescence Immune electron microscopy Western blots
Functional analysis	Microinjection of antibodies into cells Expression of antibodies in cells

sites" placed next to the amino-terminal tail of histone H3 and the DNA, respectively. The model illustrates that an IgG will recognize only a small portion of the nucleosomal surface, because the antibody-binding site comprises an area of 500–1000 $\text{Å}^{2,8}$ and the cross-sectional area of the nucleosome core is about 8000 Å^2. As a consequence, antibody binding is affected by changes only within the localized region in the binding site, and nucleosomal changes that are solely distal to the binding site do not affect antibody interactions. The model also illustrates that several antibodies can simultaneously interact with a nucleosome, provided that they are specific to different antigenic sites. Obviously, antibody accessibility to

[8] S. Jones and J. M. Thornton, *Proc. Natl. Acad. Sci. USA* **93,** 13 (1996).

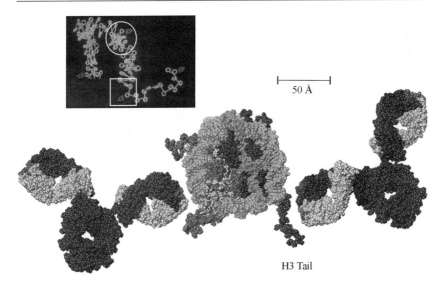

50 Å

H3 Tail

FIG. 1. Molecular model of the interaction of antibodies with a nucleosome. Structures shown are based on coordinates from the Protein Data Bank for nucleosome core particle (1KX5), an antibody Fc fragment (1DN2), and an antibody–peptide complex represented by the Fab of antibody PC283 bound to a peptide derived from a hepatitis B virus surface antigen (1KC5). Antibody heavy and light chains are blue and yellow, respectively. A peptide bound in the antibody-binding site is shown in red, and illustrates the location and relative size of an active site. Nucleosomal DNA is cyan, the H3 tail is black, and the histones are represented by different colors. The model illustrates the size relationship between an antibody and a nucleosome core. The *inset* illustrates the size relationship between antibody probes (in red) and chromatin (model provided by C. Woodcock). A representative compacted and decompacted region is denoted by a circle and square, respectively; note that antigens within these regions exhibit differential accessibility to antibody probes. (See color insert.)

some nucleosomal sites is reduced in the context of chromatin, which contains regions with varying degrees of compactness (Fig. 1, inset).

The major histone antigenic determinants in chromatin are the amino termini of the histone cores. Because these protrude beyond the periphery of the nucleosome, they are readily available for interaction with antibodies. In contrast, the globular regions of the core histones are relatively inaccessible to antibody. Thus, antigenic determinants that are fully accessible in free histones may not react with antibody while embedded in the nucleosome, owing to steric hindrance by DNA or neighboring histones. It has been estimated that less than 5% of the antigenic sites of free histones are available in nucleosomes.[9]

[9] D. Goldblatt and M. Bustin, *Biochemistry* **14,** 1689 (1975).

The model in Fig. 1 illustrates some points to consider when planning or interpreting immunochemical studies involving structural modifications in the tail. The most basic consideration is that although such changes are indeed immunodetectable, even the longest histone tail (histone H3) cannot accommodate more than one antibody at any time. Thus, two antibodies that target modifications at two different sites in the tail cannot simultaneously bind because of steric hindrance. However, a single antibody may recognize two or more modifications if these are localized within the same epitope.

Immunogens

The choice of the immunogen used to elicit antibodies is the single most important factor that determines the utility of the resulting serum. Immunization with a pure, defined immunogen ensures a specific immune response that yields antiserum with chromatin-specific antibodies. Monoclonal approaches could, in principle, circumvent the need to use defined immunogens; however, the identity of an epitope recognized by the antibodies is difficult to define. Consequently, the production of monoclonals against complex mixtures of nuclear extracts, or even against chromatin, failed to produce a battery of useful antibodies. We describe some immunogens used to elicit antisera for chromatin analysis. Protocols for preparing and using these immunogens have been described in this series and elsewhere.[3,4,10]

Synthetic Histone Peptides

Synthetic peptides are presently the immunogen of choice for producing antibodies to histones, histone variants, or posttranslationally modified histone regions. The advantages of using synthetic peptides for producing specific immunochemical reagents are obvious. First, large quantities of defined peptides can be synthesized at relatively low cost, thereby circumventing the need for cumbersome biochemical purification of "native" immunogens. Another unique advantage of this approach is the ability to produce antibodies to specific posttranslational modifications. The exquisite specificity of these antibodies was the key factor in identifying the biological importance of specific posttranslational modifications in the histone tails.

For immunization, the synthetic peptides must first be coupled to large and highly immunogenic carriers, by approaches that have been described.[11] All the procedures require affinity purification of the antibodies on resins

[10] B. D. Stollar, *Methods Cell Biol.* **18,** 105 (1978).

[11] E. Harlow and D. Lane, "Antibodies." Cold Spring Harbor Laboratory Press, Cold Spring Harbor, NY, 2002.

containing either the peptide or peptide conjugated to a carrier different from the one used for immunization. Several chapters in this volume provide details on the procedures for obtaining pure antibodies that target specific posttranslational modification in the histone tails. Examples of the use of these antibodies, especially for immunoprecipitation and analysis of posttranslational histone modification at the single gene level, are described below. The specificity of the antibodies is due to the fact that the antibodies can recognize as few as 6–10 amino acids, a point that is discussed in detail below. The exquisite sensitivity of antibody to the fine structure of chromatin is exemplified by a study using antibodies to the peptide corresponding to the dimethylated H3 Lys-9 tail.[12] Antibodies were prepared to both this single tail peptide and to a branched peptide consisting of four tail peptides. Only the latter antibody recognized heterochromatin, presumably because of the use of an immunogen with multiple and spatially proximate methylated Lys-9 residues.

Most commercial antibodies against nonmodified histones are elicited by immunizing with synthetic peptides, usually against sequences derived from the amino-terminal tails. Although these antibodies have proved useful for chromatin analysis, an antibody against full-length, purified histones is more suitable and produces stronger signals for certain applications.

Histone Fractions

Core histones are poor immunogens and do not elicit a significant immune response. Serum with an adequate titer of histone-specific antibodies can be elicited by immunizing with purified core histone–RNA complexes.[10] When eliciting antibodies to purified histones, it is extremely important to use highly purified histones that do not contain even small amounts of highly immunogenic contaminants that could potentially elicit a disproportionately strong immune response.

Only a small fraction of the antibodies elicited by purified histone–RNA complexes reacts with chromatin or nucleosomes.[9] As evident from Fig. 1, access to some of the antigenic determinants in isolated histones is sterically hindered by the DNA or by other histones. Nevertheless, because these sera have been elicited against a defined histone, they are an excellent source of specific antibodies. Furthermore, because the mixture of antibodies within sera is directed to a multiplicity of antigenic determinants, the signal produced in Western analysis is often stronger than that obtained with single peptide-elicited antibodies, which target a more localized region with perhaps limited accessibility. Sera and antibodies

---

[12] C. Maison *et al.*, *Nat. Genet.* **30,** 329 (2002).

produced by immunizing with histone–RNA complexes have proved useful for chromatin analysis. Our use of these antibodies in a variety of studies including Western analysis, immunohistochemistry, immunofluorescence, immune electron microscopy, supershift assays, and microinjection has already been described.[3]

Nucleosomes

Antisera to nucleosomes react well with mono- and oligonucleosomes and linker histones, weakly with DNA and purified core histones, and not at all with known nonhistone proteins such as HMGs.[13] Digestion of nucleosomes with proteases results in essentially complete loss of antigenicity whereas DNase I digestion results in partial loss of antigenicity. The interpretation of the results is obvious in light of Fig. 1. In nucleosomes, the histones are organized into octamers whose topography is distinct from that of isolated histones. Consequently, numerous antibodies will be elicited against epitopes composed of surface regions near histone–histone and histone–DNA interfaces. It is well documented that double-stranded DNA is a poor immunogen and that few antibodies would thus be directed against histone-free DNA regions. DNase I digestion nicks the DNA but does not break histone–DNA contacts. Because the DNA remains attached to the histone octamer, those epitopes composed of histone–DNA contacts are intact and are still recognized by the antibodies. In contrast, trypsin digestion cleaves the histones, thereby destroying epitopes composed of histones or histone–DNA surfaces. Such proteolysis would thus be expected to greatly reduce antigenicity.

Interestingly, antisera elicited against purified nucleosomes exhibit significant species specificity.[13] The specificity could be due to the sequence variability of the histone amino-terminal tails, to distinct patterns of posttranslational modifications, or to both effects. By using affinity chromatography, the sera could be a source of specific antibodies. However, the inability to define the epitopes recognized by these antibodies seriously limits their use for chromatin studies.

Purified and Dehistonized Chromatin

Antisera elicited against chromatin or "dehistonized chromatin" display remarkable specificity for the immunogen and distinguish between preparations obtained from various cell types. The sera even distinguish between chromatin preparations obtained from a specific cell type at different developmental stages or with different proliferation rates.[14] However, the

[13] C. S. Tahourdin and M. Bustin, *Biochemistry* **19,** 4387 (1980).

antigenic sites recognized by the antibodies have not been clearly identified. For dehistonized chromatin, the specificity of the sera is attributed to tissue-specific nonhistone proteins that bind tightly to DNA whereas for chromatin preparations, the specificity could be due either to nonhistone proteins or to posttranslational modifications in the histone tails. As mentioned above, the sera can serve as analytical tools to detect chromatin differences between closely related cells. However, the information gleaned from such studies is rather descriptive and thus far has not provided new insights into chromatin structure.

Nonhistone Proteins

Antibodies are an important tool for elucidating the mechanism whereby nonhistone proteins modify the structure and function of chromatin. The choice of immunogen is guided by the same criteria as described above; the most important consideration is rigorous purification and characterization of the protein used as immunogen. Synthetic peptides corresponding to an amino acid sequence deduced from genomic information circumvent the need for protein purification and enable the production of antibodies to modified amino acids in these proteins.

Studies with HMG proteins are a good example of the versatility of immunochemical techniques for analysis of the role of nonhistone proteins in chromatin structure and function.[15,16] Thus:

- Immunofluorescence, confocal immunofluorescence, immune electron microscopy, and antibody microinjections into living cells have been used to demonstrate that HMGN proteins are associated with transcriptionally active chromatin.[17,18]
- Microinjection of antibodies to HMGB1 into *Pleurodeles waltlii* oocytes suggested a role for HMGB1 in transcription from lampbrush chromosomes[19] and immunofluorescence indicated that part of the proteins are found in the cytoplasm.[20]
- Mobility supershift assays with specific antibodies demonstrated that in chromatin, nucleosomes form complexes containing two molecules of either HMGN1 or HMGN2. Furthermore, the nucleosomes containing HMGN proteins are clustered into domains, in which

[14] L. Hnilica, "Chromosomal Nonhistone Proteins." CRC Press, Boca Raton, FL, 1983.
[15] Y. V. Postnikov and M. Bustin, *Methods Enzymol.* **304**, 133 (1999).
[16] Y. V. Postnikov and M. Bustin, *Methods Mol. Biol.* **119**, 303 (1999).
[17] L. Einck and M. Bustin, *Proc. Nat. Acad. Sci. USA* **80**, 6735 (1983).
[18] R. Hock, F. Wilde, U. Scheer, and M. Bustin, *EMBO J.* **17**, 6992 (1998).
[19] U. Scheer, J. Sommerville, and M. Bustin, *J. Cell Sci.* **40**, 1 (1979).
[20] M. Bustin and N. Neihart, *Cell* **16**, 181 (1979).

each domain contains an average of six contiguous HMGN-containing nucleosomes. This approach, described in detail elsewhere, is quite informative for analysis of the organization of nonhistones in chromatin.[15]

- Immunofluorescence techniques were used to demonstrate the organization of HMGB and HMGA in mitotic and polythene chromosomes, to show that during mitosis HMGN proteins are phosphorylated and not associated with chromosomes, and to study the intranuclear organization of acetylated HMGN proteins.

- Chromatin immunoprecipitation (ChIP), which is widely used for analysis of histone posttranslational modification and for studies on the binding of regulatory regions to specific genomic sequences, was also used in studies with HMG proteins. ChIP analysis of non-histone proteins and regulatory factors requires cross-linking the proteins to chromatin before chromatin fragmentation. This is a critical step in the procedure because overextensive cross-linking may reduce antigenicity. Additional examples of the use of antibodies to study the role of HMG proteins in chromatin and details of the protocols can be found elsewhere.[15] These protocols can be easily adapted to any chromatin-binding protein of interest.

Nucleic Acids

A large repertoire of antibodies specific to nucleic acids and modified bases is available and can be used for chromatin analysis. The fundamental approaches and technical considerations for using these antibodies are similar to those discussed above.

Antibodies

Preparation

Procedures that maximize antibody efficacy and minimize batch-to-batch variation have been described.[11] For studies with chromatin, it is advantageous to prepare antibodies in two different species, especially if immunolocalization studies are planned. In our laboratory we routinely prepare antibodies in both rabbits and goats. Repeated bleedings allow accumulation of large quantities of sera (approximately 500 ml from rabbits and more than 1 liter from goats). The spectrum of epitopes recognized can vary from bleed to bleed. Thus, after ensuring that each bleed has an adequate titer, we pool the various bleeds, characterize the pooled serum, and then store it lyophilized. Under dry and cold conditions the lyophilized sera can be stored for long periods (we recover full activity after 20 years of storage). The serum is reconstituted by adding H_2O to the original volume, and then centrifuged to remove insoluble residue.

Suka et al.[7] describe the production of antibodies highly specific for distinct acetylation sites in yeast histone tails. The specificity of the sera was determined by examining yeast strains bearing specific point mutations, an extremely stringent test. Interestingly, the sera seem to be of sufficiently high titer and specificity to be used directly, without additional purification. However, in most cases, antibodies need to be purified by passage through affinity columns containing the immunogen. The content of IgG in the serum is on the order of 100 mg/ml and the content of specific antibodies, which varies widely, is less than 1% of the total IgG. For anti-histone and anti-HMG proteins, we recover about 0.2 mg of affinity-pure antibodies per 1 ml of original serum. These solutions are stable for many years when stored frozen.

Monoclonal antibodies can be an excellent choice, as high quantities of these highly specific reagents can be prepared. However, caution must be exercised because these are directed against a single epitope, which may fortuitously resemble an epitope in a heterologous antigen, leading to cross-reaction. Sera from patients suffering from autoimmune diseases have served as a useful source of specific reagents.[21–23] However, their routine use is limited by their limited availability, low amounts, and by the labor–intensive process for deciphering their specificity. For most studies, the antibody of choice is usually obtained by affinity purification from sera elicited either by "native" proteins or by specific peptides.

Specificity and Affinity

The binding of antibody to an antigen is a noncovalent, bimolecular reaction that can be measured and expressed in quantitative terms. A single bleed of polyclonal antibodies typically contains a spectrum of IgG molecules (although only a few may predominate in terms of quantity or affinity), directed against multiple epitopes on the immunogen. The dissociation constants for these IgGs can vary from 10^{-5}/mol to 10^{-12}/mol. These values need to be taken into consideration during affinity purification procedures. Normally only the highest affinity IgG molecules are recovered from the affinity columns. If the purification procedure is especially stringent, fewer IgG molecules directed toward a more limited number of epitopes

[21] R. W. Burlingame, *Clin. Lab. Med.* **17**, 367 (1997).
[22] M. Monestier, *Methods* **11**, 36 (1997).
[23] B. L. Kotzin, J. A. Lafferty, J. P. Portanova, R. L. Rubin, and E. M. Tan, *J. Immunol.* **133**, 2554 (1984).

would be obtained. This would yield an antibody product whose narrow specificity might approach that of a monoclonal antibody, whose possible cross-reactivity problems have been mentioned above. Stringent affinity purification may thus yield higher affinity yet lower specificity.

The question of specificity is relevant to practical immunochemical applications. An antibody that reacts only with the test antigen by Western analysis may react nonspecifically in immunofluorescence where other cellular components with somewhat lower affinity are concentrated in an organelle. These considerations need to be taken into account in studies with posttranslationally modified peptides. For example, in testing nuclear extracts, we found that several antibodies to acetylated peptides of HMGN proteins, which reacted with acetylated HMGN proteins with apparent high specificity, cross-reacted strongly with acetylated, but not unacetylated, core histones (our unpublished data). In these nuclear extracts the amount of core histones is approximately 100-fold higher than the amount of HMGN proteins. The weak cross-reaction with the acetyl-lysine residue rendered these antibodies unsuitable for measuring this modification in HMGN proteins. In the preparation of antibodies to posttranslationally modified proteins, it is customary to remove antibodies against nonmodified peptide by affinity chromatography. In our experience, overpurification of antibodies intended to recognize a specific modification may result in targeting the modified residue to the exclusion of the surrounding residues. Such antibodies may then recognize such a limited structural target in unrelated proteins, reducing their specificity.

It is important to consider the immunochemical implications of post-translational modifications in structural terms. For example, the amino-terminal tail of histone H3 is shown in Fig. 2. An antibody raised against a peptide containing phosphorylated Ser-10, and that distinguishes between the phosphorylated and nonphosphorylated protein, must recognize unique structural features aside from the phosphorylated Ser-10, in order to minimize cross-reactivity with other proteins that contain phosphoserine. The makeup of the epitope, for example, the relative contributions of the phosphoserine and neighboring residues, dictates the antibody affinity toward proteins with a phosphoserine and/or the residue neighbors. In any event, higher antibody concentrations can compensate for lower affinity, and antibody will then bind to nonphosphorylated H3 and to phosphorylated serines in other proteins. It is therefore extremely important to use appropriate concentrations of antibodies and antigens to obtain meaningful results. The most stringent control is to use point mutants that lack the modified residues, controls that have been done with yeast. The size of an epitope is in the range of 500–1000 Å.[2,8] This can cover a span of as few as 6–21 amino acids, depending on the surface topography of the

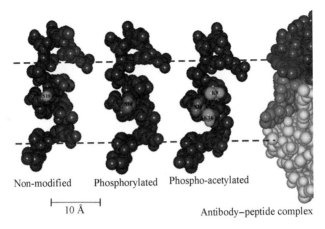

Non-modified   Phosphorylated   Phospho-acetylated

├──── 10 Å ────┤   Antibody–peptide complex

Fig. 2. Models of an antibody–peptide interaction, and a fragment of the amino-terminal tail of histone H3 with posttranslational modifications. For illustrative purposes, residues 1 to 20 are arbitrarily shown in an α-helical conformation. *From left to right:* the unmodified fragment with the oxygen of Ser-10 denoted by cyan; the fragment phosphorylated at Ser-10, with the phospho moiety denoted by red; the fragment phosphorylated at Ser-10 and acetylated at Lys-9 and Lys-14 (the acetyl moiety is denoted green). The representative antibody active site is rightmost, and is composed of regions from heavy chain (blue) and light chain (yellow). The antigen (violet) is a peptide derived from the hepatitis B virus surface. Note the complementarity between the residues in the binding surfaces of the antigen and the antibody, which results in an exact fit. The distance between the dashed lines denotes the approximate size of an epitope. (See color insert.)

epitope and the strength of the contacts with the antibody. Changes in the epitope contact residues can considerably change both the affinity and specificity of the IgG for its intended target. Indeed, it has been reported that acetylation of Lys-9 and Lys-14 abolishes the ability of the anti-phospho H3S10 antibody to recognize the phosphorylated epitope.[24] The steric models in Fig. 2 illustrate how significant changes may occur in the vicinity of the phosphorylated residue on acetylating the neighboring lysines. Insertion of a few atoms within this region may significantly disrupt critical epitope–IgG contacts if they alter the positioning of phosphate to hinder its accessibility to the IgG.

The steric considerations mentioned above for posttranslational modifications also pertain to macromolecular interactions. This is exemplified by the interaction of a nonhistone with chromatin, in the HMG Box–DNA complex (Fig. 3A). It is obvious that potential epitopes proximate to

[24] A. L. Clayton, S. Rose, M. J. Barratt, and L. C. Mahadevan, *EMBO J.* **19**, 3714 (2000).

A   B

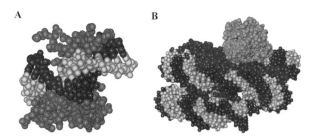

Fig. 3. Effect of macromolecular interactions on accessibility of antigenic binding sites in chromatin. (A) Model of an HMGBox–DNA complex, based on coordinates of Sry (1J46). Regions in the interface between protein (violet) and DNA (black and yellow) are inaccessible to antibody. (B) Globular region of histone H1 (1GHC) positioned on the nucleosome core. Note that complex formation blocks antibody access to binding surfaces on histone H1 (green) and nucleosomal DNA (black and yellow) and proteins (blue). (See color insert.)

the HMG–DNA interface are not fully accessible to antibody binding. Similarly, nucleosomal binding of histone H1 (Fig. 3B) will block antibody access to epitopes derived from the macromolecular interface. Measurements of antibody accessibility can thus be a useful probe to map chromatin-binding sites.

Concluding Remarks

The exquisite specificity of antibodies and their relative ease of preparation render them extremely useful reagents for chromatin analysis. They allow studies of specific components in the dynamic structure of a living cell. To take full advantage of immunochemical approaches and avoid pitfalls, one must carefully consider the basic features of the antibody reaction with chromatin subunits, with their components, with nonhistones such as HMGs, and with regulatory factors that constantly change and remodel the structure of the chromatin fiber. The most important considerations are as follows: choice of immunogen, purification of antibody, and precautions in the experimental conditions to ensure that the immunochemical reaction specifically reflects the quantity and state of the intended antigen.

[15]   Generation and Characterization of Antibodies Directed Against Di-Modified Histones, and Comments on Antibody and Epitope Recognition

By PETER CHEUNG

In eukaryotic cells, all nuclear processes that require access to the DNA template must contend with the fact that genomic DNA is packaged into chromatin *in vivo*. The fundamental unit of chromatin is the nucleosome and the definition of which is 146 bp of DNA wrapped around a histone octamer core comprising two copies each of histones H2A, H2B, H3, and H4.[1] Histone proteins are highly positively charged due to the high content of lysine and arginine residues. In fact, many of these residues, particularly those on the histone N-terminal tails, are subjected to post-translational modifications such as acetylation, methylation, and ubiquitylation.[2] In addition, a number of serine residues on histones are subjected to phosphorylation as well. Because of the intimate contacts between histones and DNA, these histone modifications can alter chromatin structure and in turn play important regulatory roles in many DNA-templated processes. It was almost 40 years ago that Allfrey *et al.* proposed that histone acetylation serves to facilitate transcription[3]; however, it is not until recent years that the biological roles of the different modifications on specific histone residues are beginning to be elucidated. Just as a few examples, acetylation of H4 at lysines 5 and 12 was found to be associated with the histone deposition process, acetylation of H3 at lysines 9 and 14 as well as H4 at lysine 16 has been correlated to gene activation, and more recently, methylation of lysine 9 of H3 has been shown to direct formation of heterochromatin leading to gene silencing.[1,4] As research into the functional roles of histone modification continues to grow, we are beginning to understand that there is a complex array of modified residues on the histones. Moreover, evidence suggests that some histone modifications occur in combinatorial fashions in association with specific nuclear processes.[5] Given the large number of potential combinations of modifications that can occur at the level of the nucleosome, this has led to the proposal that histone modifications collectively may form a histone or epigenetic code that specifies

[1] A. P. Wolffe, *in* "Chromatin: Structure and Function." Academic, San Diego, 1998.
[2] K. E. van Holde, *in* "Chromatin." Springer, New York, 1988.
[3] V. G. Allfrey, R. Faulkner, and A. E. Mirsky, *Proc. Natl. Acad. Sci. USA* **51,** 786 (1964).
[4] Y. Zhang and D. Reinberg, *Genes Dev.* **15,** 2343 (2001).
[5] P. Cheung, C. D. Allis, and P. Sassone-Corsi, *Cell* **103,** 263 (2000).

distinct biological responses to the different cellular pathways that converge on histones.[6–8] In this context, histone modifications not only function to modulate chromatin structure but also extend the amount of information encoded by the chromatin-bound genome.

The recent advances in our understanding of histone modifications is in part facilitated by the development and widespread uses of site- and modification-specific histone antibodies. Early efforts were focused on generating acetyl histone–specific antibodies to study the links between histone acetylation and a variety of nuclear processes. For example, by immunoblotting analyses, histones acetylated at specific sites are linked to histone deposition, transcription activation, and cell cycle progression.[9–11] The acetyl histone antibodies are also amenable to fluorescent microscopy studies useful for examining the nuclear localization of the modified histones. Indeed, such studies revealed that histones on the transcriptionally silent inactive X chromosome in female mammal cells are hypoacetylated compared to those on the autosomes, thus strengthening the link between histone acetylation and transcription activity.[12,13] An anti-acetyl lysine antibody was also used in immunoselection studies that showed that acetylated histones are directly associated with transcriptionally active genes.[14] Refinement of this experimental approach has led to development of the chromatin immunoprecipitation (ChIP) assay, which is now widely used to map direct association of modified histones or specific transcription factors to known genes and genomic regions.

Following the successes of synthesizing acetyl histone antibodies, a variety of antibodies against histones phosphorylated or methylated at different sites have also been generated. These reagents have proven to be invaluable for dissecting the diverse biological functions associated with these histone modifications. By far, the majority of the modified-histone antibodies that have been made are directed against single modifications on the histone molecules. In addition, antibodies to di-modified H3 molecules have also been developed to examine the significance of specific combinations of histone modifications. To date, two separate di-modified

[6] B. D. Strahl and C. D. Allis, *Nature* **403**, 41 (2000).

[7] B. M. Turner, *Bioessays* **22**, 836 (2000).

[8] T. Jenuwein and C. D. Allis, *Science* **293**, 1074 (2001).

[9] R. Lin, J. W. Leone, R. G. Cook, and C. D. Allis, *J. Cell Biol.* **108**, 1577 (1989).

[10] T. R. Hebbes, A. W. Thorne, and C. Crane-Robinson, *EMBO J.* **7**, 1395 (1988).

[11] B. M. Turner, *Exp. Cell Res.* **182**, 206 (1989).

[12] P. Jeppesen and B. M. Turner, *Cell* **74**, 281 (1993).

[13] B. A. Boggs, B. Connors, R. E. Sobel, A. C. Chinault, and C. D. Allis, *Chromosoma* **105**, 303 (1996).

[14] C. Crane-Robinson, T. R. Hebbes, A. L. Clayton, and A. W. Thorne, *Methods* **12**, 48 (1997).

H3 antibodies have been generated: one specifically recognizes H3 phosphorylated at serine 10 (Ser10) and acetylated at lysine 14 (Lys14), the other specifically recognizes H3 acetylated at Lys9 and phosphorylated at Ser10.[15,16] Both antibodies have been used to show that these combinations of phosphorylation/acetylation (Phos/Ac) occur *in vivo* and are induced in association with transcriptional activation of the immediate-early genes. These studies not only illustrate that combinations of histone modifications can be coupled, but also that antibodies are useful tools for determining which combinations of modifications occur in the cell. In this chapter, I will describe the approach and methods used in generating and purifying the polyclonal Phos Ser10/Ac Lys14 di-modified H3 antibody (hereafter Phos/Ac H3 Ab) from rabbit serum. While this particular affinity purification scheme may not be necessary for all antisera, the overall protocol can serve as a general guide for the generation and purification of antibodies against modified histones.

Description of Strategy and Method

Peptide Synthesis, Coupling to Carrier Protein, and Rabbit Immunization

Peptides synthesized with the desired modifications on specific residues are now routinely used for generating modification- and site-specific histone antibodies. Description of peptide synthesis, coupling to carrier proteins, and immunization of rabbits can be found in earlier publications.[9,17] To raise an antiserum against the Phos/Ac H3 epitope, we synthesized a peptide corresponding to amino acid residues 7–20 of the human H3 molecule with a phosphate group on Ser10 (the number refers to the amino acid residue on the full-length H3 molecule) and an acetyl group on Lys14. We have chosen this combination because H3 Ser10 phosphorylation has previously been shown to correlate with activation of the immediate-early genes,[18] and because H3 Lys14 is one of the preferred sites of acetylation by transcription-associated histone acetyltransferases such as GCN5, PCAF, and p300.[19,20] In addition, peptides corresponding to the

---

[15] P. Cheung, K. G. Tanner, W. L. Cheung, P. Sassone-Corsi, J. M. Denu, and C. D. Allis, *Mol. Cell* **5,** 905 (2000).

[16] A. L. Clayton, S. Rose, M. J. Barratt, and L. C. Mahadevan, *EMBO J.* **19,** 3714 (2000).

[17] D. A. White, N. D. Belyaev, and B. M. Turner, *Methods* **19,** 417 (1999).

[18] L. C. Mahadevan, A. C. Willis, and M. J. Barratt, *Cell* **65,** 775 (1991).

[19] M. H. Kuo, J. E. Brownell, R. E. Sobel, T. A. Ranalli, R. G. Cook, D. G. Edmondson, S. Y. Roth, and C. D. Allis, *Nature* **383,** 269 (1996).

[20] R. L. Schiltz, C. A. Mizzen, A. Vassilev, R. G. Cook, C. D. Allis, and Y. Nakatani, *J. Biol. Chem.* **274,** 1189 (1999).

unmodified, the Ser10 phosphorylated, and the Lys14 acetylated forms of H3 were also synthesized and used as controls. The Ser10 Phos/Lys14 Ac di-modified peptide was synthesized with a cysteine residue added to the C-terminal end for sulfhydryl coupling purposes. This peptide was coupled to keyhole limpet hemocyanin (KLH) as carrier and then injected into rabbits. Sera from the injected rabbits were collected over regular intervals and evaluated by enzyme-linked immunosorbent assay (ELISA).

ELISA

To determine whether the rabbit serum contained antibodies specific for the di-modified H3, reactivity to the original Phos/Ac H3 peptide used for injections, as well as to the unmodified (unmod), Ser10 phosphorylated (Phos S10), Lys14 acetylated (Ac K14) H3 control peptides was assayed by ELISA. A brief protocol for ELISA is as follows:

1. Dilute each peptide to the desired concentrations (ranging from 0 to 25 ng/ml) in 0.05 *M* sodium carbonate buffer, pH 9.6.
2. Dispense 200 μl of the diluted peptide solutions into the appropriate wells on the ELISA plate and incubate at 37° overnight.
3. Wash the wells twice with PBS-T (1 × PBS, pH 7.4, 0.05% Tween 20), add 200 μl PBS-T containing 1% BSA to each well, and incubate at 37° for 1 h to block the uncoated areas of the ELISA wells with BSA.
4. Wash the wells twice with PBS-T, add 200 μl of diluted rabbit serum (in PBS-T) to each well and incubate at 37° for 2 h. Initial tests should be done using a range of dilution of the rabbit serum (from 1:500 to 1:2000 for this Phos/Ac H3 antiserum).
5. Wash the wells twice with PBS-T, add 200 μl of diluted (1:5000) HRP-conjugated goat anti-rabbit IgG secondary antibody to each well, and incubate at 37° for 2 h.
6. Wash the wells twice with PBS-T, add 100 μl of OPD substrate (Sigma Fast: P-9187) per well, and let the reactions develop for 30 min.
7. At the end of the incubation period, stop reactions by adding 50 μl of 1 *M* H_2SO_4 and assay the binding of rabbit antibody to the plate-bound test peptides by measuring the absorbance of the reactions at 492 nm using a microtiter plate reader.

Figure 1A shows a typical example of the ELISA results obtained using a 1:1000 dilution of one of the immunized rabbit bleeds. The results show that the rabbit serum reacted to all four peptides tested and the same results were found with different bleeds as well as with the

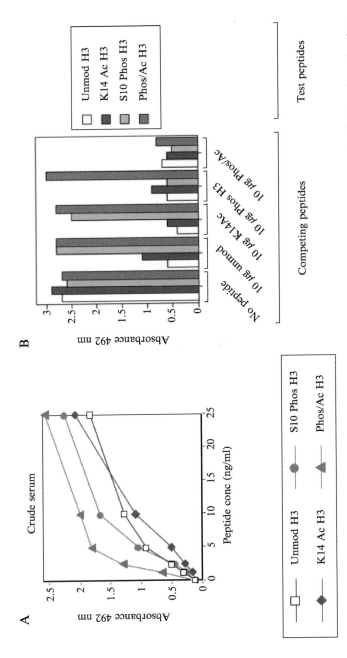

Fig. 1. ELISA analyses of the crude serum from a rabbit immunized with H3 peptides phosphorylated at Ser10 and acetylated at Lys14. (A) The crude serum was tested for reactivity to the four indicated test peptides by ELISA. Binding of antibodies to the test peptides was measured by a colorimetric assay and quantified by absorbance at 492 nm. (B) Parallel samples of crude serum were either mock-treated or incubated with 10 µg of the indicated competing peptides prior to being tested by ELISA. The addition of different competing peptides led to selective loss of reactivity of the antibody mixture present in the crude serum.

sera from a second immunized rabbit. To test whether the serum contained one population of antibody that reacted to all four peptides or whether there was a mixture of antibodies each reacting to the different peptides, peptide competition assays were done by adding various amounts of competing peptides to the serum prior to addition to the ELISA plate wells in step 4. Figure 1B shows one example of the peptide competition assays in which 10 μg of each competing peptides was added to crude serum aliquots prior to dilution and addition to the ELISA plates. Addition of excess unmodified H3 competing peptide selectively removed reactivity of the rabbit serum to the unmodified test peptide. Addition of the Lys14 acetylated H3 peptide as competitor removed reactivity of the serum to both the unmodified and acetylated-alone peptide. Addition of the Ser10 phosphorylated competing peptide blocked reactivity of the serum to the unmodified as well as singly modified test peptides but did not diminish reactivity of the serum to the phosphorylated/acetylated test peptide. In fact, only the original di-modified peptide used to immunize the rabbits was able to successfully compete away recognition of the phosphorylated/acetylated test peptide by the serum. These results together suggest that the crude rabbit serum contained a mixture of antibodies that reacted to the unmodified, the singly modified as well as the di-modified forms of H3. Therefore, affinity purification is required to isolate the fraction of antibodies that specifically recognizes the di-modified form of the H3.

Ammonium Sulfate Precipitation

To begin the antibody purification process, the serum was first mixed with ammonium sulfate to preferentially precipitate the immunoglobulins. This procedure separates the immunoglobulins from serum albumin and plasma proteins that otherwise could interfere with the affinity purification process. A brief description of the ammonium precipitation protocol is as follows:

1. Prepare a stock of saturated (approximately 4.1 M) ammonium sulfate solution, filter through Whatman paper to remove undissolved crystals, and pH to 7.6 before use.
2. Dilute 10 ml of serum in an equal volume of 1× PBS containing 1 mM PMSF.
3. Spin the diluted serum at 3000g for 10 min to remove particulate matters.
4. Transfer the supernatant/serum to a clean beaker and stir solution gently using a stir bar over a magnetic stirrer.

5. In a dropwise fashion, slowly add an equal volume of saturated ammonium sulfate to the diluted serum. A final concentration of 50% ammonium sulfate is often used for precipitation of immunoglobulins; however, one should start to see a white precipitate forming when the percentage of the ammonium sulfate in the solution reaches >30%.
6. Continue to stir the 50% ammonium sulfate/serum solution for 10 min, transfer the mixture to clean centrifuge tube, and spin at 5000g for 10 min.
7. Remove the supernatant and resuspend the precipitated pellet in 10 ml of 1× PBS. Repeat the precipitation process (steps 4–7) to remove any residual albumin left from the first round of precipitation.
8. The final immunoglobulin fraction in 1× PBS can be stored at 4° or can be directly applied to the peptide column for affinity purification.

Affinity Purification Strategy and Protocol

Initial attempts of a subtractive approach involving serial passages of the crude serum through peptide columns containing singly acetylated or singly phosphorylated H3 peptides resulted in non-specific loss of the di-modified H3-specific antibodies over the peptide columns. Also, single passage of the crude serum over a peptide column containing the di-modified form of H3 did not effectively separate the antibody populations that recognize the mono- and di-modified forms of H3. A final scheme based on peptide competition was devised and found to be effective for isolating the fraction of antibodies that specifically recognizes the Phos S10/Ac K14 H3 peptide. This strategy involves adding an excess of the singly modified (Phos S10 or Ac K14) H3 peptides to the serum to block binding of the antibodies specific for the singly phosphorylated or singly acetylated forms of H3 to the column-bound di-modified H3 peptides. The protocol for this purification scheme is as follows:

1. Prepare a peptide column by coupling 1 mg of the desired peptide (in this case the Phos S10/Ac K14 peptide) to 1 ml of SulfoLink Coupling Gel (Pierce, cat# 20401) as per manufacturer's instructions and transfer the peptide-coupled gel to an Econo-Pac chromatography column (BioRad, cat# 732-1010).
2. Wash the peptide column sequentially with:

    a. 10 ml of 10 mM Tris, pH 7.5
    b. 10 ml of 100 mM glycine, pH 2.5

c. 10 ml of 10 mM Tris, pH 8.8

d. 10 ml of 100 mM triethanolamine, pH 11.5

e. 10 ml of 10 mM Tris, pH 7.5

3. Add 1 mg each of the Phos S10 and Ac K14 H3 peptides to the 10 ml of precipitated immunoglobulins in PBS (see step 8 of ammonium sulfate precipitation protocol) and incubate with occasional mixing at room temperature for 15 min.

4. Apply the immunoglobulin plus peptide mixture (step 3) over the Phos S10/Ac K14 peptide column (step 1) and allow mixture to drip through the column by gravity. Collect flow through and re-apply to column one more time.

5. Wash peptide column with 3 cycles of:

a. 10 ml 1 M NaCl, 50 mM sodium acetate, pH 4.0

b. 10 ml of 1 M NaCl, 20 mM Tris, pH 8.0

6. Elute antibodies bound to peptide column with 10 ml of 100 mM glycine, pH 2.5. Collect the eluate in four 2.5-ml fractions in separate tubes each containing 0.5 ml of 1 M Tris, pH 8.0.

7. Wash peptide column with 10 ml of 10 mM Tris, pH 8.8.

(One can omit steps 8 and 9 if the desired antibody is known to elute in glycine.)

8. Elute with 10 ml of 100 mM triethanolamine, pH 11.5. Collect eluate in a tube containing 2 ml of 1 M Tris, pH 8.0.

9. Wash peptide column with 10 ml of 10 mM Tris, pH 8.8.

10. Elute one last time with 5 ml of 100 mM glycine, pH 2.5. Collect eluate in a tube with 1 ml of 1 M Tris, pH 8.0.

11. Test the collected fractions for antibody reactivity by ELISA (see earlier protocol).

12. Dialyze the fractions containing the desired antibody against 1× PBS until the fractions are equilibrated with 1× PBS, divide into small aliquots and store at −80°.

Notes About This Protocol

1. This protocol is based on a standard antibody purification scheme that uses cycling of pH's to elute the column-bound antibodies.[21] Most antibodies are eluted by the low pH glycine elution buffer;

---

[21] E. Harlow and D. Lane, in "Antibodies: A Laboratory Manual." CSHL Press, Cold Spring Harbor, 1988.

however, some antibodies are only eluted by high pH. Therefore, to be thorough, steps 8 and 9 are included in this protocol.

2. It is important to collect the eluate from the first glycine elution step (step 6) in separate small volume fractions because the profile of the antibodies eluted changes over the course of the elution step. By this scheme, the majority of antibodies were found to be eluted in the first two 2.5-ml aliquots collected (elution fraction 1a and b, see Fig. 2). Fraction 1b contained the bulk of the Phos/Ac H3-specific antibody. An even higher specificity fraction was eluted from the second glycine wash (step 10, elution fraction 2 in Fig. 2); however, the antibody yield from this round of elution was significantly less.

3. The eluates were collected in tubes containing 1 M Tris, pH 8, to immediately neutralize the extreme pH's of the glycine or triethanolamine buffers and minimize damage to the eluted antibodies.

4. In this case, the purified Phos/Ac H3 antibody was eluted from the first and second rounds of glycine elution (steps 6 and 10). No detectable antibody was eluted by the triethanolamine buffer (data not shown).

Verification of the Specificity of the Phos/Ac H3 Antibody

In addition to testing the purified antibody by ELISA, the specificity of histone modification antibodies should also be tested by western blot analyses coupled with peptide competition assays. In the case of this di-modified H3 antibody, previous studies have shown that H3 phosphorylation in mammalian cells is induced upon epidermal growth factor (EGF) stimulation and correlates with transcription activation of the immediate early genes.[18] Indeed, we developed this particular Phos/Ac H3 antibody to test whether EGF also induced accumulation of the Ser10-phosphorylated and Lys14-acetylated form of H3. Western blot analyses of histones extracted from EGF-stimulated tissue culture cells showed that the purified antibody specifically detected a band corresponding to histone H3 and the level of this modified form of H3 is induced upon EGF stimulation (see Fig. 3). A similar induction profile was seen using a previously characterized antibody specific for the phosphorylated form of H3 (Phos H3, see Fig. 3). To confirm that the species of H3 detected by the purified antibody was indeed di-modified, excess amounts of the same panel of modified H3 peptides used in the ELISA tests were incubated with the purified antibody for 15 min at room temperature prior to its use in western blot analyses. As shown in Fig. 3, only the Phos/Ac H3 peptide blocked detection of the H3 band by the purified antibody, whereas the phosphorylated H3

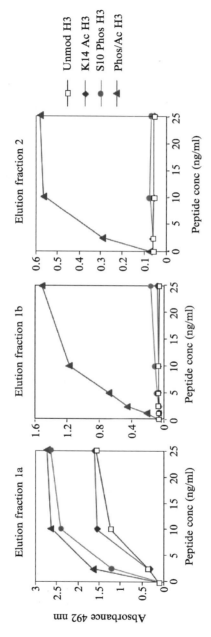

Fig. 2. Antibody specificities of the eluate collected from the affinity purification procedure. As indicated in the protocol, fractions of the eluate from the glycine elution steps were tested by ELISA. Fractions 1a and b were the first and second 2.5-ml eluate fractions of the first 10-ml glycine elution step. Fraction 2 was from the second glycine elution step. These ELISA analyses show that fractions 1b and 2 were specifically enriched for antibodies that recognize the Phos/Ac H3 epitope.

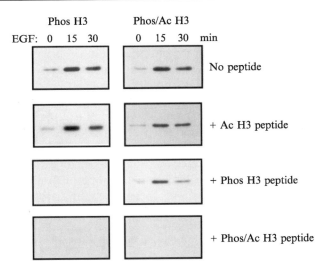

Fig. 3. Further analysis of the specificity of the purified antibody by western blots. Histones from mock-treated or EGF-treated cells were separated on SDS polyacrylamide gels, transferred to PVDF membranes, and probed using a Phos H3 specific antibody or with the affinity-purified Phos/Ac H3 antibody. Various competing peptides were added to the antibodies prior to immunoblotting to test the antibodies' specificities.

peptide was sufficient to block recognition of the H3 band by the Phos H3 antibody. These peptide competition assays therefore further confirm the specificity of the purified Phos/Ac H3 antibody.

Density of Histone Modifications, Epitope Recognition, and the Potential of Masking or Occlusion

While it has been known for decades that histone proteins are subjected to a variety of post-translational modifications, the extent and functions of these modifications are far from being fully understood. In fact, new sites of modifications on core and variant histones are still being identified today. The growing catalog of these modified histone residues indicates that some regions on the N-terminal tails of H3 and H4 are particularly rich in modified residues. For example, within the first five amino acids of H4 (Ser-Gly-Arg-Gly-Lys), it is known that Ser1 can be phosphorylated, Arg3 can be methylated, and Lys5 can be acetylated.[2,22,23] Similarly, on

[22] H. Wang, Z. Q. Huang, L. Xia, Q. Feng, H. Erdjument-Bromage, B. D. Strahl, S. D. Briggs, C. D. Allis, J. Wong, P. Tempst, and Y. Zhang, *Science* **293**, 853 (2001).

[23] B. D. Strahl, S. D. Briggs, C. J. Brame, J. A. Caldwell, S. S. Koh, H. Ma, R. G. Cook, J. Shabanowitz, D. F. Hunt, M. R. Stallcup, and C. D. Allis, *Curr. Biol.* **11**, 996 (2001).

H3 from Lys9 to Lys 14 (Lys-Ser-Thr-Gly-Gly-Lys), Lys9 can be acetylated or methylated, Ser10 can be phosphorylated, Thr11 has recently been shown to be phosphorylated during mitosis, and Lys14 is well documented to be acetylated in association with transcription activation.[1,24] At present, the extent of multiple modifications that occur concurrently at these modification-dense regions is not known. Based on antibody studies, we know that H3 phosphorylated at Ser10 and acetylated at Lys14, as well as those acetylated at Lys9 and phosphorylated at Ser10 do exist[15,16]; however, other combinations of histone modifications within these regions have not been directly tested. The fact that multiple modifications can occur on adjacent or nearby residues raises a potential problem in using histone modification-specific antibodies. As documented by Clayton *et al.*, the authors of that study generated an antibody that specifically recognizes the H3 phosphorylated at Ser10, but recognition of the Phos H3 epitope by that antibody is totally abolished when the H3 molecules are also acetylated at Lys9 or Lys14.[16] Therefore, some of the histone modification-specific antibodies may be susceptible to epitope disruption or occlusion when additional modifications are present on adjacent or nearby residues. ELISA tests of another antibody specific for the Ser10 phosphorylated H3 (previously described in Hendzel *et al.*[25]) showed that this antibody can recognize peptides that are singly phosphorylated at Ser10 and those that are phosphorylated at Ser10 and acetylated at Lys14 equally well (see Fig. 4). However, it is unable to recognize the Phos Ser10 epitope when the adjacent Lys9 is also acetylated. On the other hand, recognition of the intended epitope of Phos Ser10/Ac Lys14 by the affinity-purified antibody described in this chapter is not affected at all by the additional acetylation at Lys9. Therefore, the problem of epitope disruption varies from antibody to antibody and from site to site. With a wide variety of histone modification-specific antibodies currently being used by many laboratories, it is difficult to know which ones are potentially affected by epitope occlusion. It is simply not feasible to test the specificity of every one of these antibodies against all possible combinations of modifications on surrounding residues. Moreover, there is no doubt that modified residues on histone molecules are yet to be discovered and therefore cannot be accounted for. In spite of potential limitations, histone modification-specific antibodies remain as some of the most useful tools for chromatin research. With better characterization of new antibodies, the use of multiple sources of antibodies, and further analyses of histone modification clusters, we can

[24] U. Preuss, G. Landsberg, and K. H. Scheidtmann, *Nucleic Acids Res.* **31**, 878

[25] M. J. Hendzel, Y. Wei, M. A. Mancini, A. Van Hooser, T. Ranalli, B. R. Brinkley, D. P. Bazett-Jones, and C. D. Allis, *Chromosoma* **106**, 348 (1997).

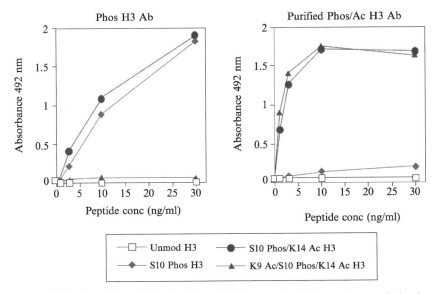

FIG. 4. The Phos H3 and Phos/Ac H3 antibodies were tested for epitope occlusion by ELISA tests. ELISA was performed as in Fig. 1 and 2 with the addition of a new test peptide that corresponds to tri-modified H3 (acetylated at Lys9, phosphorylated at Ser10, and acetylated at Lys14). The additional acetyl group on Lys9 inhibited recognition of the Phos H3 epitope by the Phos H3 antibody but had no effect on the recognition of the Phos Ser10/ Ac Lys14 epitope by the affinity-purified antibody.

minimize potential issues with epitope disruption. At minimum, by being more aware, we can be more careful and make better judgments regarding data interpretation.

The fact that multiple modifications within close proximity can affect the recognition of single modifications by specific antibodies points to an interesting phenomenon that may actually have biological significance. Recent studies have shown that modified histone residues can act as binding motifs and mediate histone-protein interactions. For example, acetyl lysines have been shown to bind the bromodomain motif and H3 methylated at Lys9 can bind to the chromodomain of the heterochromatin protein HP1.[4,26] The addition of other modifications on adjacent or nearby residues potentially can disrupt binding of nuclear factors to singly modified histones and could represent a mechanism for regulating modification-dependent binding of nuclear proteins to chromatin.[5] Indeed, recent data suggest that binding of HP1 to Lys9 methylated H3 is precluded by

[26] C. Dhalluin, J. E. Carlson, L. Zeng, C. He, A. K. Aggarwal, and M. M. Zhou, *Nature* **399,** 491 (1999).

the phosphorylation of the adjacent Ser10 residue (C. D. Allis, personal communication; Jaskelioff and Peterson[27]). Therefore, it may be useful to generate antibodies specific for the combination of H3 methylated Lys9 and phosphorylated at Ser10 to probe for their existence and occurrences *in vivo.* With more and more histone residues found to be modified, and some of them being clustered together, the generation of antibodies against specific combinations of histone modifications will further our understanding of the complex code of modifications present on the histone molecules.

Acknowledgments

I would like to thank C. David Allis for frequent discussions and scientific support. Portion of this work was done in the Allis lab. I would also like to thank Craig Mizzen for excellent advice on antibody purification protocols. Finally, I would like to acknowledge Elsevier for permission to adapt some of the figures originally published in Mol. Cell 5, 905–915 (2000) for this work.

[27] M. Jaskelioff and C. L. Peterson, *Nat. Cell Biol.* **5,** 395 (2003).

[16] Generation and Characterization of Methyl-Lysine Histone Antibodies

By Laura Perez-Burgos, Antoine H. F. M. Peters, Susanne Opravil, Monika Kauer, Karl Mechtler, and Thomas Jenuwein

Histones are among the most conserved proteins throughout evolution and responsible for the compaction of DNA into the higher-order polymer, known as chromatin. In addition to their structural roles, histones play important functions in the control of gene expression by regulating access to the underlying nucleosomal template. Histone amino-termini (tails) protrude from the nucleosome core and are subject to a variety of post-translational modifications, including acetylation (on lysine residues), phosphorylation (on serine and threonine residues), methylation (on lysine and arginine residues), ubiquitination (on lysine residues), and ADP-ribosylation (on glutamic acid residues).[1] The various histone tail modifications have been proposed to constitute a "histone code" that could significantly extend the information potential of the genetic information.[2–4]

[1] K. E. van Holde, "Chromatin." Springer, New York, 1988.

Over the last three years, histone lysine methylation (K-methylation) has emerged as a central modification that can discriminate chromatin transitions in most of the epigenetic paradigms. These include the formation of silent chromatin domains at pericentric heterochromatin, decoration of the inactive X chromosome, Polycomb- and Trithorax-mediated transcriptional memory, and regulation of promoter activity at euchromatic positions.[5] There are at least five methylatable lysine positions in the histone tails of H3 (K4, K9, K27, and K36) and H4 (K20)[1,6] (Fig. 1A); another one exists in the histone-fold domain of H3 (K79).[7–10] Of those, H3-K4, H3-K36, and H3-K79 have been primarily associated with transcriptional stimulation and an activated chromatin state, whereas H3-K9, H3-K27, and H4-K20 are linked with gene repression and silenced/condensed chromatin domains.[11–13] Particularly for the "repressive" lysine methylation marks, distinct subnuclear compartments can be visualized, thereby allowing the analysis of the relative distribution of silenced chromatin regions by indirect immunofluorescence (IF). Differences in the ratio between active and repressed chromatin states have been proposed to correlate with the developmental potential of progenitor cells, to indicate the plasticity of stem cells and to be a diagnostic marker for the subnuclear transformation that is a frequent hallmark of tumor cells.[14,15]

Based on these implications and broad application in many experimental systems, methyl-lysine histone antibodies have been in high demand, both from research laboratories and from commercial sources. However, although the currently available methyl-lysine histone antibodies have been useful tools to probe distinct methylation marks, they need further development

[2] B. D. Strahl and C. D. Allis, *Nature* **403**, 41 (2000).

[3] B. M. Turner, *Bioessays* **22**, 836 (2000).

[4] T. Jenuwein and C. D. Allis, *Science* **293**, 1074 (2001).

[5] M. Lachner and T. Jenuwein, *Curr. Opin. Cell Biol.* **14**, 286 (2002).

[6] Y. Zhang and D. Reinberg, *Genes Dev.* **15**, 2343 (2001).

[7] Q. Feng, H. Wang, H. H. Ng, H. Erdjument-Bromage, P. Tempst, K. Struhl, and Y. Zhang, *Curr. Biol.* **12**, 1052 (2002).

[8] N. Lacoste, R. T. Utley, J. M. Hunter, G. G. Poirier, and J. Cote, *J. Biol. Chem.* **277**, 30421 (2002).

[9] H. H. Ng, Q. Feng, H. Wang, H. Erdjument-Bromage, P. Tempst, Y. Zhang, and K. Struhl, *Genes Dev.* **16**, 1518 (2002).

[10] F. van Leeuwen, P. R. Gafken, and D. E. Gottschling, *Cell* **109**, 745 (2002).

[11] J. C. Rice, K. Nishioka, K. Sarma, R. Steward, D. Reinberg, and C. D. Allis, *Genes Dev.* **16**, 2225 (2002).

[12] T. Kouzarides, *Curr. Opin. Genet. Dev.* **12**, 198 (2002).

[13] A. Vaquero, A. Loyola, and D. Reinberg, "The Constantly Changing Face of Chromatin." Science's SAGE KE: http://sageke.sciencemag.org/cgi/content/full/sageke;2003/14/re4, 2003.

[14] M. A. Surani, *Nature* **414**, 122 (2001).

[15] P. A. Jones and S. B. Baylin, *Nat. Rev. Genet.* **3**, 415 (2002).

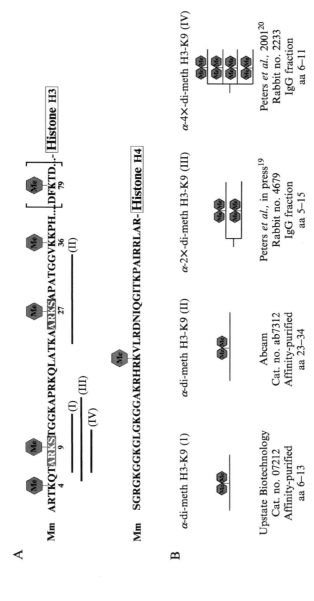

Fig. 1. Methyl-lysine positions in histone amino-termini and peptide design. (A) Schematic representation of methylatable lysine positions in the histone H3 and H4 amino-termini. The H3 and H4 sequences are indicated for the murine proteins. (B) Diagram of linear and branched peptides used to generate H3-K9 di-methylation–specific antibodies. The amino acid (aa) numbers correspond to the positions within the H3 N-terminal tail shown in panel A.

and careful controls. This is because histone lysine methylation has the additional complexity that lysine residues can be mono-, di-, or tri-methylated,[1,16,17] with significant differences in biological output.[18,19] Second, the H3-K9 and H3-K27 positions are embedded within the same amino acid sequence "-ARKS-" (see Fig. 1A), thereby allowing for considerable cross-reactivities of the respective methyl-lysine histone antibodies.

In this chapter, we summarize our experience in the development and characterization of rabbit polyclonal antibodies that are directed against the methylated H3-K9 position. We compare the specificity and subnuclear localization of several methyl H3-K9 "specific" antibodies that were generated in research laboratories[20–22] or are available from commercial sources (Upstate Biotech (UBI), USA and Abcam, UK). We also provide protocols for peptide design, rabbit immunizations, and quality controls of methyl-lysine histone antibodies, followed by their *in vivo* characterization using indirect IF of inter- and metaphase chromatin in wild-type (wt) and mutant mouse cells that are deficient for the Suv39h histone methyltransferases (HMTases). Our comparative analyses indicate significant discrepancies in the specificity and avidity of the available methyl-lysine histone antibodies and highlight the need for extensive quality controls, such that experimental data can be correctly interpreted despite the exquisite complexity of histone lysine methylation.

Results

Peptide Design and Antibody Quality Controls

The most critical step in the generation of a specific antiserum is peptide (antigen) design. Peptide antigens can differ in length (6–20 amino acids, aa), configuration (linear versus branched peptide), and methylation

[16] W. K. Paik and S. Kim, *Science* **174**, 114 (1971).

[17] J. H. Waterborg, *J. Biol. Chem.* **268**, 4918 (1993).

[18] H. Santos-Rosa, R. Schneider, A. J. Bannister, J. Sherriff, B. E. Bernstein, N. C. Emre, S. L. Schreiber, J. Mellor, and T. Kouzarides, *Nature* **419**, 407 (2002).

[19] A. H. Peters, S. Kubicek, L. Perez-Burgos, S. Opravil, K. Mechtler, A. Kohlmaier, A. Beyer, M. Tachibana, Y. Shinkai, and T. Jenuwein, *Mol. Cell,* in press.

[20] A. H. Peters, D. O'Carroll, H. Scherthan, K. Mechtler, S. Sauer, C. Schofer, K. Weipoltshammer, M. Pagani, M. Lachner, A. Kohlmaier, S. Opravil, M. Doyle, M. Sibilia, and T. Jenuwein, *Cell* **107**, 323 (2001).

[21] B. A. Boggs, P. Cheung, E. Heard, D. L. Spector, A. C. Chinault, and C. D. Allis, *Nat. Genet.* **30**, 73 (2002) (published online: 10 December 2001).

[22] I. G. Cowell, R. Aucott, S. K. Mahadevaiah, P. S. Burgoyne, N. Huskisson, S. Bongiorni, G. Prantera, L. Fanti, S. Pimpinelli, R. Wu, D. M. Gilbert, W. Shi, R. Fundele, H. Morrison, P. Jeppesen, and P. B. Singh, *Chromosoma* **111**, 22 (2002).

state (mono-, di-, or tri-) of the lysine of interest. To illustrate how different peptide designs can affect antibody specificity, we have compared four rabbit polyclonal antisera raised against various H3-K9 di-methylated peptides (see Fig. 1B). These peptides comprise a linear octamer (aa 6–13) (UBI #07-212),[21] a linear 12-mer (aa 23–34) (Abcam #ab7312), a 2-branched 11-mer (aa 5–15),[19] and a 4-branched hexamer (aa 6–11).[20]

To assess the specificity of a given antiserum, three types of quality controls are available. These comprise enzyme-linked immunosorbent assays (ELISAS), peptide spotting analyses (dot blots), and protein blots with recombinant and nuclear histones. Whereas ELISAS or peptide and histone blots are used for the initial characterization, additional protein blots with whole nuclear extracts allow the detection of possible cross-reactivities with other epitopes that may be present on non-histone molecules. We have developed a comprehensive peptide blot analysis with a panel of 19 linear peptides spanning the histone H3 and H4 amino-termini and which are specific for unmodified or mono-, di-, or tri-methylated H3-K4, H3-K9, H3-K27, H3-K36, and H4-K20. A serial dilution of these peptides (50, 10, 2, 0.4 pmol) is spotted onto a Hybond-P membrane and binding of an antibody is visualized by enhanced chemiluminescence. The antibody titer and avidity are determined by scoring binding to decreasing peptide amounts. The four antisera described in this section were each probed at three different dilutions, ranging from 1:200 to 1:2500.

The UBI antibody #07-212 displays a high affinity toward di-methylated H3-K9, but also shows minor cross-reactivities (i.e., only when ≥50 pmol peptides are spotted) with di-methylated H3-K4 and tri-methylated H3-K27 (see Fig. 2). This profile is slightly different from a "golden bleed" batch, which lacks the minor cross-reactivity toward tri-methylated H3-K27 (D. Allis, personal communication, data not shown). The Abcam #ab7312 antibody also has a high preference for di-methylated H3-K9 and to a minor extent cross-reacts with di-methylated H3-K36 and H4-K20 positions, and with tri-methylated H3-K27. This is a surprising result, since the used antigen did not even comprise the H3-K9 residue but was designed to generate an α-dimeth H3-K27 antibody. The detected cross-reactivities illustrate the difficulty to discriminate the highly similar H3-K9 and H3-K27 positions. For each of these commercially available antibodies, the most specific results were obtained at dilutions 1:1000 (UBI #07-212) and 1:500 (Abcam #ab7312).

In an attempt to increase specificity and antigenicity, we developed a novel antiserum by using a 2-branched peptide that encompasses the di-methylated H3-K9 position by five amino acids on either side. This α-2×-dimeth H3-K9 antibody (#4679) has both a high titer (as shown by its use at a 1:2500 dilution) and is exclusively specific for di-methylated H3-K9,

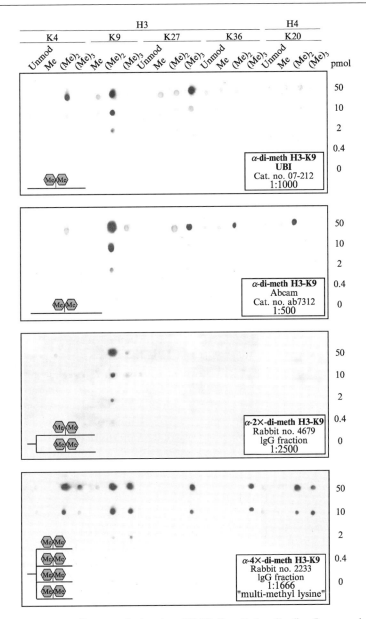

FIG. 2. *In vitro* quality control of various H3-K9 di-methyl antibodies. Immuno-dot blots showing reactivity of antibodies raised against linear, two- and four-branched H3-K9 di-methylated peptide antigens (lower left corners). Antibodies were probed at various dilutions, with the most specific reactivity and dilution shown. Dot blots contain 0, 0.4, 2, 10, and 50 pmol of linear H3 (aa 1–20; aa 19–34; aa 25–45) peptides, unmodified or mono-, di- and tri-methylated at the K4, K9, K27, or K36 positions. In addition, a linear H4 (aa 12–31) peptide, differentially methylated at the K20 position, was also used.

without cross-reacting with H3-K27 or any other tested position (see Fig. 2). Using a similar strategy, we had previously designed a 4-branched peptide spanning the di-methylated H3-K9 position by only 2–3 amino acids on either side. The rationale for this peptide was to mimic the potential *in vivo* clustering of methylated histone tails at pericentric heterochromatin, which may be organized in a more compact nucleosome structure. The resulting α-4$\times$-dimeth H3-K9 antibody (#2233) has indeed been very useful to detect histone methylation at pericentric heterochromatin in an Suv39h-dependent manner.[20,23] At that time, dot blot analysis with only a limited peptide panel, including unmodified and di-methylated H3-K9 and H3-K27 peptides, categorized this antiserum as being specific for the di-methylated H3-K9 position (see Fig. 2).

Using the complete peptide panel in dot blot analyses, however, reveals that the α-4$\times$-dimeth H3-K9 antibody has highest affinity for tri-methylated H3-K9 and significantly cross-reacts (i.e., with $\leq$10 pmol of spotted peptides) with di- and tri-methylated states of all tested lysine positions (see Fig. 2), even including H1-K26 di- and tri-methylated peptides (data not shown). The promiscuity of this antibody, which may well extend beyond histone epitopes, can be attributed to the relatively short peptide (only 6 aa) used to present the antigen. Despite the potential to recognize many di- and tri-methylated lysine positions, this antiserum does not uniformly stain interphase chromatin but selectively decorates silenced chromatin regions at pericentric heterochromatin and on the inactive X chromosome (see Figs. 3 and 4). This staining pattern has been interpreted to be a consequence of the α-4$\times$-dimeth H3-K9 antibody as having a preferential affinity toward a high concentration or density of methylated lysine residues, rather than detecting more isolated methyl epitopes.[20,23] Based on its promiscuity and potential to interact with multiple methylated lysine positions, we would like to rename the α-4$\times$-dimeth H3-K9 antibody (# 2233) as a "multi-methyl lysine" antiserum.

Protocol A: Generation of an H3-K9 tri-Methyl Specific Antiserum (#4861)

Peptide Synthesis. A 2-branched peptide with the sequence 2$\times$ (QTARK(Me)$_3$ STGGKA)-1-K-cys comprising amino acids 5–15 of histone H3 was synthesized in a ABI 433 Peptide Synthesizer (Applied Biosystems).

---

[23] C. Maison, D. Bailly, A. H. Peters, J. P. Quivy, D. Roche, A. Taddei, M. Lachner, T. Jenuwein, and G. Almouzni, *Nat. Genet.* **30**, 329 (2002).

Di-methyl H3-K9 antibodies

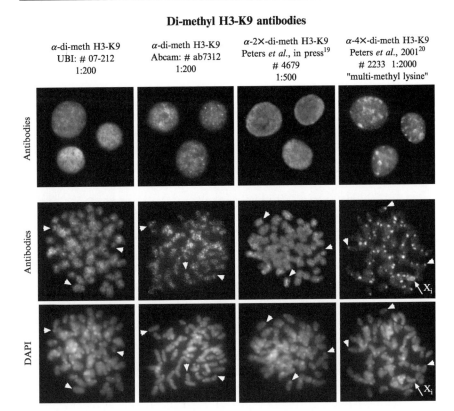

FIG. 3. Immunofluorescence analysis for H3-K9 di-methylation in mouse chromatin. Indirect IF of interphase and mitotic chromatin of wt female MEFs stained with the indicated α-di-meth H3-K9 antibodies. Heterochromatic regions of the acrocentric mouse chromosomes are under-represented for H3-K9 di-methylation (see arrowheads in the mitotic chromosome spreads) but enriched for H3-K9 methylation with the "multi-methyl lysine" antibody.[20] The inactive X chromosome (see arrow X_i) is also decorated by the "multi-methyl lysine" antibody.[24]

1. Inject protected amino acids (MultiSynTech GmbH, Germany) in reverse order at 1 mM each, as synthesis takes place from C-terminal to N-terminal end. The first amino acid is a cysteine (for coupling and purification purposes) that is already bound to the resin TCP Pepchem (Goldammer & Clausen, cat. no. PC-01-0114). Two hundred and fifty milligrams of Pepchem of low substitution

[24] A. H. Peters, J. E. Mermoud, D. O'Carroll, M. Pagani, D. Schweizer, N. Brockdorff, and T. Jenuwein, *Nat. Genet.* **30**, 77 (2002) (published online: 10 December 2001).

Tri-methyl H3-K9 and H3-K27 antibodies

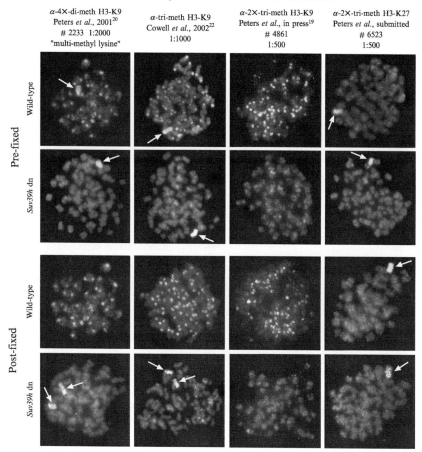

FIG. 4. Immunofluorescence analysis for H3-K9 tri-methylation on mitotic chromosomes. Indirect IF of mitotic chromosomes of wt and *Suv39h* dn female MEFs stained with the indicated antibodies under prefixed (top panels) and post-fixed (bottom panels) conditions. Pericentric heterochromatin is significantly enriched for H3-K9 tri-methylation in an *Suv39h*-dependent manner. By contrast, H3-K27 tri-methylation at the inactive X chromosome (see arrows) is *Suv39h* independent. Both the "multi-methyl lysine"[20] and Cowell *et al.*[22] antibodies decorate constitutive (pericentric) and facultative (X_i) heterochromatin. Some *Suv39h* dn cells are hypo-tetraploid[20,24] and therefore contain two X_i.

grade (0.16 mmol/g) are used for the entire synthesis. The peptide synthesis is monitored by conductivity measurement.

2. The second amino acid is a DI-Fmoc-lysine-OH (Bachem, Switzerland, cat. no. B-1610), which is used to drive the synthesis of a 2-branched peptide via its α- and ε-amino groups.

3. A tri-methylated Fmoc-Lysine-(Me)$_3$-OH chloride (Bachem, Switzerland, cat. no. B-2685) is incorporated at the H3-K9 position.

4. Release and deprotect the completed peptide by treating the resin with 11 ml of 90% TFA (v/v), 0.05% tri-ethylsilan (v/v), 0.05% H$_2$O for 60 min. Isolate the peptide by rinsing and filtering the resin (MultySynTech GmbH, cat. no. V100PE120) with TFA.

5. Precipitate the peptide by adding it dropwise to a 35-ml cold solution of 60% ether (v/v), 40% heptane (v/v). Store for 2 h at $-80°$. Collect the precipitate by centrifugation and wash three times with 40 ml ether. Lyophilize the peptide under vacuum.

6. Dissolve the peptide in 5 ml of 0.1% (v/v) TFA in water. Inject into a HPLC (Applied Biosystems) and purify by reverse-phase chromatography on a C18 column (Phenomenex, lunar 5C18) using a 45-min, 2–45% gradient of acetonitrile. The positive fractions are identified by mass spectroscopy in linear and reflector mode (Bruker Reflex III MALDI-TOF).

7. Freeze the peptide at $-80°$ and lyophilize for 2 days. The yield obtained for the 2-branched H3-K9 tri-methyl peptide was 66 mg.

Rabbit Immunizations. Screen pre-immune sera from 6–8 rabbits by dot blots (see later) with unmodified and methylated histone peptides. Proceed with the 2 rabbits showing the lowest background signals.

1. Combine 3 mg of peptide with 3 mg keyhole limpet hemocyanin (KLH) (Calbiotech, cat. no. 3744805) that has been activated through a maleimide group (Sulfo-GMBS; Pierce, cat. no. 22324) and is dissolved in 320 μl of 250 mM NaCl and 20 mM HEPES, pH 7.3. Add 340 μl 1 M boric acid, pH 6.0, and 340 μl 6 M guanidine chloride. Rinse with argon to remove residual oxygen and incubate for 1 h at RT. To assess the coupling efficiency, combine 5 μl of KLH-peptide and 5 μl of Ellman's reagent (Sigma, cat. no. D-8130) in 1 ml of 2 M HEPES, pH 7.3; efficient coupling is indicated by a decreased signal at 412 nm in a UV-spectrophotometer against a peptide-only blank.

2. Ship the KLH-peptide (1 ml at 6 mg/ml) at $4°$ to an antibody company (e.g., Gramsch Laboratories, Schwabhausen, Germany).

3. On day 1, 400 μg of KLH-peptide (i.e., 200 μg peptide) in complete Freud's adjuvant (CFA) are injected subcutaneously into the lymph nodes and the neck of rabbits.

4. On day 21, the animals are boosted subcutaneously in the shoulder and the neck with 400 μg of KLH-peptide in incomplete Freud's adjuvant (IFA).

5. On day 35, the first bleed is withdrawn (5 ml) and analyzed by dot blots.
6. Responding rabbits are boosted intramuscularly (rear thigh) with 200 μg of KLH-peptide in PBS on day 42–45, followed by additional boosts at subsequent timepoints.
7. On day 63, the second bleed is obtained (20 ml) and additional bleeds (20–25 ml each) can be withdrawn every other week. The rabbits can be maintained (more boosts are required) or exsanguinated (on day 85–90; preferred procedure), yielding around 80–90 ml of crude serum.
8. To preserve the serum, fresh 10% (w/v) sodium azide is added to a final concentration of 0.02% (w/v). The serum is then snap-frozen in liquid N_2 and kept at $-80°$ for long-term storage in 4-ml aliquots. A 0.5-ml working aliquot is kept at $4°$.

Generation of an IgG Fraction. In most cases, it is advised to generate an IgG fraction from the crude serum, which will improve the signal to background ratio in IF and chromatin immunoprecipitation (ChIP) analyses. Although the protein A incubation can sometimes weaken antibody avidity, a high-titer serum will almost always gain in specificity after an IgG fractionation.

1. To de-lipify the crude serum, mix 4 ml with an equal volume of Freon (Aldrich, cat no. 24050-8). Vortex and centrifuge three times for 10 min at 3300 rpm at $4°$. Transfer supernatant containing serum to fresh Freon after each wash. Filter the supernatant through a 0.22-μm pore-sized filter and add 5 M NaCl to a final concentration of 1 M.
2. Apply the serum to a Protein A-Poros® resin column (Applied Biosystems) and elute antibodies in two steps, using 1.5 M MgCl$_2$, 20 mM sodium acetate, pH 5, followed by 0.1 M glycine, pH 2.45.
3. Immediately neutralize eluted fractions with 0.1 volume of 2 M HEPES, pH 7.9, to avoid degradation of the antibodies. Pool positive fractions (identified by O.D. at 280 nm).
4. Dialyse the ≈2 ml eluate for 3 h at $4°$ against 2 l of PBS, followed by an overnight incubation in fresh buffer. After dialysis, the volume increases to 5 ml.
5. Determine antibody concentration at 200–400 nm by UV scan (for antibody #4861, it was 1.3 mg/ml).
6. Preserve and store the dialyzed IgG fraction as described for the crude serum.

Dot Blot Analysis. ELISA and dot blots can be used equivalently to probe antibody specificities. We prefer dot blots, simply because the same procedure can be extended to the analysis of nuclear proteins by western blotting.

1. Rinse a Hybond-P membrane (cat. no. RPN303F, Amersham Biosciences) for 10 s in methanol, wash with water, and soak in PBS until use. Identify the correct side of the membrane (characterized by a higher surface tension) by spotting 2 μl of PBS on a corner: the drop sits on the surface for a few seconds before it is absorbed.

2. Cover a glass or Plexiglas plate with parafilm and place a sheet of PBS-soaked Whatman paper on top. Place the Hybond-P membrane on the Whatman paper and allow all excess liquid to run off before proceeding to spot. Mark the membrane with blue or black ballpoint dots and cut the upper left corner for orientation.

3. Spot 2 μl each of 0, 0.2, 1, 5, and 25 μM of peptide solution in PBS, with a minimum of 1 cm between dots; peptides are spotted on the x-axis and serial dilutions on the y-axis. After spotting, place the membrane between two sheets of Whatman paper and allow it to air-dry overnight.

4. On the next day, rinse the membrane for 10 s in methanol, wash in PBS and block for 1 h in 3% (w/v) BSA in PBS (blocking solution).

5. Incubate for 1 h at RT with the appropriate primary antibody dilution in 3 ml of blocking buffer, by placing the membrane face-down on a parafilm-covered glass or Plexiglas plate.

6. Wash five times for 10 min in 1× TBS-T (50 mM Tris-HCl, pH 8.0, 100 mM NaCl, 0.1% (v/v) Tween 20).

7. Dilute the peroxidase-conjugated anti-rabbit secondary antibody (Jackson Immunoresearch Laboratories, cat. no. 111-035-144) 1:2000 in 5% (w/v) non-fat milk in PBS and incubate with the membrane for 1 h.

8. Wash five times for 10 min in 1× TBS-T (50 mM Tris-HCl, pH 8.0, 100 mM NaCl, 0.1% (v/v) Tween 20).

9. Develop using a dual ECL system (Amersham Biosciences) for 4 min and expose to scientific imaging film (Kodak X-Omat Blue XB-1, cat. no. 165-1454).

Using the previous protocol, we have generated a highly specific, high-titer IgG fraction for an α-2×-trimeth H3-K9 (# 4861) antibody. In a similar manner, we also succeeded to generate highly selective α-2×-mono-meth H3-K9 (# 4858) and α-2×-dimeth H3-K9 (# 4679) antibodies. All of these antisera show negligable (i.e., only very weak signals with ≥50 pmol spotted peptides) or no cross-reactivities with other methylated lysine positions (see Table I). These antisera have further been controlled in IF and ChIP experiments in wt and mutant (i.e., *Suv39h* dn) cells.[19]

TABLE I

SUMMARY OF SPECIFICITIES FOR METHYL-LYSINE ANTIBODIES[a]

Antigen		Abcam Mono linear (aa 23–34) ab9045	Abcam Di linear (aa 23–34) ab7312	Abcam Tri linear (aa 5–11) ab8898	UBI Di linear (aa 6–13) 07-212	Peters et al.[19] Mono 2× (aa 5–15) 4858	Peters et al.[19] Di 2× (aa 5–15) 4679	Peters et al.[19] Tri 2× (aa 5–15) 4861	Peters et al.[20] Di 4× (aa 6–11) 2233	Cowell et al.[22] Tri linear (aa 1–20)	Peters et al.[19] Tri 2× (aa 22–33) 6523
											H3-K27
H3-K4 (aa 1–20)	Unmodified	–	–	–	–	–	–	–	–	–	–
	Mono	+	–	–	–	–	–	–	–	–	–
	Di	–	–	–	+	–	–	–	++	–	–
	Tri	–	–	+	–	–	–	–	+	–	–
H3-K9 (aa 1–20)	Unmodified	–	–	–	–	–	–	–	–	–	–
	Mono	+++	+++	–	–	+++	–	–	–	–	–
	Di	–	–	–	+++	–	+++	–	++	+	–
	Tri	–	–	+++	–	–	–	++++	+++	++++	–
H3-K27 (aa 19–34)	Unmodified	–	–	–	–	–	–	–	–	–	–
	Mono	+	–	–	–	–	–	–	–	–	–
	Di	+	–	–	–	–	–	+	++	++	–
	Tri	+	+	–	–	–	–	–	–	+	+++
H3-K36 (aa 25–45)	Unmodified	+	–	–	–	–	–	–	–	–	–
	Mono	+	+	+	–	–	–	–	++	–	–
	Di	+	+	–	–	–	–	–	–	+	–
	Tri	+	–	–	–	–	–	–	–	–	–
H4-K20 (aa 12–31)	Unmodified	–	–	–	–	–	–	–	–	–	–
	Mono	–	–	–	–	–	–	+	++	+++	+
	Di	–	+	–	–	–	–	–	++	–	–
	Tri	–	–	–	–	–	–	–	–	–	–
Fraction		Affinity	Affinity	Affinity	Affinity	IgG	IgG	IgG	IgG	Serum	IgG
Concentration (mg/ml)		0.35	0.38	0.09	n.a.	1.7	1.7	1.3	1	n.a.	1.1
Dilution		1:1000	1:500	1:500	1:1000	1:2500	1:2500	1:2500	1:1666	1:1500	1:1500

[a] Each indicated antibody has been tested at various dilutions on peptide dot blots as shown in Fig. 2. The quality control of the α-2×-trimeth H3-K27 antibody (#6523) is included for comparison. The relative affinity of an antibody for a given methylated lysine residue is determined by its reactivity toward a peptide dilution series (see Fig. 2) and indicated by the number of +, ++++ represents the highest antibody affinity (detecting 0.4 pmol peptide), whereas + indicates very low-affinity (detecting ≥50 pmol peptide). Most of the antibodies (ab7312, UBI 07-212, #4858, #4679, #4861, #2233, Cowell et al., and #6523) have been tested by IF of mouse inter- and metaphase chromatin (see Figs. 3 and 4). Based on these IF analyses, we observed that minor cross-reactivities (+) can sometimes be ignored, since they do not result in significantly altered staining patterns. However, we do not know how these minor cross-reactivities would affect readout in ChIP experiments.
n.a.: not available.

In Vivo *Characterization of H3-K9 Methylation Patterns*

Although ELISA or dot blot analyses provide the first level of quality control for methyl-lysine histone antibodies, they are by no means sufficient to evaluate antibody specificities in a physiological setting. For this, IF analyses of inter- and metaphase chromatin offer a valuable platform to discriminate the subnuclear distribution of distinct methylation marks at constitutive (e.g., pericentric) and facultative (e.g., the inactive X chromosome) heterochromatin and in euchromatin. Particularly in mouse cells, constitutive heterochromatin can readily be visualized at one end of all 20 acrocentric chromosomes (except the Y chromosome) by intercalation of the fluorochrome 4′,6-diamidino-2-phenylindole (DAPI) at pericentric A/T-rich satellite repeats.

We have tested all four α-dimeth H3-K9 antibodies described in Fig. 2 by IF on interphase and mitotic chromatin present in female mouse embryonic fibroblasts (MEFs) (see Fig. 3). The antibody from UBI (#07-212) and our new 2-branched antiserum (#4679) display a rather uniform staining of interphase chromatin (upper panel) and strongly decorate the arms of mitotic chromosomes (middle panel). Although the Abcam antibody (#ab7312) also broadly stains interphase chromatin, it reveals a more punctate pattern along the chromsomal arms. None of these three different antisera show enriched staining at DAPI-dense regions, but instead are under-represented at pericentric heterochromatin of the mitotic chromosomes (see arrowheads in Fig. 3). Since all three antisera have highest affinity for di-methylated H3-K9 (see Fig. 2 and Table I), these data suggest H3-K9 di-methylation to be a very general mark for eukaryotic chromatin. Moreover, since all mitotic chromosomes are labeled, these results also indicate H3-K9 di-methylation to be present at the inactive X chromosome (X_i),[21,25] although it does not appear as a prominent mark for the X_i; (see later).

By contrast, the "multi-methyl lysine" antibody (# 2233) strongly decorates pericentric heterochromatin both in interphase chromatin and on mitotic chromosomes[20] and is under-represented along the chromosomal arms. In addition, it selectively stains the X_i; (see arrows in Fig. 3).[24] This surprisingly distinct staining pattern is likely to be a consequence of the promiscuity of the "multi-methyl lysine" antibody, but which of the many recognized positions (see Fig. 2) could discriminate pericentric heterochromatin and the X_i? Since H3-K4 and H3-K36 are mainly linked

[25] E. Heard, C. Rougeulle, D. Arnaud, P. Avner, C. D. Allis, and D. L. Spector, *Cell* **107**, 727 (2001).

with transcriptional activation at euchromatic positions,[26] H3-K9 tri-methylation[22] and H4-K20[11,27,28] or H3-K27 methylation[29,30] remain as possible candidates.

H3-K9 and H3-K27 Tri-Methylation Can Discriminate Constitutive and Facultative Heterochromatin

We repeated the above-mentioned IF analyses of mitotic chromosomes from wt female MEFs with a recently developed H3-K9 tri-methyl antibody[22] and our new α-2$\times$-trimeth H3-K9 antiserum (#4861). Intriguingly, the Cowell et al. antibody[22] displayed a very similar staining pattern as the "multi-methyl lysine" (#2233) antibody, with prominent signals at both pericentric heterochromatin and the X_i (see Fig. 4, top panel). By contrast, the α-2$\times$-trimeth H3-K9 antiserum (#4861) almost exclusively decorates pericentric regions but fails to detect the X_i. Since the Cowell et al. antibody shows stronger cross-reactivities with tri-methylated H3-K27 peptides as compared to antiserum #4861 (see Table I), H3-K27 tri-methylation could be a "diagnostic" mark for the X_i. We, therefore, generated another antibody that is highly specific for the tri-methylated H3-K27 position, by following a similar stategy as outlined earlier under protocol A. This novel α-2$\times$-trimeth H3-K27 antiserum (#6523) has successfully been used to detect the X_i[30] and does not cross-react with the methylated H3-K9 position (see Table I). Importantly, it also does not decorate pericentric heterochromatin (see Fig. 4, top panel), indicating that tri-methylation of two separate lysine positions (H3-K9 versus H3-K27) can discriminate constitutive from facultative heterochromatin in mammalian cells. Using similar assays, an H4-K20 tri-methyl antibody (UBI, #07-463) is also not generating broad signals, but selectively stains pericentric heterochromatin on mitotic chromosomes (M. Lachner, J. Bone, and T. Jenuwein, unpublished).

[26] M. Hampsey and D. Reinberg, Cell 113, 429 (2003).

[27] K. Nishioka, J. C. Rice, K. Sarma, H. Erdjument-Bromage, J. Werner, Y. Wang, S. Chuikov, P. Valenzuela, P. Tempst, R. Steward, J. T. Lis, C. D. Allis, and D. Reinberg, Mol. Cell 9, 1201 (2002).

[28] J. Fang, Q. Feng, C. S. Ketel, H. Wang, R. Cao, L. Xia, H. Erdjument-Bromage, P. Tempst, J. A. Simon, and Y. Zhang, Curr. Biol. 12, 1086 (2002).

[29] K. Plath, J. Fang, S. K. Mlynarczyk-Evans, R. Cao, K. A. Worringer, H. Wang, C. C. de la Cruz, A. P. Otte, B. Panning, and Y. Zhang, Science 300, 131 (2003).

[30] J. Silva, W. Mak, I. Zvetkova, R. Appanah, T. B. Nesterova, Z. Webster, A. H. Peters, T. Jenuwein, A. P. Otte, and N. Brockdorff, Dev. Cell 4, 481 (2003).

The Ultimate Control for Methyl-Lysine Histone Antibodies

The most stringent test for any antibody specificity is to analyze its reactivity in a wt and mutant background. To illustrate this important control, we repeated the IF analyses with mitotic chromosomes derived from female MEFs that are double null (dn) for both Suv39h HMTases.[20] In *Suv39h* dn chromosomes, both the "multi-methyl lysine" and the Cowell *et al.* antibodies loose pericentric staining but gain decoration in the arms. However, signals on the putative X_i remain present (see arrows in Fig. 4). IF analyses with the highly selective H3-K9 (#4861) and H3-K27 (#6523) tri-methylation antibodies revealed the "true" *in vivo* specificity of the Suv39h HMTases, whereas in *Suv39h* dn cells H3-K9 tri-methylation at pericentric heterochromatin is selectively abrogated, H3-K27 tri-methylation at the X_i is unaltered. Thus, the Suv39h HMTases are H3-K9 tri-methylating enzymes with a strong preference for pericentric heterochromatin, whereas Suv39h-unrelated HMTases are responsible for H3-K27 tri-methylation of the X_i.[19,30] These results also suggest that the less-specific "multi-methyl lysine" and Cowell *et al.* antisera contain a significant fraction of H3-K27 tri-methylation antibodies.

All of the above-mentioned IF analyses have been performed under fixed conditions (pre-fixed). For IF as well as for ChIP, fixation may alter epitope presentation which could affect staining and immunoprecipitation patterns. However, cells can also be analyzed under native conditions and fixed *after* the secondary antibody has been washed off (post-fixed). To assess the influence of fixation on epitope presentation and antibody recognition, we repeated the earlier analyses in wt and *Suv39h* dn backgrounds under post-fixed conditions (see Fig. 4, lower panels). For all four antibodies tested, we did not observe significant differences between pre-fixed and post-fixed conditions, with the exception that the "multi-methyl lysine" and Cowell *et al.* antibodies have a slightly reduced potential to decorate the X_i under more native conditions.

Protocol B: Analysis of Histone Lysine Methylation by Immunofluorescence of Mitotic Chromosomes

Mouse cells are particularly useful for IF analyses with methyl-lysine histone antibodies, as their nuclei contain well-defined euchromatic and heterochromatic domains that can be distinguished with the fluorochrome DAPI. Antibodies should preferably be tested in wt and mutant cells that are deficient for a given HMTase with known target specificities.

1. Prepare $10\times$ KCM solution: 1.2 M KCl, 200 mM NaCl, 100 mM Tris, pH 8, EDTA 5 mM. Adjust pH to 7.5 and autoclave. Dilute to $1\times$ KCM with mono Q H$_2$O.

2. Grow cells to 70–80% confluency in a maxi dish and arrest mitotic cells at metaphase with 0.2 μg colchicine (Sigma, cat. no. C-9754) per ml for 1–2 h at 37°.

3. Perform mitotic shake-off or trypsinization, centrifuge cells for 5 min at 1100 rpm, and resuspend in a small volume of growth medium.

4. Add 3 volumes of RBS hypotonic solution (10 mM Tris-HCl, pH 7.5, 10 mM NaCl, 5 mM MgCl$_2$) and incubate for 15 min at 37°. Place cells on ice for a maximum of 1 h until centrifugation.

5. Add 0.1 volume of 3% (v/v) Tween 20 in RBS to cells just prior to centrifugation.

6. Centrifuge 100 μl of cell suspension onto Superfrost slides using a Shandon Cytospin3 (Shandon, UK) for 10 min at 2000 rpm. To check cell density and presence of mitotic chromosomes, analyze one test slide by DAPI staining and fluorescence microscopy.

7. After centrifugation, transfer all slides directly in KCM-T solution (10 mM Tris, pH 8.0, 120 mM KCl, 20 mM NaCl, 0.5 mM EDTA, 0.1% (v/v) Tween 20). Extract with KCM-T containing 0.1% (v/v) Triton X-100 for 10 min at RT and subsequently wash the slides three times for 10 min in KCM-T.

8. Fix in freshly prepared 2% (v/v) para-formaldehyde in PBS, pH 7.4, for 10 min at RT and subsequently wash three times for 10 min in KCM-T.

9. Wash slides three times for 10 min at RT in KCM-BT (KCM-T containing 0.25% (w/v) BSA).

10. Block slides at RT for 30 min with 100 μl of blocking buffer (KCM-T containing 2.5% BSA (w/v) and 10% goat serum).

11. Incubate 1 h at RT or overnight at 4° with 100 μl of primary antibody (e.g., α-2$\times$-tri-methyl H3-K9 (#4861) IgG fraction, 1.3 mg/ml) using dilutions from 1:500 to 1:2000 in blocking buffer.

12. Subsequently or on the next day, wash the slides three times for 10 min in KCM-BT.

13. Add 100 μl of a secondary antibody diluted 1:500 in blocking buffer (e.g., goat α-rabbit IgG, Alexa Fluor488, Molecular Probes, cat. no. A-11034). Incubate for 1 h in the dark.

14. Wash slides three times in KCM-T, followed by a last wash in PBS.

15. Embed in Vectashield containing 1.5 μg DAPI (Vector Laboratories, cat. no. H-1200) per ml and store at 4° until analysis.

16. Images are taken with an Axioplan 2 fluorescence microscope (Zeiss, Germany).

Immunofluorescence Labeling Under Native Conditions. Both inter-phase chromatin and mitotic chromosome spreads can also be stained under native conditions (post-fixation). The above-mentioned protocols are followed as described, but the fixation step (= step 8) is performed *after* the secondary antibody is washed off (i.e., after step 15). Before embedding in Vectashield containing DAPI (Vector Laboratories, cat. no. H-1200), three washes of 10 min are performed with KCM-T followed by a PBS-only wash.

Discussion

The above-mentioned examples underscore the technically complex analysis of the functional significance for mono-, di-, and tri-methylation of histone lysine positions *in vivo*. Although it is anticipated that highly specific antibodies for each methylatable histone lysine position will be developed that can discriminate every distinct methylation state, only the minority of the currently available methyl-lysine histone antibodies fulfill all the quality criteria and specificity controls outlined in this chapter. It is therefore highly advised that any methyl-lysine histone antibody should be rigorously tested *in vitro* and *in vivo* before being used in experimental settings. Similarly, antibody companies such as Upstate Biotechnology, USA, and Abcam, UK, would do an invaluable service to the scientific community if they would very stringently indicate strength and weaknesses of several of their methyl-lysine histone antibodies and, if required, stop commercializing less-specific antibodies or sub-optimal batches.

We have taken the difficulty to discriminate between the H3-K9 and H3-K27 positions as a quality criterion to evaluate nine different antisera raised against the methylated H3-K9 residue (see Table I). For each antiserum, dot blot analyses comprising the complete peptide panel as shown in Fig. 2 were performed. The relative affinity of an antibody for a given methylated lysine residue was scored using reactivity toward 0.4 pmol peptide (highest affinity; $++++$) up to 50 pmol peptide (very low affinity; $+$). These *in vitro* data have then been extended and compared to *in vivo* analyses by performing IF of mouse inter- and metaphase chromatin. Although we have not done extensive IF with all antibodies, unspecific or perturbed staining patterns became apparent if antibodies displayed cross-reactivities in peptide dot blots ranging from high affinity ($++++$) to moderate/low affinity ($++$). By contrast, minor cross-reactivities ($+$) can sometimes be ignored, as they did not result in significantly altered staining patterns (particularly for IF on mitotic chromosomes). However, we do not know how these minor cross-reactivities may affect read-out of ChIP experiments.

In addition to the problems outlined earlier, there are even more technical and functional complexities in analyzing the specificity of methyl-lysine histone antibodies. For example, differences in peptide length appear to affect antibody reactivities in dot blots (see Chapter 17) again highlighting the need to control antibody specificity beyond peptide spotting or ELISA assays. Moreover, it is also largely unexplored how additional modifications would affect epitope recognition, as for example, by the simultaneous presence of H3-K9 methylation and H3-S10 phosphorylation. Intriguingly, every methylatable histone lysine position is neighborored by either a serine or threonine residue (D. Allis, personal communication), thereby allowing a phosphorylation event to potentially modulate or even mask a methylated epitope. We have tested the reactivity of the α-2×-dimeth H3-K9 antibody (#4676) toward a doubly modified (H3-K9 di-methylation and H3-S10 phosphorylation) peptide, without observing a reduction in epitope recognition (data not shown), and are currently extending these analyses for the mono- and tri-methyl–specific H3-K9 antisera. Similarly, it is also possible that a doubly methylated H3 amino terminus (e.g., H3-K9 and H3-K27 methylation) may affect epitope presentation and antibody reactivity. Particularly for multiple histone tail modifications, extensive *in vivo* analyses are required, ideally comprising mass spectrometry of chromatin fragments that have been immunoprecipitated with the respective antibodies or which were prepared during different stages of the cell cycle. Finally, antibody development in different host species (e.g., rabbit, chicken, mouse, lama) will provide another important tool to analyze the presence of multiple modification marks by colocalization or coimmunoprecipitation experiments. The availability of highly selective methyl-lysine histone antibodies from different species will, for example, prove advantageous for understanding the possible combinatorial contribution of H3-K9 and H3-K27 tri-methylation during Polycomb-mediated transcriptional memory.[31–34]

[31] R. Cao, L. Wang, H. Wang, L. Xia, H. Erdjument-Bromage, P. Tempst, R. S. Jones, and Y. Zhang, *Science* **298**, 1039 (2002).

[32] B. Czermin, R. Melfi, D. McCabe, V. Seitz, A. Imhof, and V. Pirrotta, *Cell* **111**, 185 (2002).

[33] J. Muller, C. M. Hart, N. J. Francis, M. L. Vargas, A. Sengupta, B. Wild, E. L. Miller, M. B. O'Connor, R. E. Kingston, and J. A. Simon, *Cell* **111**, 197 (2002).

[34] A. Kuzmichev, K. Nishioka, H. Erdjument-Bromage, P. Tempst, and D. Reinberg, *Genes Dev.* **16**, 2893 (2002).

Important Considerations and Technical Advice

Based on our experience, we would like to highlight the following steps in the generation and characterization of methyl-lysine histone antibodies:

1. *Golden Rule:* obtain a high-titer, position-specific antiserum that has extensively been characterized in *in vitro* and *in vivo* analyses and which ideally has also been controlled in wt and mutant backgrounds.
2. To obtain a specific antiserum, the peptide length is crucial and should be equal or greater than 10 amino acids, with the lysine of interest centered in the middle of the sequence.
3. When branched peptides (preferably 2-branched) are used, nearly every rabbit produces an immune response by yielding a high-titer antiserum; in contrast, only a fraction of immunized rabbits respond to linear peptide antigens and sometimes weakly.
4. A high-titer, position-specific antiserum (which can be used at dilutions ≥1:2000) is a hallmark of a robust antibody that can be used with confidence in assays such as protein blot, IF, and ChIP.
5. The specificity of a crude antiserum can be further improved by performing an IgG fractionation. Although this procedure can sometimes weaken the avidity of the antibody (particularly for ChIP), it increases the signal to background ratio. We recommend that a gain in specificity should be favored over a stronger output.
6. We have observed that pre-absorbing a less-specific antiserum against peptides with which it cross-reacts, followed by affinity purification, considerably weakens the avidity of the antibody without a significant improvement in specificity. We therefore recommend that new rabbits should be immunized rather than applying extensive purification schemes that only yield suboptimal antibody preparations. As an alternative to elaborate purification, peptide competition may be used in some assays.

The fast-moving histone methylation field has generated many novel insights that may well underpin basic mechanisms for epigenetic control. However, many questions remain unanswered, such as, for example, whether differences between mono-, di-, and tri-methylation of selective lysine residues in the histone amino-termini will be of functional importance for the organization of constitutive and facultative heterochromatin, and for the long- and short-term regulation of promoter activity. We currently have only limited knowledge about mono-methylated states, and it is unclear whether mono-methylation could specify distinct chromatin regions or whether it may be required to prepare a lysine position for subsequent di- and tri-methylation. By contrast, di-methylation states appear as more general marks that can broadly decorate eukaryotic chromatin. Intriguingly, tri-methylation states are more selectively distributed along inter- and metaphase chromatin and would be good candidates to index subchromosomal domains. For example, H3-K4 tri-methylation has been correlated with the promoter regions of fully activated genes[18,35] but is excluded from pericentric heterochromatin (data not shown). Similarly, H3-K36 methylation has also been implicated in gene control during transcriptional elongation,[26] although H3-K36 tri-methyl antibodies are currently not available. By contrast, tri-methylation of the H3-K9, H3-K27, and H4-K20 positions can be associated with repressive functions, since H3-K9 tri-methylation is highly enriched at pericentric heterochromatin[19,22] and H3-K27 tri-methylation is a prominent mark for the Xi.[19,29,30] H4-K20 di-methylation has been observed at condensed chromatin regions during mitosis,[11,27,28] and H4-K20 tri-methylation antibodies decorate pericentric heterochromatin (M. Lachner, J. Bone, and T. Jenuwein, unpublished). It is without doubt that the development of high-quality, position-specific methyl-lysine histone antibodies will provide important tools for the further decoding of the epigenetic information which is, in part, indexed by distinct methylation states of selective lysine residues in the histone amino-termini.

Acknowledgments

We would like to thank Upstate Biotechnology (UBI, Lake Placid, NY) (http://www.upstatebiotech.com/) and Abcam (Cambridge, UK) (http://www.abcam.com/) for exchange of antibodies and critical comments to the manuscript. In particular, we would like to acknowledge Judy Nisson, Rene Rice, Mary-Ann Jelinek, Thomas Jelinek, and Jim Bone from Upstate Biotechnology and Darren Harper from Abcam. We also acknowledge Gramsch Laboratories (Schwabhausen, Germany) (http://www.gramsch.de/) for excellent antibody development. We are indebted to Judd Rice and David Allis for their contribution and help in characterizing the new series of α-2×-mono-di-tri H3-K9 methyl-specific antibodies. We thank Prim Singh for contributing the "Cowell et al." antibody, and Danny Reinberg and Bryan Turner for helpful discussions. We would like to thank Mathias Madalinski for the protocol on peptide synthesis and Ines Steinmacher for help on mass spectrometry analysis. Research in the laboratory of T.J. is supported by the I.M.P. through Boehringer Ingelheim and by grants from the Vienna Economy Promotion Fund, the European Union, and the GEN-AU initiative, which is financed by the Austrian Ministry of Education, Science, and Culture.

[35] H. H. Ng, F. Robert, R. A. Young, and K. Struhl, *Mol. Cell* **11**, 709 (2003).

[17] Tips in Analyzing Antibodies Directed Against Specific Histone Tail Modifications

By Kavitha Sarma, Kenichi Nishioka, and Danny Reinberg

Histone methylation has been known to exist for over 40 years[1] but the enzymes that catalyze this reaction have remained elusive until the discovery that Suv39H1 methylates histone H3 specifically at lysine 9.[2] This discovery was followed by a bevy of papers describing other methyltransferases specific for different residues and their apparent function *in vivo*. Histones are methylated at lysine as well as arginine residues. Lysines can be mono-, di-, or tri-methylated *in vivo*. The sites for lysine methylation on histone H3 are 4, 9, 27, 36, and 79 (for reviews see Zhang and Reinberg (2001),[3] Turner (2002),[4] Bannister *et al.* (2002),[5] Lachner *et al.* (2003),[6] and Vaquero *et al.* (2003)[7]) while lysines 20 and 30 remain the sole site for methylation on H4[8,9,9a] and H2B,[10] respectively (see Fig. 1).

Studies on the effects of these modifications on gene regulation have been greatly facilitated by the production of antibodies "specific" for the modified state. The need to carefully characterize antibodies raised against methylated histone peptides stems from the observation by several laboratories that these antibodies can be promiscuous depending on several factors such as concentration, peptide context, substrate, etc. Due to this, several papers have been subject to scrutiny in recent months as the specificity of the antibodies used was questionable. The results of chromatin immunoprecipitation (ChIP) experiments became increasingly difficult to interpret

[1] K. Murray, *Biochemistry* **3**, 10 (1964).

[2] S. Rea, F. Eisenhaber, D. O'Carroll, B. D. Strahl, Z. W. Sun, M. Schmid, S. Opravil, K. Mechtler, C. P. Ponting, C. D. Allis, and T. Jenuwein, *Nature* **406**, 593 (2000).

[3] Y. Zhang and D. Reinberg, *Genes Dev.* **15**, 2343 (2001).

[4] B. M. Turner, *Cell* **111**, 285 (2002).

[5] A. J. Bannister, R. Schneider, and T. Kouzarides, *Cell* **109**, 801 (2002).

[6] M. Lachner, R. J. O'Sullivan, and T. Jenuwein, *J. Cell Sci.* **116**, 2117 (2003).

[7] A. Vaquero, A. Loyola, and D. Reinberg, published online at http://sageke.sciencemag.org/cgi/content/full/sageke;2003/14/re4

[8] K. Nishioka, J. C. Rice, K. Sarma, H. Erdjument-Bromage, J. Werner, Y. Wang, S. Chuikov, P. Valenzuela, P. Tempst, R. Steward, J. T. Lis, C. D. Allis, and D. Reinberg, *Mol. Cell* **9**, 1201 (2002).

[9] J. C. Rice, K. Nishioka, K. Sarma, R. Steward, D. Reinberg, and C. D. Allis, *Genes Dev.* **16**, 2225 (2002).

[9a] J. Fang, Q. Feng, C. S. Ketel, H. Wang, R. Cao, L. Xia, H. Erdjument-Bromage, P. Tempst, J. A. Simon, and Y. Zhang, *Curr. Biol.* **12**, 1086 (2002).

[10] K. Zhang and H. Tang, *J. Chromatogr. B* **783**, 173 (2003).

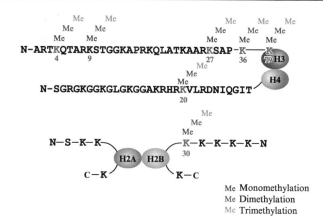

FIG. 1. A schematic representation of the various histone lysine methylation sites and their potential degrees of methylation identified to date *in vivo*. (See color insert.)

since the antibodies that had been used in these experiments were not specific to the residue or its modified state. The ideal method to test this would be to perform ChIP and then analyze the products that have been immunoprecipitated by mass spectroscopy in order to confirm that the antibody has reacted only with the modification of interest. Since this is not feasible, and the experiments required are too difficult to perform, initial characterization of the antibodies and determination of optimal concentration to be used in ChIP becomes extremely important.

In this chapter, we present several parameters to be taken into consideration and some useful hints for systematic characterization of antibodies raised against methylated histone peptides. Although we have focused on antibodies against methylated residues: H3-K4, H3-K27, and H4-K20, the methods and procedures described herein are applicable for any antibody directed against the histone tail modifications, including arginine methylation and lysine acetylation, among other modifications.

ELISA

This is the first step in antibody characterization. Analysis of the antibody with enzyme-linked immunosorbent assay (ELISA) is the traditional method for characterization. The advantage of ELISA is that of all assays described, it is the most quantitative and sensitive and the optimal concentration of the antibody to be used in experiments can be ascertained, but see later. An example of antibody characterization using ELISA is shown in the case of polyclonal di-/tri-methyl K27 antibody (see Fig. 2). At higher concentrations (1:300), the antibody is able to recognize both the di- and

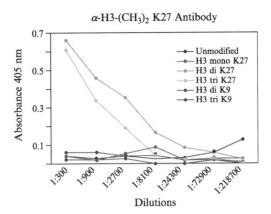

FIG. 2. ELISA with polyclonal antibodies against di-/tri-methyl K27 (D.R.) shows almost equal reactivity against both di- and tri-methyl K27 peptides at 1:300 antibody dilution. But using 1:2700 dilution, the antibody is three times more preferential to di-methyl K27 compared to tri-methyl K27.

tri-methyl K27 peptides with equal efficiency, but at lower concentrations (1:2700 dilution), the antibody is three times more preferential to the tri-methyl peptide than to the di-methyl peptide. The ELISA should be followed by dot blot analyses. If monoclonal antibodies are to be screened, it is easier to perform dot blot analysis first to isolate specific clones and then proceed to ELISA to ascertain optimal concentrations for the antibody.

Dot Blot Analysis

Care should be taken that equimolar amounts of peptide are loaded onto the blots and this must be quantified using Ellman's reagent. Quantification that involves staining of the membrane is ill-advised since all peptides do not stain equally well with Coomassie blue or Ponceau.

Serum can be checked before affinity purification by dot blots. Equimolar amounts of peptide should be blotted on nitrocellulose membrane and processed in a manner similar to that used for western blot analysis followed by detection using enhanced chemiluminescence (ECL). Initial characterization should be performed with unmodified, mono-, di-, and tri-methyl peptides, preferably in the same or very similar peptide context. This provides an initial gauge of the immune response generated by the antigen. It has been observed that, sometimes, a di-methyl peptide causes a strong anti-monomethyl response (see Fig. 3A and B). Once it is confirmed that the signal is due to a di-methyl peptide, several other di-methyl peptides should be checked for cross-reactivity to ensure that the antibody recognizes the di-methyl peptide in a residue-specific context and not the

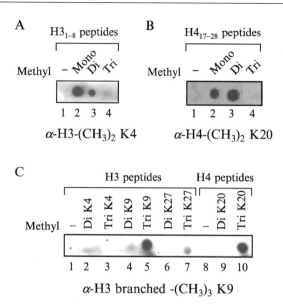

FIG. 3. Dot blot analyses of various antibodies raised against methylated histone peptides. (A) Polyclonal antibodies raised against di-methyl H3 lysine K4 peptide (Upstate 07-030) analyzed using unmodified H3 (lane 1), mono-methyl K4 (lane 2), di-methyl K4 (lane 3), and tri-methyl K4 (lane 4). (B) Monoclonal antibodies raised against di-methyl H4K20 peptide (D.R.) tested with unmodified H4 (lane 1), mono-methyl K20 (lane 2), di-methyl K20 (lane 3), and tri-methyl K20 (lane 4). (C) Polyclonal antibodies raised against the branched tri-methyl H3-K9 peptide (T.J.) analyzed for specificity using di- and tri-methyl peptides in the context of different methylated residues. The methylation site and status of the peptides are as indicated above the panels and the antibody used in each experiment is indicated below each panel.

di-methyl moiety alone. Recognition of a tri-methyl moiety without residue specificity was observed with antibodies raised against the branched tri-methyl K9, where the antibody was able to recognize tri-methyl K9, K27, and K20 (see Fig. 3C). An example of desirable optimal conditions is presented in Fig. 2, where we have tested anti-di-methyl H4-K20 polyclonal antibodies (Upstate 07-367). Dot blot analysis with K20 peptides with various degrees of modification shows antibody reactivity with the di-methyl K20 peptide exclusively (see Fig. 4A). Other di-methyl peptides were not recognized, confirming that this antibody is specific to di-methyl K20 (see Fig. 4A). To further strengthen this observation, competition experiments with the H4 peptides show that the antibody is competed out with the di-methyl peptide alone and is not affected by the mono- or the tri-methyl peptides (see Fig. 4B). Western blot analysis showed a strong signal with the native histone H4 polypeptide, but no reactivity with the recombinant H4 (see Fig. 4C, left panel). The signal obtained with native

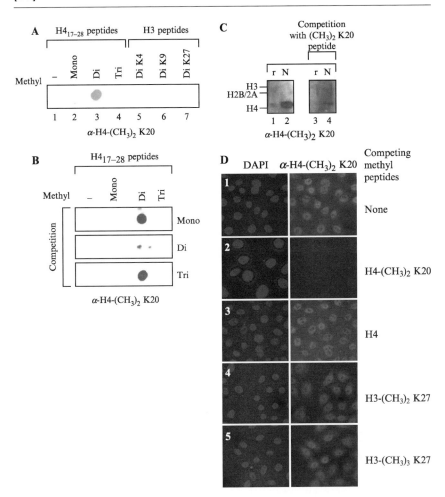

FIG. 4. Characterization of polyclonal antibodies raised against di-methyl H4-K20 antibodies (Upstate 07-367) (A) Dot blot analysis to check antibody specificity using methyl peptides as indicated above the panel. (B) Competition dot blot performed in the presence of 1 μg/ml of various peptides as indicated to the right. The di-methyl signal is lost only in the middle panel where di-methyl H4-K20 is the competing peptide. (C) Western blot analysis of the antibody using recombinant nucleosomes labeled r (left panel, lane 1) and native oligonucleosomes isolated from HeLa cells labeled N (left panel, lane 2) shows reactivity with native H4 and not with recombinant H4. Competition with di-methyl H4-K20 peptide resulted in loss of signal on native oligonucleosomes (compare lane 2 on left and right panels). The antibody dilution used was the same as in the dot blot analyses and competition dot blots. (D) Immunofluorescent staining of HeLa cells with these antibodies at 1:100 dilution show a prominent nuclear signal (panel 1) as is expected for antibodies raised against histones. This signal is completely obliterated when the antibody is used in the presence of di-methyl K20 peptide (panel 2). There is no change in signal when unmodified H4 peptide (panel 3) or non-specific peptides like di- and tri-methyl K27 peptides (panels 4 and 5) are used. (See color insert.)

histones was competed out when the western blot was performed in the presence of the di-methyl peptide (see Fig. 4C, right panel). Immunofluorescence studies also showed localization of the antibody to the nucleus and this was lost on addition of the di-methyl K20 peptide at the time of staining (see Fig. 4D, panel 2). Addition of non-specific peptides, in this case di-methyl or tri-methyl H3-K27, did not affect staining (see Fig. 4D, panels 4 and 5). Similar results were obtained using peptides containing mono- and tri-methyl H4-K20 (data not shown).

Some problems that have been encountered in our laboratory during the characterization of antibodies using dot blots are as follows. While the antibody may be extremely specific for a given antigen, when the context of the peptide is changed, the antibody is then able to recognize even the unmodified peptide. This was seen in the case of antibodies against methylated H3-K9, where, in the context of residues 4–15, the antibodies remained specific for di- and tri-methyl K9, but the antibodies were found to react with unmodified H3 peptide of a longer length, that is, residues 4–32 (see Fig. 5A and B).

It is also very important to take into consideration the context of the peptide when characterizing antibodies against the methylated H3-K9 and H3-K27 residues. Since both of these residues are contained within a very similar context, that is, $TKQTARK_9S$ for K9 and $TKQTAARK_{27}S$ for K27, it is very likely that the antibodies will cross-react with the different peptides. It becomes crucial then, to check every K9 antibody prepared for cross-reactivity with K27 peptides and vice versa. Several initial reports regarding X chromosome inactivation in mammals focused on H3-K9 methylation as the early event in X inactivation.[11,12] The paper by Heard et al.[11] shows that di-methylation at K9 is present on the inactive X chromosome; this is a noteworthy point because the antibodies used in this case were the anti-dimethyl K9 "Golden Bunny" antibody from David Allis' lab and the anti-dimethyl K9 antibodies from Upstate Biotechnology. Both of these have been extensively characterized and have been shown to be dimethyl K9 specific and show no cross-reactivity with other modifications. But recently it has been shown without doubt that the prominent mark for the establishment of the inactive X chromosome is tri-methylation at H3-K27.[13,14]

[11] E. Heard, C. Rougeulle, D. Arnaud, P. Avner, C. D. Allis, and D. L. Spector, Cell 107, 727 (2001).

[12] A. H. Peters, J. E. Mermoud, D. O'Carroll, M. Pagani, D. Schweizer, N. Brockdorff, and T. Jenuwein, Nat. Genet. 30, 77 (2002).

[13] K. Plath, J. Fang, S. K. Mlynarczyk-Evans, R. Cao, K. A. Worringer, H. Wang, C. C. de la Cruz, A. P. Otte, B. Panning, and Y. Zhang, Science 300, 131 (2003).

[14] J. Silva, W. Mak, I. Zvetkova, R. Appanah, T. B. Nesterova, Z. Webster, A. H. Peters, T. Jenuwein, A. P. Otte, and N. Brockdorff, Dev. Cell 4, 481 (2003).

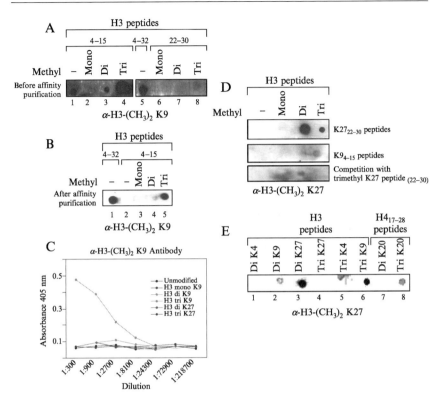

Fig. 5. (A) Polyclonal antibodies raised against di-methyl H3-K9 before affinity purification showed strong reactivity with the tri-methyl K9 peptide within the same context (lane 4) and weak reactivity with either unmodified H3 within the same context (lane 1), di-methyl K9 (lane 3), or the longer unmodified H3 peptide (lane 5). There was no cross-reactivity seen with mono-methyl K9, mono-, di-, or tri-methyl K27 (lanes 2, 6, 7, and 8, respectively). (B) After affinity purification on the di-methyl H3K9 column, the antibody is able to recognize predominantly tri-methyl K9 (lane 5) and almost no reactivity is seen with di-methyl K9 (lane 4). A strong signal is still seen with the longer unmodified H3 (lane 2) while the shorter peptide showed no cross-reactivity. This antibody was used at a dilution of 1:1000. (C) ELISA analysis with the same antibody after affinity purification also confirmed that it recognizes the tri-methyl K9 modification but not di-methyl K9, K27 peptides with different degrees of methylation or the shorter unmodified H3 peptide. (D) Polyclonal antibodies raised against the di-methyl H3-K27 showed specific reactivity with di- and tri-methyl K27 (top panel), but not against methyl K9 peptides (middle panel). Competition with the tri-methyl K27 peptide completely eliminates the signal with di-methyl K27 peptide also (bottom panel). (E) Polyclonal antibodies raised against di-/tri-methyl H3-K27 are unable to recognize or react to an insignificant level with di- and tri-methyl K4 and K20 peptides, as indicated above the panel. (See color insert.)

Therefore, while it is still likely that methylation of K9 may play a role in this process (for example, in maintenance, rather than establishment, of X-inactivation), the major modification associated with the establishment of inactive X chromosome is methylation of K27 catalyzed by the PRC2 complex or the enhancer of Zeste complex.[15–18] This was not seen earlier, because antibodies raised against methylated H3-K9 cross-reacted with methylated H3-K27.

The antibody must always be checked before and after affinity purification, since the specificity or preference for the substrate may be altered. This was seen in the case of the di-methyl H3-K9 antibody where before affinity purification, the antibody could recognize both the di- and tri-methyl forms of K9, with a preference for the tri-methylated form (see Fig. 5A), but after affinity purification using bound di-methyl K9 peptide, the antibody was now found to be tri-methyl K9 specific and no longer showed cross-reactivity with the di-methylated form (see Fig. 5B); although cross-reactivity was still seen with the longer form of the H3 unmodified peptide, amino acids 4–32. The same result was obtained when the antibody was checked by ELISA (see Fig. 5C).

Affinity purification must be performed with the antigen that was used for generating the immune response. In most cases, pre-clearing the antibody using peptides showing lower reactivity in order to enrich for a specific antibody, seems to result in complete loss of signal. An example of this is the case of the methyl H3-K27 antibody which recognizes both the di- and tri-methylated forms of K27 (see Fig. 5D and E). This antibody was affinity purified using the di-methyl K27 peptide. Theoretically, in a competition experiment with tri-methyl peptide, only the tri-methyl signal should be reduced without any effect on the di-methyl signal. But as shown in Fig. 5D, the di-methylated signal is also greatly reduced, suggesting that the epitope recognized by the antibody is common to the di-as well as tri-methylated H3-K27 peptide. The effect observed is essentially equivalent to having pre-cleared this antibody on a tri-methyl peptide column.

Western Blot Analysis

After a particular antibody has been analyzed by either or both of the above-mentioned techniques, it must next be tested by western blot. This

[15] A. Kuzmichev, K. Nishioka, H. Erdjument-Bromage, P. Tempst, and D. Reinberg, *Genes Dev.* **16,** 2893 (2002).

[16] R. Cao, L. Wang, H. Wang, L. Xia, H. Erdjument-Bromage, P. Tempst, R. S. Jones, and Y. Zhang, *Science* **298,** 1039 (2002).

[17] B. Czermin, R. Melfi, D. McCabe, V. Seitz, A. Imhof, and V. Pirrotta, *Cell* **111,** 185 (2002).

[18] J. Muller, C. M. Hart, N. J. Francis, M. L. Vargas, A. Sengupta, B. Wild, E. L. Miller, M. B. O'Connor, R. E. Kingston, and J. A. Simon, *Cell* **111,** 197 (2002).

must be tested with recombinant and native histones run side by side on an SDS-PAGE. A good antibody should not show any reactivity with the recombinant histone polypeptide (see Fig. 4C).

Sometimes, an antibody that does not exhibit reactivity with the unmodified peptide during dot blot (Fig. 6A) or ELISA analyses will give strong positive results with the recombinant histone polypeptide in western blots (see Fig. 6B).

Another example is shown in Fig. 6C, in which the polyclonal antibody generated against histone tri-methyl H4-K20 showed no reactivity with other tri-methyl peptides during dot blot analysis. When it was then tested by western blot analysis, it showed no reactivity with the recombinant H4 or H3 polypeptides. But there was a strong signal seen with native H4 polypeptides and, surprisingly, with native H3 polypeptides (see Fig. 6D). This could mean that there may be some other modification on the native H3 that is recognized by the antibody, one that we have not tested or one that is as yet unknown. Thus, during characterization of these antibodies, one cannot rule out the possibility that the antibody may recognize some modification in native histones that has not been tested or identified.

An ideal set of controls to include in the western blot characterization is to use a series of recombinant histone polypeptides having specific modifications that were chemically incorporated, using the technology described by Loyola et al. in this series or that described by the Peterson laboratory.[19]

Immunofluorescence

In order to test if the antibody is able to recognize its target in vivo, it becomes essential to stain cells with this antibody. Optimal concentrations must be determined by titration of the antibody. The knowledge that the inactive X chromosome is enriched in H3-K27 tri-methylation facilitates characterization of good antibodies as the inactive X signal can be identified during staining along with the loss of such a signal obtained with competition using the tri-methyl peptide. For example, in Fig. 7 in which monoclonal antibodies raised against di-/tri-methyl H3-K27 were raised, the inactive X chromosome is detected as a strong signal in mouse embryonic fibroblasts (MEFs). The specificity of these antibodies in vivo was confirmed by competition experiments performed with methylated H3-K9 peptides. This characterization can become difficult, however, when the exact function of the modification is unknown. One can detect heterochromatic or euchromatic localization depending on the modification. The most

[19] M. A. Shrogen-Knaak, C. J. Fry, and C. L. Peterson, J. Biol. Chem. 278, 15744 (2003).

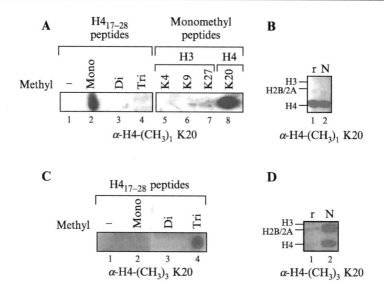

Fig. 6. (A) Dot blot analysis of polyclonal antibodies raised against mono-methyl H4-K20 (Abcam ab9051) shows very high specificity to the mono-methyl K20 peptide (lanes 2 and 8) and does not recognize unmodified H4, di-methyl K20, or tri-methyl K20 peptides (lanes 1, 3, and 4). They do not react with mono-methyl K4, K9 or K27 peptides (lanes 5, 6, and 7). (B) Antibodies raised against mono-methyl H4-K20 (ab9051) were analyzed by western blot analysis using recombinant nucleosomes (r) and native oligonucleosomes (N) from HeLa cells. They recognized both unmodified and native forms of H4 equally well (lanes 1 and 2). (C) Polyclonal antibodies raised against tri-methyl H4-K20 (D.R.) were found to be extremely specific for the tri-methyl K20 (lane 4) and showed no reactivity with either unmodified H4, or mono- or di-methyl K20 (lanes 1, 2, and 3, respectively) by dot blot. (D) Western blot analysis with antibodies raised against tri-methyl H4-K20 (D.R.) shows a strong signal with the native H4 and H3 (lane 2) but no cross-reactivity with recombinant nucleosomes (lane 1).

general protocol to characterize this is to detect nuclear staining in cells and loss of such signal upon competition with the specific antigen, but not with a series of peptides containing other modifications.

Immunofluorescence is also the best method to test the *in vivo* specificity of the antibody. To test this, competition experiments must be performed with the antigen, as well as the unmodified peptide within the same context, and also with modified peptides within different contexts. Only if the signal is lost with the antigen alone can the antibody be used with confidence in *in vivo* experiments (as shown in Figs. 4D and 7). A list of antibodies that have been tested in our laboratory by all the methods described earlier has been summarized in Table I along with their advantages and shortcomings for various experiments.

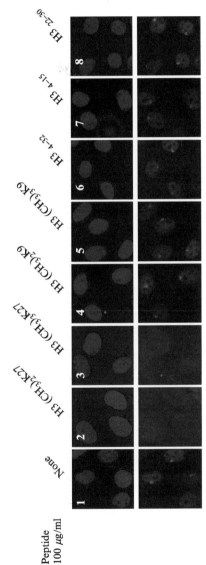

Fig. 7. Immunostaining of female mouse embryonic fibroblasts (MEFs) with monoclonal di-/tri-methyl K27 antibodies. Panel 1 shows staining of MEFs with di-/tri-methyl K27 antibody with a prominent signal that correlated with the inactive X chromosome in independent experiments (for details refer to the chapter by J. Chaumeil *et al.* in this volume). Panels 2 and 3 show the result of antibody competition with the di- and tri-methyl K27 peptides where the signal on the inactive X is lost. Competition with di- and tri-methyl K9 (panels 4 and 5) and unmodified H3 peptides of different lengths (panels 6–8) do not affect the signal obtained with this antibody, demonstrating the specificity of these antibodies *in vivo*. The concentration of peptide used for competition was 100 μg/ml, as indicated. (See color insert.)

TABLE I

Antibody	Upstate Biotechnology	Abcam	D.R.	Others
H3-K4				
Mono-methyl	N/A	Highly specific to mono-methyl K4 only	N/A	N/A
Di-methyl	Methyl H3-K4 specific but reacts with mono- and di-K4	Methyl H3-K4 specific but reacts with mono- and di-K4	N/A	N/A
Tri-methyl	N/A	Highly specific to tri-methyl K4 only	Highly specific to tri-methyl K4 only	N/A
H3-K27				
Mono-methyl	N/A	N/A	N/A	N/A
Di-methyl	Highly specific to di-methyl K27 only	N/A	Methyl H3-K27 specific but reacts with di- and tri-methyl K27	N/A
Tri-methyl	N/A	N/A	Methyl H3-K27 specific but reacts with di- and tri-methyl K27	Highly specific to tri-methyl K27 only (Thomas Jenuwein's laboratory)
H4-K20				
Mono-methyl	N/A	Mono-methyl K20 specific in dot blots but cross-reacts with recombinant H4 in western	N/A	N/A
Di-methyl	Highly specific to di-methyl K20 only	N/A	Specific to methyl K20 but cross-reacts with mono-methyl K20	N/A
Tri-methyl	Highly specific in dot blot analysis but shows slight reactivity to trimethyl K9 in western blot	Shows cross-reactivity with di-methyl K20	Shows cross-reactivity with mono- and di- methyl K20	N/A

Note: Some antibodies that are now available in various companies have been marked N/A since they were not in circulation at the time this manuscript was prepared.

Concluding Remarks

Even after antibodies have been characterized by the techniques described herein, one can perform additional experiments to confirm the antibody specificity. Another system that can be used for characterization of some of these antibodies is *Saccharomyces cerevisiae*. This becomes especially useful when the modification to be studied exists in yeast. A yeast strain in which the H3 and H4 chromosomal copies have been deleted and the only source of H3 and H4 is through a plasmid-borne copy, can be used to introduce point mutations.[20] For example, H3-K4 can be mutated and histones can be extracted from this strain and used for analysis of the methyl K4 modifications. Unfortunately, not all methyl lysine modifications observed in higher eukaryotes exist in *S. cerevisiae* such as H3-K9, H3-K27, and H4-K20.

It has become evident that all studies that address the functional relevance of the different methylated states of histones should be undertaken only after very careful analysis of the antibody being used. The antibodies have to be characterized by as many methods as possible and with as many controls as possible. The number of parameters that affect the specificity of the antibody are too many and any result that is derived from ChIP and IF experiments must be interpreted with extreme caution.

Materials and Methods

Antibodies and Peptides

Polyclonal antibodies were obtained from the following sources: di-methyl H3K4 (Upstate 07-030), branched tri-methyl H3-K9 (Thomas Jenuwein), di-methyl H4-K20 (Upstate 07-367), mono-methyl H4-K20 (Abcam ab9051). Polyclonal antibodies against di-methyl H3-K27, di-methyl H3-K9 and tri-methyl H4-K20 were generated in D.R.'s laboratory. Monoclonal antibodes against di-methyl H3-K27 and di-methyl H4-K20 were generated by Bios Chile.

The contexts of the various human histone peptides synthesized (Global Peptide Services) were as follows: H3-K4 (amino acids 1–8), H3-K9 (amino acids 4–15 and 4–32), H3-K27 (amino acids 22–30), and H4-K20 (amino acids 17–28).

[20] W. Zhang, J. R. Bone, D. G. Edmondson, B. M. Turner, and S. Y. Roth, *EMBO J.* **17**, 3155 (1998).

Dot Blot and Western Blot

For dot blot, load 1 μl of 0.5 mM peptide solution on nitrocellulose membrane and dry completely. Block with 3% milk in TTBS (10 mM Tris, 200 mM NaCl, and 0.05% Tween 20, pH 7.9). Wash and incubate in primary antibody at 1:1000 dilution (or as directed by the manufacturer) for 2 h at room temperature. Wash with TTBS and incubate in 1:5000 dilution of secondary antibody conjugated to horseradish peroxidase (HRP) for 40 min. Wash thoroughly three times, 10 min each time and develop using ECL. Cover membrane with plastic wrap and expose to film for various times, ranging from 2 s to 1 min. For western blot, load 2 μg of recombinant and native core histones onto a 15% SDS-PAGE, transfer to nitrocellulose membrane and stain with Ponceau to ensure efficient transfer. Proceed as described earlier for dot blot. For competition experiments, add peptide to a final concentration of 1 μg/ml to the primary antibody dilution.

ELISA

Peptide solution was prepared at a concentration of 5 μg/ml in PBS and 50 μl were added to 96-well plates (Costar) and left overnight at room temperature to coat the plates with antigen. All subsequent procedures were performed at room temperature. The plates were then washed with PBST (1$\times$ PBS with 0.05% Tween 20) twice, for 5 min each time and incubated in blocking solution (5% bovine serum albumin in PBST) for 30 min. Plates were then washed with PBST twice for 10 min each time and 200 μl primary antibody was added at the following dilutions: 1:300, 1:900, 1:2700, 1:8100, 1:24300, 1:72900, and 1:218700. A non-specific polyclonal or monoclonal antibody was used as a control at 1:300 dilution. The plates were incubated with primary antibody for 2 h. The plates were then washed three times with PBST for 10 min each time. Fifty microliters of secondary antibody conjugated to alkaline phosphatase were added at a dilution of 1:5000 and incubated for 2 h. The plates were washed thoroughly with PBST three times with vortexing the last time. Fifty microliters of developing solution (Sigma Fast p-Nitrophenyl Phosphate tablet sets) were added to each well and color was allowed to develop. The reaction was stopped by adding 50 μl of 3 N NAOH. The plates were read at 405 nm on an ELISA plate reader (Tecan).

Immunofluorescence

HeLa cells were grown overnight at 37° on cover slips. All procedures after this were performed at room temperature. Cells were fixed with 3.7%

formaldehyde in PBS for 10 min, washed twice in PBS, and then permeabilized with 0.1% Triton X-100 in PBS for 10 min. Cells can then be processed immediately for staining or can be stored in PBS for a few hours. Cells were incubated with blocking solution (3% bovine serum albumin in PBS) for 30 min, washed, and then primary antibody was added at appropriate concentrations (1:30 to 1:200 dilution). The optimal concentration is determined by titration. After 1 h in solution containing primary antibody, coverslips were washed in PBS and incubated with secondary antibody conjugated to rhodamine (Santa Cruz) at 1:100 dilution for 20 min. Cover slips were washed thoroughly in PBS, three times for 10 min each time before mounting with Vectashield mounting medium with DAPI.

For immunofluorescent staining of MEFs, refer to chapter by J. Chaumeil *et al.* in this volume.

Acknowledgments

We thank Dr. Edith Heard for providing the figure with immunostaining of MEFs and Dr. Lynne Vales for critical reading of the manuscript. We also thank Upstate Biotechnology and Dr. Thomas Jenuwein for kindly providing us with antibodies for testing and members of the Reinberg laboratory for helpful discussions. This work was supported by grants to D.R. (NIH GM37120) and the Howard Hughes Medical School.

[18]   Histone Methylation: Recognizing the Methyl Mark

By ANDREW J. BANNISTER and TONY KOUZARIDES

Histone N-terminal tails are subject to a variety of covalent modifications that ultimately affect chromatin structure.[1,2] One such modification is *N*-methylation, which occurs on Lys (K) and Arg (R) amino acids. The enzymes performing these modifications are the histone *N*-methyltransferases (HMTs) that catalyze the transfer of a methyl group from *s*-adenosylmethionine (SAM) to the ε-amino groups of lysine and/or arginine residues within histones. Many of the methylated sites within the histone N-termini have now been mapped (see Fig. 1).

[1] S. L. Berger, *Curr. Opin. Genet. Dev.* **12**, 142 (2002).
[2] B. D. Strahl and C. D. Allis, *Nature (London)* **403**, 41 (2000).

Known sites of methylation in H4 and H3

FIG. 1. Known sites of methylation within mammalian histone H3 and histone H4 N-terminal tails. Note that lysines may be mono-, di-, or tri-methylated, whereas arginines may be mono-, symmetric di-, or asymmetric di-methylated (see Fig. 2). (See color insert.)

HMTs can be divided into two groups: (i) histone lysine N-methyltransferases and (ii) histone arginine N-methyltransferases.

Histone Lysine N-*Methyltransferases*

All of the enzymes known to methylate lysines within histone N-terminal tails contain a conserved methyltransferase domain termed an SET [Su(var)3-9, Enhancer-of-zeste, Trithorax] domain.[3] Comparison of various SET domain proteins has allowed them to be classified into four subfamilies.[4] Recent evidence now strongly implicates these enzymes in human disease, especially cancer.[5,6]

In vivo methylated lysines may be found in the mono-, di-, or tri-methylated state (see Fig. 2A). It is becoming clear that these states of methylation have differing effects with respect to chromatin structure and transcription. For instance, di-methylation of lysine 4 of histone H3 (K4H3) is found at inactive genes as well as at active genes, whereas tri-methylated K4H3 is found predominantly at active genes. Thus, high levels of tri-methyl K4H3 appear to "mark" active genes.[7]

The recent solving of the crystal structure of the SET7/9 methyltransferase revealed a clue as to how the different methyl states of lysine are regulated.[8] This enzyme specifically mono-methylates K4H3. It is the

[3] M. Lachner and T. Jenuwein, *Curr. Opin. Cell Biol.* **14,** 286 (2002).

[4] T. Kouzarides, *Curr. Opin. Gen. Dev.* **12,** 198 (2002).

[5] S. Varambally, S. M. Dhanasekaran, M. Zhou, T. R. Barrette, C. Kumar-Sinha, M. G. Sanda, D. Ghosh, K. J. Pienta, R. G. Sewalt, A. P. Otte, M. A. Rubin, and A. M. Chinnaiyan, *Nature (London)* **419,** 572 (2002).

[6] R. Schneider, A. J. Bannister, and T. Kouzarides, *Trends Biochem. Sci.* **27,** 396 (2002).

[7] H. Santos-Rosa, R. Schneider, A. J. Bannister, J. Sherriff, B. E. Bernstein, N. C. Emre, S. L. Schreiber, J. Mellor, and T. Kouzarides, *Nature (London)* **419,** 407 (2002).

[8] B. Xiao, C. Jing, J. R. Wilson, P. A. Walker, N. Vasisht, G. Kelly, S. Howell, I. A. Taylor, G. M. Blackburn, and S. J. Gamblin, *Nature (London)* 421, **652** (2003).

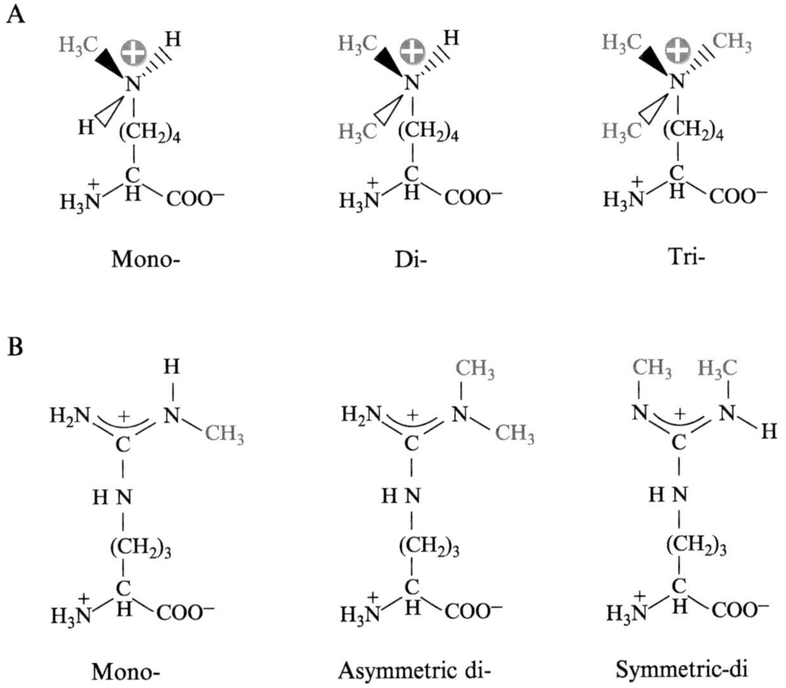

A

Mono-   Di-   Tri-

B

Mono-   Asymmetric di-   Symmetric-di

FIG. 2. Chemical structures of methylated states of lysine (A) and arginine (B). (See color insert.)

positioning of a tyrosine within the active site of the enzyme that prevents the site accommodating di-methylated K4H3, which confers the specificity. Indeed, if the tyrosine is mutated, the active site can now accommodate di-methyl lysine and the mutated SET7/9 is now able to di- and tri-methylate K4H3. Perhaps *in vivo* an allosteric change within the SET7/9 catalytic site regulates this "switching" activity. Such a change could be introduced, for example, by post-translational modification of SET7/9 or, alternatively, protein binding partners may have an effect. Understanding the function and interplay of these modifications requires the generation of specific antibodies that are capable of distinguishing the various methylated forms *in vivo*.[7,9]

[9] C. Maison, D. Bailly, A. H. Peters, J. P. Quivy, D. Roche, A. Taddei, M. Lachner, T. Jenuwein, and G. Almouzni, *Nat. Genet.* **30,** 329 (2002).

Histone Arginine N-*Methyltransferases*

Enzymes that methylate arginines in histones are referred to a histone PRMTs (protein arginine methyltransferases). These enzymes are divided into two classes, Type I and Type II, depending on whether they catalyze assymmetric or symmetric dimethylation of arginines (see Fig. 2B). To date, the best characterized histone PRMTs (PRMT1 and PRMT4 [CARM1]) both belong to the Type I PRMTs.[10,11] Little is currently known concerning how arginine methylation regulates chromatin structure but it is strongly implicated, at least, because many histone PRMTs are known transcriptional activators.[12,13] For instance, at the estrogen-regulated pS2 promoter addition of hormone induces chromatin remodeling and transcriptional activation.[14] These events are tightly linked to the recruitment of CARM1 to the pS2 promoter and the subsequent appearance of R17H3 methylation.[15,16]

Perhaps the most powerful tools in the study of histone methylation are site-specific antibodies capable of discriminating between the different methylated forms. These can then be used in numerous analyses aimed at unraveling the complexities of histone methylation. Of course, analysis of the HMTs themselves is often desirable. Here we also describe various methods and protocols that are useful in the study of the enzymatic activity associated with histone methyltransferases.

Methods

Raising of Polyclonal Rabbit Antisera that Recognize Methylated Histones

Choice of Inoculating Antigen

Generally speaking there are two possibilities concerning the choice of antigen:

---

[10] Y. Zhang and D. Reinberg, *Genes Dev.* **15,** 2343 (2001).

[11] J. M. Aletta, T. R. Cimato, and M. J. Ettinger, *Trends Biochem. Sci.* **23,** 89 (1998).

[12] M. R. Stallcup, *Oncogene* **20,** 3014 (2001).

[13] D. Chen, H. Ma, H. Hong, S. S. Koh, S. M. Huang, B. T. Schurter, D. W. Aswad, and M. R. Stallcup, *Science* **284,** 2174 (1999).

[14] G. F. Sewack, T. W. Ellis, and U Hansen, *Mol. Cell. Biol.* **21,** 1404 (2001).

[15] U. M. Bauer, S. Daujat, S. J. Nielsen, K. Nightingale, and T. Kouzarides, *EMBO Rep.* **3,** 39 (2002).

[16] S. Daujat, U.-M. Bauer, V. Shah, B. Turner, S. Berger, and T. Kouzarides, *Curr. Biol.* **12,** 2090 (2002).

1. Generation of site-specific antibodies: If the actual site of methylation within a histone has been determined (e.g., by mass spectroscopy), a sequence-specific peptide may be synthesized that contains a methylated lysine or arginine at the appropriate position.

2. Generation of pan–anti-methylated amino acid antibodies: A more general anti-methylated amino acid antibody may be desired that is capable of recognizing a broad spectrum of proteins that contain methylated arginines or lysines.

Synthesis of Peptides Containing Methylated Lysines or Arginines

Once the site of methylation within a histone has been determined (e.g., by mass spectroscopy), the investigator can chemically synthesize a peptide containing a methylated lysine or arginine at the appropriate position. For this approach the investigator needs to decide to what degree the lysine or arginine is methylated (see Fig. 2). In other words, for a lysine residue is it mono-, di-, or tri-methylated, and for an arginine residue is it mono-, symmetric di-, or asymmetric di-methylated? Without additional information (e.g., from mass spectroscopy data) this is a very difficult decision for histone lysines because they exist in all three states *in vivo*. As a first approach, synthesizing a peptide containing di-methyl lysine at the appropriate position is a good starting point. For histone arginines the situation is perhaps a little simpler; to date all the methylated arginines in histone tails are asymmetric di-methylated.

Another consideration is the length of the peptide to be immunized. We typically synthesize peptides that contain three or four specific amino acids on each side of the methylated residue. The exact length actually depends on the results from an analysis of the sequence by the Protean epitope prediction package that identifies optimal immunogenic response. However, we have also synthesized relatively long peptides (16 amino acids) that have been used successfully to generate specific anti-methylated histone antibodies. We generally incorporate a Cys residue at the C-terminal end of the peptide that is used for coupling to KHL cyanin prior to immunization and is also used for coupling the peptide to a column matrix to aid affinity purification of antibodies.

Peptide Synthesis

Peptides are synthesized using Fmoc chemistry[17] on an Applied Biosystems 9050+ or Pioneer automatic peptide synthesizer with customized

[17] E. Atherton and R. C. Sheppard, eds., "Solid Phase Peptide Synthesis." IRL Press (Oxford University) 1989.

protocols. Basically, *N*-a-Fmoc amino acids, with side chain protection where necessary, are coupled in the specified order to a polystyrene resin with polyethylene glycol units attached as spacers (PEG-PS). The resin is obtained with the C-terminal amino acid already attached via a cleavable ester bond with a typical substitution of 0.2 mmol/g and the peptides are synthesized from C- to N-terminus. The amino acid side chain protections used are *t*-butyl for ser, thr, glu, and asp; trityl for cys, asn, gln, and his; 2,2,5,7,8-pentamethylchroman-6-sulfonyl (pmc) for arg; *t*-butoxycarbonyl for lys and trp. For peptides containing side chain methylated lysines no protecting groups are used except in the case of monomethyl where *t*-butoxycarbonyl is used. Reagents are purchased from Bachem (UK).

A 4-fold excess of amino acid to resin substitution is used and coupling is performed using HBTU (benzotriazol-N,N,N^1,N^1-tetramethyluronium hexafluorophosphate) chemistry—1 equivalent of HBTU, 1-hydroxybenzotriazole (HOBt), and 2 equivalents of *N*-methylmorpholine are dissolved with the protected amino acid in dimethylformamide (DMF) and added to the DMF solvated resin where the previous amino acid has been *N*-a-deprotected (Fmoc removed with 20% piperidine solution in DMF) exposing a free amino terminus. The final Fmoc group is removed and the resin is washed with methanol and dichloromethane and then dried under high vacuum.

The peptide is released from the resin (500 mg) during a 2-h incubation with a cocktail (50 ml) of trifluoroacetic acid (Tfa) 92.5%, phenol 2%, water 2%, ethanedithiol 2%, anisole 1%, and triisobutylsilane 1%. The peptide is then concentrated and ether precipitated and collected by centrifugation, washed three times in ether and then vacuum dried for 30 min.

Characterization of the peptide is performed using reversed-phase HPLC on a C18 silica wide-pore column (15 cm × 2.1 mm) with an elution gradient of 10–50 acetonitrile containing 0.1% Tfa over 40 min. Further purification (where necessary) is performed on a 25 cm × 22 mm column using the same conditions. The molecular weight of the peptide is confirmed by mass spectrometry performed on a VG Quattro triple quadrupole instrument with electrospray ionisation. We find that peptides with a purity of >80% make good immunogens and work well in peptide competition assays (see later).

The above-mentioned approach is very useful for generating site-specific, context-dependent antibodies. However, for many applications an investigator may require an antibody that generally recognizes methylated lysines/arginines within *different* amino acid contexts. One way to generate such an antibody is to inoculate rabbits with a chemically synthesized peptide that contains a methylated lysine/arginine within the context of randomized amino acids. To do this the peptide synthesis is altered such that the process randomly incorporates amino acids except at the methylated position.

An alternate approach to generate a general pan-methylated lysine antibody is to chemically methylate lysines within proteins *in vitro* and then use this substrate to inoculate with. Because it contains methylated lysines in a wide variety of sequence contexts, the immune response often generates antibodies that generally recognize methylated lysines in a variety of contexts.

Chemical Methylation of Lysine Residues

A very efficient way of generating *in vitro* proteins containing methylated lysines is via reductive methylation with formaldehyde. Only lysine residues are modified and the main product is ε-*N,N*-di-methyllysine.[18] If radioactive formaldehyde is used (e.g., ^{14}C-formaldehyde), then the radionuclide is incorporated in the methyl groups. The use of radioactive formaldehyde is of course optional but it does allow the reaction efficiency to be easily determined. Obviously, if the aim of the reaction is to generate a substrate for injection in animals, then radioactive formaldehyde should not be used. To prevent deleterious effects of formaldehyde its concentration is kept low and borohydride is used in excess. Any protein substrate may be methylated in this approach, though we typically methylate histone proteins *in vitro*.

1. Dissolve 1 mg of purified histone H3 (Roche) in 500 ml of 200 mM borate buffer (pH 9.0) and place on ice.
2. To the chilled histone solution add 300 μg of NaBH$_4$ and continue to keep on ice.
3. 1 ml of ^{14}C-formaldehyde (0.83 MBq; 56 mCi/mmol, NEN) is then added at 5-min intervals on ice, six times in total. Subsequent to the last addition the reaction is allowed to proceed on ice for another 15 min.
4. Dialyze the reaction products against two changes of 2 L of PBS over a 12-h period in a Mini Dialysis Unit (Pierce, 10,000 MWCO).

The relative amount of radiolabel (i.e., CH$_3$ groups) incorporated can be determined by SDS-PAGE and autoradiography. Alternatively, if histones have been used as the substrate, a filter binding assay may be used (see later).

[18] G. E. Means and R. E. Feeney, *Biochemistry* **7,** 2192 (1968).

Inoculating Rabbits

Peptides containing methylated residues are usually coupled to KLH carrier protein using standard techniques. However, if the immunogen is a chemically methylated protein, then it is used directly. A typical immunization schedule is outlined later. Before initiating the immunization, pre-immune bleeds are taken from the rabbit to identify existing immune responses. The first immunization (week 0) injection contains 200 μg of immunogen per rabbit emulsified in complete Freund's adjuvant and this is then followed by two further immunizations (weeks 3 and 6) of 100 μg of immunogen per rabbit in incomplete Freund's adjuvant. Our injection procedure typically involves four subcutaneous site injections per rabbit per immunization. Two bleeds at weeks 7 and 9 are then taken, followed by a further immunization in incomplete Freund's adjuvant. Two further bleeds are taken at weeks 11 and 13 and finally exsanguination of the rabbit occurs at week 14. The blood is centrifuged to isolate the sera. The level and specificity of immune response is then determined using an ELISA plate assay.

ELISA Plate Assay Protocol

This assay is a relatively quick and easy way to screen bleeds for a positive immune response.

Individual Covalink microtitre plates (Nunc) are coated with 100 μl of 1 μg/ml peptide solution (in PBS) per well, covered and incubated overnight at 4°. The plates are then washed three times with 100 μl PBS, 0.1% Tween 20 (PBS-T) per well ensuring rigorous expulsion of liquids out of the wells at the end of each wash. The wells are subsequently filled with 100 μl blocking buffer (PBS-T with 5% (w/v) BSA) and incubated for 1 h at 37° before being emptied.

A 1/500 dilution of the antibody serum is made in blocking buffer. Two hundred microliters of this dilution are placed in the first well of the plate and 100 μl are withdrawn and placed into the second well containing 100 μl PBS-T. This serial dilution is performed up to eight times to give a final serum concentration of 1:64,000 in well 8. The plates are then covered and incubated for 1 h at 37° before being washed three times with PBS-T again ensuring vigorous expulsion of the liquids from the wells at the end of each wash. One hundred microliters of goat anti-rabbit IgG alkaline phosphatase linked antibody conjugate (1:1000 dilution in blocking buffer) are then added to each well and the plates are covered and incubated for an additional 1 h at 37°.

During the secondary antibody incubation development buffer is prepared by the addition of 0.051 g $MgCl_2$ to 53.625 ml diethanolamine. Purified water is then added to a final volume of 450 ml. The pH is adjusted to 9.8 with dilute HCl and finally the buffer is topped up to 500 ml with water. This buffer may then be stored at 4° for 2–3 days. To generate substrate, one tablet of p-nitrophenyl phosphate (pNPP: Sigma) is added to 5 ml of development buffer and mixed by gentle rotation.

After the secondary antibody incubation the plates are washed three times with PBS-T and then twice in PBS. Fifty microliters of substrate are added per well and incubated at room temperature until an appreciable color gradient appears. The reaction is stopped by the addition of 50 μl of 1 M NaOH per well. The optical density (at 405 nm) of the reaction in each well is then determined by placing the ELISA plate into an optical plate reader. Selected bleeds with the highest ELISA titre values are then affinity purified.

Affinity Purification of Antibody

Peptides bound to a solid phase can be used to affinity purify antibodies from total serum. In addition, they can be used to test other proteins for their ability to specifically bind the methylated peptides. When performing binding assays of either sort it is essential to make a control peptide that has the same amino acid sequence as the experimental (methylated) peptide but which is itself not methylated. Antibodies or other proteins binding only to the experimental peptide are the desired end products.

We design and synthesize our peptides to contain a C-terminal Cys residue such that we can use this to link them to SulfoLink coupling gel (Pierce, cat. no. 20401) that binds specifically to free sulfhydryls. All references to "gel" refer to a 50% (v/v) slurry of SulfoLink bead in the appropriate buffer.

1. The peptide is dissolved in binding buffer (50 mM Tris, 5 mM EDTA, pH 8.5).
2. Equilibrate the SulfoLink gel in binding buffer by washing 1 ml of gel in 50 ml buffer.
3. 1 mg of peptide from step 1 (in a volume no greater than 1 ml) is added to 1 ml of SulfoLink gel. This suspension is slowly rotated for 1 h at room temperature.
4. The SulfoLink gel is then washed three times in 50 ml of binding buffer.

5. Non-specific binding sites on the gel are blocked by adding 1 ml of 50 mM cysteine (in binding buffer) to 1 ml of gel and then slowly rotating the suspension for 1 h at room temperature.

6. The gel is washed once with 50 ml 1 M NaCl and once with 50 ml binding buffer. The peptide bound gel is then stored in binding buffer containing 0.05% sodium azide.

The next step is to apply serum (identified in ELISA as having high titre values) to specific immobilized peptides in order to bind out specific antibodies. We find the following protocol usually gives good consistent results and it may be applied to a batch purification method or to a column procedure.

1. Spin serum at 12,000g for 10 min at room temperature and recover the cleared serum into a fresh tube. Add EDTA to a final concentration of 5 mM.

2. Equilibrate the peptide-bound SulfoLink gel in TBS/5 mM EDTA (50 mM Tris-HCl, pH 8.0, 150 mM NaCl, 5 mM EDTA).

3. Incubate 500 μl of equilibrated peptide-bound SulfoLink gel with 5 ml of serum/5 mM EDTA by rotating slowly for 90 min at room temperature.

4. The bound IgG is then washed by subjecting it to 1 M NaCl and then eluted using 0.1 M glycine solution, pH 2.75 (2 ml glycine solution per 0.5 ml SulfoLink). The eluted IgG (in 2 ml of glycine solution) is immediately added to a tube containing 100 μl ice-cold 1 M Tris, pH 8.8. This elution step can be repeated one to two times to ensure complete elution of the IgG from the peptide. Ensure that the final pH of the neutralized antibody solution is about pH 7.0.

5. The affinity-purified antibody is then dialyzed against three changes of 1 l PBS, and its final concentration is determined using a BCA protein assay employing an IgG protein standard curve.

We prefer to add BSA to the purified antibodies to a final concentration of 1% (w/v) instead of adding glycerol. We have found that this often gives lower background when used to probe western blots. For long-term storage the antibody solution is aliquotted and stored at $-80°$. For short-term storage sodium azide is added to 0.04% (w/v) final concentration and then stored at 4°. Another ELISA assay is then performed with the antibody to ensure its functionality and to determine the post-purification titre. It can then be further tested by western blot, etc.

Characterization of Rabbit Polyclonal Anti-Methylated Histone Antibodies

Is My Antibody Specific?

Once an antibody has been purified, it is imperative to fully characterize its specificity. There are two common approaches that can be taken to achieve this: (i) an ELISA assay or (ii) a western blot approach can be performed in the presence or absence of varying concentrations of different peptides. The first decision here concerns the source of substrate. In an ELISA assay the wells are usually coated with the specific peptide or protein that the antibody was raised against. In a western blot, the blotted protein is typically purified histones from calf thymus (Sigma) since these are a good source of many different methylated sites within the histone N-terminal tails. In both the ELISA and western blot approaches, competition peptides should be included to test the specificity of the antibody. These competition peptides should include as wide a range as possible of similar, but not identical, epitopes. For instance, if the antibody to be characterized was raised against a H3 peptide (amino acids 1–8) tri-methylated at lysine 4, the competition peptides should include a non-methylated H3 peptide (1–8) as well as the mono-, di-, and tri-methylated derivatives. Only the tri-methylated peptide should block the binding of the antibody to its substrate. An example of a typical result from a western blot approach is shown in Fig. 3. In addition to these peptides, a number of other tri-methylated histone peptides should also be tested to determine the absolute specificity of the antibody.

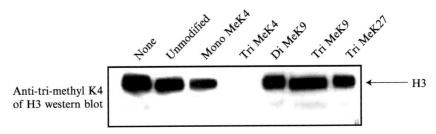

FIG. 3. Determining antibody specificity using peptide competition. Claf thymus histones were loaded into a single wide well of a 20% SDS-PAG and resolved. They were then western blotted to nitrocellulose. Strips were then cut from the blot and separately probed with an anti-tri-methyl K4 of H3 antibody (available from www.abcam.com) in the presence (1 µg/ml) or absence of various peptides as indicated.

Histones that have been expressed in and purified from bacteria also serve as excellent controls in assays designed to test the specificity of anti-methylated histone antibodies. Because bacteria lack the methyltransferases that modify histones, the histones are unmethylated and therefore should not be recognized by the antibodies.

Characterization by ELISA Assay

Our typical ELISA protocol was described earlier. To characterize the specificity of a particular antibody using this approach we include peptide competitors over a wide concentration range.

Characterization by Western Blotting

Histones are resolved in a standard 20% (w/v) SDS-PAG and then blotted to nitrocellulose. We find that histones blot more efficiently in a simple phosphate buffer (20 mM Na phosphate; prepare a stock solution of 1 M Na phosphate, pH 6.7—this is a 50× stock made by mixing approximately 1000 ml 1 M NaH$_2$PO$_4$ with 850 ml 1 M Na$_2$HPO$_4$, the exact quantities being dependent on pH). Using a Biorad miniblot apparatus blot for 35–40 min at 0.8 A. The nitrocellulose membrane is then recovered from the apparatus and transferred to block solution (1× TBS, 5% [w/v] BSA, 0.5% [v/v] Tween 20) and incubated for 1 h at room temperature. While the blocking is taking place, the anti-methylated histone antibody, at an appropriate concentration, is added to block solution in a series of tubes. Competitor peptides are then be added to the tubes in order to test the specificity of the antibody. Typically peptides are used at a final concentration of 1 μg/ml and include a variety of methylated, as well as non-methylated, peptides encompassing the amino acid backbone of the inoculating peptide (see Fig. 3). Peptides and antibody mixes are incubated at room temperature for 1 h.

The blot is probed in blocking solution containing antibody plus/minus peptide competitor at 4° overnight. Following this incubation the blot is briefly washed in blocking solution and then probed at room temperature with an appropriate HRP-conjugated secondary antibody in blocking solution. After washing, bound antibody is detected using a chemoluminescent kit, such as the ECL kit available from Amersham.

Using Anti-Methylated Histone Antibodies

Western Blots

An obvious application for the use of anti-methylated histone antibodies is the characterization of substrates by western blot analysis. The substrate protein may include (i) histones extracted from various cellular sources or (ii) the methylated histone product of an *in vitro* methylation assay (Fig. 4). Our protocol for this procedure is described earlier.

Immunofluorescence

An investigator often wishes to ask whether a particular histone methylation mark is enriched in specific structures within the genome, in specific cells, or at specific points of the cell-cycle. One way to address these issues is by using immunofluorescence. The main decisions here regard which fixation and permeabilization techniques to use. The following is our general protocol that should serve as a reasonable starting point.

1. Cells are grown under the appropriate condition in 25-cm^2 slide flasks (NUNC).
2. Aspirate growth medium and wash the cells twice with PBS.
3. The cells are then fixed in 2 ml of 4% (v/v) formaldehyde in PBS for 20 min at room temperature. Alternatively, paraformaldeyde may be used if formaldehyde fails to give reasonable results.
4. Wash cells three times (5 min each) with PBS.
5. Block/permeabilize the cells with 3% (w/v) BSA, 0.6% (v/v) Triton X-100 in PBS for 15 min at room temperature. An alternative detergent to consider is 0.2% (v/v) Tween 20, though we find that this often fails to fully permeabilize the cells.
6. Wash three times (5 min each) in PBS.
7. Incubate with secondary antibody (conjugated to an appropriate flour) in 3% (w/v) BSA in PBS for 1 h in the dark.
8. Wash once in PBS.
9. Hoechst 33258 in PBS is added to a final concentration of 1 μg/ml and incubated for 5 min at room temperature in the dark.
10. The cells are finally washed four times in PBS, each for 5 min.
11. The flask is then opened using the tool supplied by the manufacturer and the resulting slide is mounted using standard techniques.

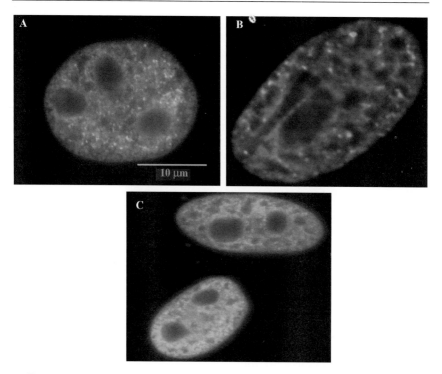

FIG. 4. Immunoflourescence of mammalian cells using anti-methylated K4 of histone H3 antibodies. (A) Indian muntjac fibroblast cell stained with anti-monomethylated K4 of H3 antibody (green) and DAPI (red). (B) Indian muntjac fibroblast cell stained with anti-dimethylated K4 of H3 antibody (green) and DAPI (red). (C) Indian muntjac fibroblast cells stained with anti-trimethylated K4 of H3 antibody (green). (See color insert.)

The slides are then visualized under a microscope, typically a confocal microscope, capable of irradiating the mounted cells with the appropriate wavelength light in order to stimulate the flour attached to the secondary antibody.

Chromatin Immunoprecipitation Analysis

One of the most important questions concerning chromatin research is where in the genome do particular covalent modifications of histones occur. In order to answer this type of question a chromatin immunoprecipitation (ChIP) may be performed. Chromatin is prepared from cells of interest that have been subjected to "gentle" formaldehyde cross-linking. If the researcher is interested in a chromatin-bound factor, such as a transcription factor, then the cross-linking is necessary to link the factor to the chromatin. If the investigator wishes to analyze covalent modification of a histone tail, then the cross-linking is optional, though we find it is not deleterious to the procedure and therefore recommend that it is performed.

Once the chromatin has been prepared it is necessary to reduce it to smaller fragments. This essentially serves two purposes: (i) the immunoprecipitation becomes more reproducible and (ii) the "resolution" of the technique is dramatically increased. Purified antibodies are then added to the fractionated chromatin and an immunoprecipitation is performed. Typically antibodies include those that recognize modified histones, transcription factors, or other chromatin-associated factors. Following the immunoprecipitation, the abundance of specific sequences within the bound fraction is quantitatively determined.

The following ChIP protocols have been used successfully by our laboratory to study the role of histone methylation in gene transcription.[7,15,16,19]

Harvesting the Cells

Formaldehyde is added dropwise directly to the cell growth media for a final concentration of 0.75%. These are then gently swirled for 10 min at room temperature. The cells are then rinsed twice with cold PBS and scraped into 5 ml of cold PBS. The cells are then pelleted by spinning at 1000g for 2 min at room temperature. It is now important to directly proceed to nuclease digestion/sonication. Do not freeze the cells at this point.

Micrococcal Nuclease Digestion or Sonication?

There are two commonly used procedures for breaking high molecular weight chromatin into smaller fragments. One uses an endonuclease, typically micrococcal nuclease, to specifically digest the linker DNA between nucleosomes. Using this approach it is possible to derive a preparation of predominantly mononucleosomes; for this approach begin the protocol at step 1. An alternative approach is to use mechanical shearing by sonication to break down the chromatin. This results in random cleavage of the DNA and conditions should be found that give rise to an approximately 700 bp average length of sonicated chromatin—for sonication begin the protocol at step 9.

[19] S. J. Nielsen, R. Schneider, U. M. Bauer, A. J. Bannister, A. Morrison, D. O'Carroll, R. Firestein, M. Cleary, T. Jenuwein, R. E. Herrera, and T. Kouzarides, *Nature (London)* **412**, 561 (2001).

1. Remove the supernatant from the pelleted cells and resuspend in 10 ml cold NI buffer (15 mM Tris-HCl, pH 7.5, 60 mM KCl, 0.5 mM DTT, 15 mM NaCl, 300 mM sucrose) containing 5 mM MgCl$_2$. Then add 10 ml cold NI buffer containing 5 mM MgCl$_2$ and 1% (v/v) NP-40, and mix by inverting and leave on ice for 10 min.

2. Pellet the nuclei at 2000g for 5 min at 4°.

3. Remove the supernatant and resuspend nuclei in 5 ml cold NI buffer containing 5 mM MgCl$_2$ and then pellet as in step 2.

4. Remove the supernatant and resuspend nuclei in 5 ml cold NI buffer containing 1 mM CaCl$_2$, and then pellet nuclei as in step 2.

5. Remove the supernatant and resuspend nuclei in 1 ml cold NI buffer containing 1 mM CaCl$_2$ and transfer to a fresh 1.5 ml Eppendorf tube. Remove 50 μl—this serves as an undigested control.

6. Add the nuclei from step 5 to 400 μl lysis solution (1% [w/v] SDS, 50 mM Tris-HCl, pH 8.0, 20 mM EDTA) and then add 375 units of S7 micrococcal nuclease (15 U/μl in H$_2$O) to each sample. The chromatin is incubated with the nuclease on ice and the kinetics of digestion are followed by analyzing the products in a 1.5% agarose gel. Ideally, the digested chromatin should give rise to mono-nucleosomes with a low amount of di-nucleosomes being evident in the agarose gel.

7. The reaction is stopped by the addition of EDTA to a final concentration of 50 mM.

8. The reaction tubes are then given a brief spin (30 s at 13,000 rpm).

9. The pellet is resuspended in 0.4–1 ml (depending on amount of cells) of ChIP lysis buffer (1% [w/v] SDS, 10 mM EDTA, 50 mM Tris-HCl, pH 8.0, protease inhibitor cocktail [Roche]). If sonication is not required, proceed to step 10. Begin with this resuspension if the sonication of chromatin is to be performed. Sonicate the chromatin on ice and follow the digestion in 1.5% agarose gel. An average sonicated size of about 700 bp should be aimed for.

10. Pellet the cellular debris by spinning at 13,000g for 2 min at 4°. Then transfer the supernatant to a new tube. Remove 50 μl of each preparation and add it to 400 μl lysis solution (1% SDS [w/v], 50 mM Tris-HCl, pH 8.0, 20 mM EDTA). This sample is the input. It is used for obtaining DNA concentration for subsequent IPs (see later) and as control in final PCRs.

11. Store the remainder of the chromatin on ice at 4° (it is stable for approximately 1 week).

Checking the DNA Concentration of Inputs

Five microliters of proteinase K (20 mg/ml stock solution) are added to each sample in lysis solution (from step 10, above) and it is then heated with shaking at 65° for 4–5 h to reverse the cross-linking (the samples may be frozen here if required). The samples are then phenol:chloroform extracted with an equal volume of 1:1 phenol:choroform. Ten microliters of glycogen (5 mg/ml stock solution) are added to aid precipitation and then 1/10th volume of sodium acetate and 2 volumes of ethanol are added. The precipitations are incubated on dry ice for 10 min and the DNA is then pelleted by spinning at 13,000g for 10 min. The pellet is carefully washed in 70% (v/v) ethanol and then dried. Finally the DNA pellet is dissolved in 50 μl TE (10 mM Tris-HCl, pH 8.0, 1 mM EDTA) and its concentration is determined by absorbance at OD^{260}.

Performing Chromatin Immunoprecipitations

The protocol outlined later provides a good starting point for researchers wishing to begin ChIPs, but it should be emphasized that a number of steps will need to be empirically optimized for each system analyzed. Each amplicon to be studied requires a minimum of two independent immunoprecipitations to be performed: (1) with a specific antibody (e.g., against di-methylated K4 of histone H3) and (2) with an unrelated irrelevant antibody as a control. Each ChIP reaction should contain approximately 25 μg of chromatin-bound DNA as determined earlier.

1. Samples are adjusted to the same volume with ChIP lysis buffer and each is then diluted 1:10 with dilution buffer (20 mM Tris-HCl, pH 8.0, 150 mM NaCl, 2 mM EDTA, 1% [v/v] Triton X-100). If a high background is a problem, it is possible to use RIPA buffer here to help reduce it.

2. Add an appropriate amount of immunoprecipitation antibody. The actual amount will depend on the specific antibody characteristics, as well as factors (e.g., epitope abundance, accessibility, etc). Incubate on ice for 1 h.

3. Add 20 μl (50% [v/v] slurry) of protein A/G beads (pre-absorbed with sonicated single-stranded salmon sperm DNA at 1.5 μg/20 μl beads). Incubate overnight with rotation at 4°.

4. Pellet the immuno complex by spinning at 1000g for 30 s. Wash the pellet three times with 1 ml of wash buffer (20 mM Tris-HCl, pH 8.0, 150 mM NaCl, 2 mM EDTA, 1% [v/v] Triton X-100, 0.1% [w/v] SDS) each time spinning as earlier.

5. Wash the pellet once with 1 ml final wash buffer (20 mM Tris-HCl, pH 8.0, 500 mM NaCl, 2 mM EDTA, 1% [v/v] Triton X-100, 0.1% [w/v] SDS). Spin as earlier.

6. Elute the bound DNA by adding 450 μl of freshly made elution buffer (1% [w/v] SDS, 100 mM NaHCO$_3$) and rotating on a wheel for 15 min at room temperature. Spin down as earlier and remove the supernatant into a fresh tube.

7. Add 5 μl proteinase K (20 mg/ml) and heat with shaking at 65° for 4–5 h to reverse cross-linking (can freeze samples here if necessary).

8. The samples are then phenol:chloroform extracted with an equal volume of 1:1 phenol:choroform. Ten microliters of glycogen (5 mg/ml stock solution) are added to aid precipitation and then 1/10th volume of sodium acetate and 2 volumes of ethanol are added. The precipitations are incubated on dry ice for 10 min and the DNA is then pelleted by spinning at 13,000g for 10 min. The pellet is carefully washed in 70% (v/v) ethanol and then dried. Finally the DNA pellet is dissolved in 100 μl TE (10 mM Tris-HCl) pH 8.0, 1 mM EDTA, and stored at 4° or −20°.

Analysis of Results by Quantitative PCR

The best way to analyze the results from a ChIP assay is via real-time quantitative PCR. There are many commercially available instruments designed to perform these reactions. In each case primers against the DNA sequence of interest need to be designed and there are many computer programs to help with this. At the end of the day the actual methodology for this needs to be determined on a case-by-case basis.

Detecting Histone N-Methyltransferase Activity In Vitro

In Vitro N-Methyltransferase Assays

An investigator may wish to question whether a particular protein contains N-methyltransferase activity. A common approach to address this is to express the potential methyltransferase (assuming it has already been cloned) in bacteria and then to purify it. Typically this can be accomplished by expressing the protein of interest as a GST- or MBP-fusion protein. Sometimes, however, the potential methyltransferase activity may be present in a purified cellular extract fraction or in the immuno-pellet of an immunoprecipitation reaction. In the former case, then the extract is directly added to the N-methyltransferase assay. In the latter case, the immuno-pellet is washed into N-methyltransferase buffer and the other reaction components are then added to the tube.

A protein may be used as a substrate in this assay, though we routinely employ core histones from calf thymus (a mixture of H2A, H2B, H3, and H4, Roche) that contain pre-existing modifications found *in vivo*. Substrates expressed in and purified from bacteria should also be considered since they are most likely devoid of pre-existing methylations. This is particularly important if the reaction products are to be detected by antibodies where pre-existing modifications may mask the activity of the enzyme in question. Also it should be noted that a particular N-methyltransferase may only methylate histones when they are presented in nucleosomes. The amount of substrate required varies depending on numerous factors but in a typical assay is usually between 1 and 20 μg. The following protocol has been modified from Rea *et al.* (2000).[20]

1. The potential methyltransferase is made to 26 μl in N-methyltransferase buffer (50 mM Tris, pH 8.5, 20 mM KCl, 10 mM MgCl$_2$, 2 mM β-mercaptoethanol, 250 mM sucrose).
2. Two microliters of substrate are then added (e.g., 20 μg of core histones, Roche) and 2 μl of S-adenosyl-L-[*methyl*-^{3}H]-methionine (7.4×10^{-2} MBq; 77 Ci/mmol, NEN) as methyl donor. Radioactive SAM allows the reaction to be easily monitored. However, non-radioactive SAM may be used if required.
3. The reaction is incubated at 30° for 1 h.
4. Stop the reactions by adding SDS-PAG loading buffer and boiling.

The proteins are resolved by SDS-PAGE and blotted onto PVDF membrane. We expose the membranes to BiomaxMS (Kodak) films using a BioMax LE intensifying screen (Kodak) at −80° for 12 h. If the N-methylation is believed to reside in a lysine near the N-terminus of the protein, then the radiolabeled protein band can be excised from the PVDF membrane and sequenced by sequential Edman degradation. The amino acid residues liberated after each round of degradation are collected and counted in a liquid scintillation counter. The presence of radioactivity within a fraction indicates that the corresponding amino acid was methylated.

In our experience, the earlier protocol works well with a range of K-methyltransferases. However, if R-methyltransferases are to be analyzed, we have found that it is better to perform the assay as earlier except that PBS is used in place of N-methyltransferase buffer. We also find it

[20] S. Rea, F. Eisenhaber, D. O'Carroll, B. D. Strahl, Z. W. Sun, M. Schmid, S. Opravil, K. Mechtler, C. P. Ponting, C. D. Allis, and T. Jenuwein, *Nature (London)* **406,** 593 (2000).

better to express the arginine methyltransferases PRMT1, and CARM1 as GST fusions rather than MBP fusions. It is important to test these enzymes freshly eluted from the bead and dialyzed against PBS, 10% (v/v) glycerol.

To our knowledge there are no methyltransferase inhibitors that target specific methyltransferases. Inhibitors such as 5′-methyl-thioadenosine inhibit most, if not all, methyltransferases.

Detection of N-Methyltransferase Activity Using a Liquid Assay

If histones are used as substrate, they can be radioactively methylated and bound to filters for counting in a liquid scintillation counter. The main advantage of this technique over that of SDS-PAGE/western blotting coupled to autoradiography is that it is significantly easier and quicker to perform. However, the disadvantage is that this method does not allow the identification of which histone is methylated.

1. Spot 20 μl of a methylation reaction onto a 2-cm^2 piece of P81 paper (Whatman).
2. Wash in 50 mM carbonate buffer, pH 9.0 (3 × 10 min, 100 ml each wash) at room temperature.
3. Soak the filters in acetone for 1 min and then air-dry.
4. Scintillation counting is then performed.

Acknowledgments

We would like to thank Darren Harper, Abcam Ltd. (www.abcam.com) for suggestions and helpful advice relating to antibody production and ELISA assays, Michael Hendzel (Alberta University, Canada) for supplying immunofluorescence images, and Graham Bloomberg for advice on peptide synthesis. All our methylated peptides are synthesized by Graham Bloomberg (G.B.Bloomberg@bristol.ac.uk).

[19] Analysis of Genome-Wide Histone Acetylation State and Enzyme Binding Using DNA Microarrays

By Daniel Robyr, Siavash K. Kurdistani, and Michael Grunstein

Nuclear events such as gene transcription, DNA replication, and DNA repair must cope with local chromatin environments that may be inhibitory to these processes. Therefore, chromatin remodeling and histone post-translational modifications are an intrinsic component of the maintenance and establishment of repressive or permissive chromatin states. Histone acetylation and deacetylation play an important role in this regulation. The small genome of *Saccharomyces cerevisiae* contains numerous histone deacetylases (HDACs) including Rpd3, Hda1, Hos1, Hos2, Hos3, and Sir2. Chromatin immunoprecipitation using highly specific antibodies raised against individual lysine residues[1] has established that several of these enzymes have different histone specificities. For instance, Hda1 deacetylates histones H3 and H2B,[2] whereas Rpd3 strongly affects all core histones sites analyzed with the notable exception of H4 lysine 16.[1] In contrast, Hos2 is required for H3 and H4 deacetylation only.[3] To function at DNA regulatory elements HDAC complexes are recruited to their target promoters by specific transcriptional repressors such as Ume6 at the Rpd3-affected *INO1* gene leading to the local deacetylation of about two nucleosomes around the TATA-box.[4,5] However, Rpd3 and Hda1 also deacetylate large regions of chromatin, including promoters and open reading frames (ORFs), without apparent direct recruitment by DNA binding repressors.[6] This global deacetylation mechanism is involved in the rapid establishment of a global repressive chromatin environment once gene transcription is turned down. Finally, Hos2 primarily deacetylates histones within the coding regions of genes. Surprisingly, unlike other HDACs that are repressors, Hos2 is required directly for gene activity.[3] These observations suggest that a comprehensive understanding of HDAC functions requires the analysis of large regions of chromatin, ideally at a genome-wide level.

[1] N. Suka, Y. Suka, A. A. Carmen, J. Wu, and M. Grunstein, *Mol. Cell* **8,** 473 (2001).

[2] J. Wu, N. Suka, M. Carlson, and M. Grunstein, *Mol. Cell* **7,** 117 (2001).

[3] A. Wang, S. K. Kurdistani, and M. Grunstein, *Science* **298,** 1412 (2002).

[4] D. Kadosh and K. Struhl, *Cell* **89,** 365 (1997).

[5] S. E. Rundlett, A. A. Carmen, N. Suka, B. M. Turner, and M. Grunstein, *Nature* **392,** 831 (998).

[6] M. Vogelauer, J. Wu, N. Suka, and M. Grunstein, *Nature* **408,** 495 (2000).

HDAC locus specificity was initially probed by analyzing the influence of each HDAC deletion on genome-wide transcription using DNA micro-arrays.[7,8] *RPD3* disruption, however, led to more genes being repressed than up-regulated genome-wide, most likely as a result of indirect effects on global gene regulation. Thus, determination of HDAC function genome-wide benefits from the additional studies of acetylation (acetyla-tion arrays) and enzyme binding (binding arrays) as more direct tools to assess HDAC locus specificity. In this manner, a much more comprehensive analysis can determine the sites at which the enzyme binds, affects acetylation state, and results directly in changes in gene activity.[9,10] We describe here methods for identifying the sites of enzyme action using acet-ylation microarrays and enzyme binding. We have recently used such procedures to unravel the sites of action for Rpd3 and Hos2[3,9,10] and acetylation arrays for studies on Hda1, Hos1, Hos3, and Sir2.[10]

Acetylation microarrays use the combination of chromatin immuno-precipitation[11] and hybridization of DNA to microarray glass slides (see Fig. 1).[12,13] Highly specific antibodies are used to immunoprecipitate for-maldehyde-crosslinked chromatin fragments enriched for a given acety-lated lysine residue in cell lysates obtained from a wild-type (WT) strain and its isogenic strain disrupted for the HDAC of interest. The crosslink is reversed after chromatin immunoprecipitation allowing DNA purifica-tion, amplification by PCR, and DNA labeling with fluorophores. One fluorescent dye (e.g., Cy5) is used for DNA recovered from the strain carrying the deacetylase gene mutant, whereas a second dye (e.g., Cy3) is used to label DNA from the WT isogenic control strain. The labeled DNA probes from both strains are combined and hybridized onto DNA microarray glass slides containing either intergenic regions or ORFs or both. Glass slides are scanned for both fluorescent dyes and the normalized

[7] B. E. Bernstein, J. K. Tong, and S. L. Schreiber, *Proc. Natl. Acad. Sci. USA* **97,** 13708 (2000).

[8] T. R. Hughes, M. J. Marton, A. R. Jones, C. J. Roberts, R. Stoughton, C. D. Armour, H. A. Bennett, E. Coffey, H. Dai, Y. D. He, M. J. Kidd, A. M. King, M. R. Meyer, D. Slade, P. Y. Lum, S. B. Stepaniants, D. D. Shoemaker, D. Gachotte, K. Chakraburtty, J. Simon, M. Bard, and S. H. Friend, *Cell* **102,** 109 (2000).

[9] S. K. Kurdistani, D. Robyr, S. Tavazoie, and M. Grunstein, *Nat. Genet.* **31,** 248 (2002).

[10] D. Robyr, Y. Suka, I. Xenarios, S. K. Kurdistani, A. Wang, N. Suka, and M. Grunstein, *Cell* **109,** 437 (2002).

[11] A. Hecht, S. Strahl-Bolsinger, and M. Grunstein, *Nature* **383,** 92 (1996).

[12] V. R. Iyer, C. E. Horak, C. S. Scafe, D. Botstein, M. Snyder, and P. O. Brown, *Nature* **409,** 533 (2001).

[13] B. Ren, F. Robert, J. J. Wyrick, O. Aparicio, E. G. Jennings, I. Simon, J. Zeitlinger, J. Schreiber, N. Hannett, E. Kanin, T. L. Volkert, C. J. Wilson, S. P. Bell, and R. A. Young, *Science* **290,** 2306 (2000).

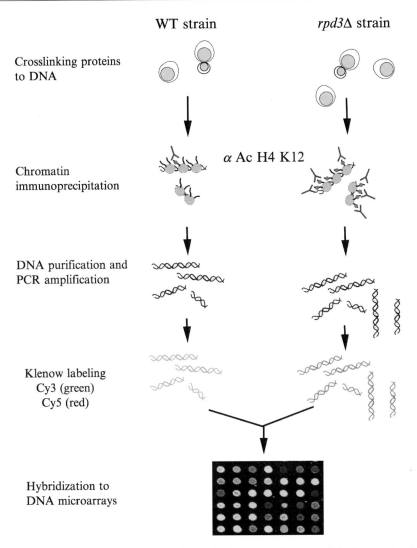

FIG. 1. Acetylation microarrays. Chromatin fragments from crosslinked mutant cells (*rpd3Δ*) and their isogenic WT counterparts were immunoprecipitated using highly specific antibodies raised against acetylated histone sites. DNA from enriched chromatin fragments was purified, amplified by PCR, and labeled with a fluorophore (Cy3 or Cy5). Probes from both sets of labeled DNA were then combined and hybridized to a DNA microarray containing either intergenic regions, ORFs, or both. For a given region on the microarray, the ratio of the normalized fluorescent intensities between the two probes indicates whether the analyzed lysine residue is hypo- or hyper-acetylated in the experiment strain. Reprinted with permission from D. Robyr and M. Grunstein, *Methods* **31,** 83–89 (2003). (See color insert.)

ratios of intensities between the deacetylase mutant and the WT probes reflect changes in acetylation in the mutant strain.

Chromatin Immunoprecipitation (ChrIP or ChIP)

1. Dilute an overnight pre-culture of yeast into 50 ml YEPD medium (2% peptone, 1% yeast extract, 2% dextrose) to $A_{600} = 0.2$ and allow the cells to reach $A_{600} \sim 1$. Histones are crosslinked to DNA *in vivo* by adding formaldehyde (Fisher) to the culture to a final concentration of 1% (w/v). The crosslinking reaction is carried out at room temperature for 15 min with constant mild agitation and is then quenched by adding 2.5 ml 2.5 M glycine (final concentration 125 mM).

2. Harvest and wash cells twice in ice-cold $1\times$ PBS (140 mM NaCl, 2.5 mM KCl, 8.1 mM Na2HPO4, 1.5 mM KH2PO4, pH 7.5) by centrifugation at $4°$ at 2800g (Beckman J2-HC, rotor JS-4.3). The cell pellet can be frozen in liquid nitrogen and stored at $-80°$ if needed.

3. Resuspend cells in 400 μl ice-cold lysis buffer (50 mM HEPES/ KOH, pH 7.5, 500 mM NaCl, 1 mM EDTA, 1% (v/v) Triton X-100, 0.1 % (v/v) SDS, and 0.1% (w/v) sodium-deoxycholate) complemented with proteases inhibitors (Complete, Roche; $100\times$ stock prepared in 500 μl H2O from 1 tablet). Transfer the resuspended cells into siliconized tubes (1.7 ml NoStick Hydrophobic Microtubes, GeneMate) and add 1 volume of acid-washed glass beads (0.45–0.52 mm diameter, Thomas Scientific).[14] Cells are lysed on an Eppendorf shaker (model 5432) between 45 and 60 min at $4°$. Puncture the bottom of the opened tubes containing the lysed cells with a red hot 25G1 needle (Becton Dickinson). Immediately close the tubes, place them on a fresh collecting 1.7-ml tube and recover the cell lysate by centrifugation (5 s in an Eppendorf tabletop centrifuge (model 5415 C) at 14,000 rpm).

4. Shear chromatin from the cell lysate by sonication down to an average fragment size of 500 bp using a Sonic dismembrator 550 (Fisher Scientific) with a Microtip model 419 from Misonix. Sonicate lysate on ice with two pulses of 15 s each (magnitude setting of 4) and a 60-s rest interval. Recover the clear cell lysate from the cell debris (pellet) by centrifugation at $4°$ in an Eppendorf centrifuge (10 min at full speed). At this point the lysate can be stored at $-80°$ for up to a month.

[14] A. Hecht, S. Strahl-Bolsinger, and M. Grunstein, *Methods Mol. Biol.* **119**, 469 (1999).

5. Immunoprecipitate acetylated chromatin fragments overnight at 4°
   on a nutator incubating 50 μl of cell lysate with 2–5 μl of a given
   antiserum (see later). Then, add a suspension of 25 μl of 50% (v/v)
   protein A sepharose CL-4B beads (Amersham-Pharmacia),
   equilibrated in lysis buffer and incubate an additional 2 h.
6. Pellet the sepharose beads for 30 s at room temperature in an
   Eppendorf centrifuge (model 5415 C) at 735g (3000 rpm). Discard
   the supernatant and wash successively for 5 min the beads on a
   nutator with 500 μl of the following solutions: twice in lysis buffer,
   once in deoxycholate buffer (10 mM Tris-HCl, pH 8.0, 0.25 M LiCl,
   0.5% (v/v) NP-40, 0.5% (w/v) sodium deoxycholate and 1 mM
   EDTA, pH 8.0), and once in TE, pH 8.0. Pellet the beads and
   discard the supernatant between each washing step as indicated
   earlier. The whole procedure is carried out at room temperature.
7. Elute the immunoprecipitated chromatin fragment from the beads
   and reverse crosslink overnight at 65° with 50 μl elution buffer
   (50 mM Tris-HCl, pH 8.0, 10 mM EDTA, pH 8.0, and 1% (v/v) SDS).
8. Add 50 μl of TE, pH 8.0, 20 μg glycogen and treat with 20 μg
   proteinase K for 2–3 h at 55°. Finally, extract DNA with 1 volume
   phenol/chloroform/isoamyl alcohol [25:24:1 (v/v)] and ethanol
   precipitate. Resuspend purified DNA into 50 μl TE, pH 8.0 and
   store at −20°.

Troubleshooting

The antibody specificity is the most critical aspect for any chromatin immunoprecipitation. The polyclonal antibodies raised against individual
acetylated lysine residues were prepared in our laboratory and are available at Upstate (http://www.upstate.com). Their specificity was verified
by ELISA assay and tested by ChrIP against histones mutated for the acet-
ylatable lysine.[1] The titration of the antibody amount required for an immunoprecipitation has to be determined experimentally for all antibodies.

Sonication conditions will determine the resolution of the chromatin
immunoprecipitation (the higher the average fragment size is, the lower
the resolution will be). Shearing efficiency will also depend on the sonicator
brand. A pilot experiment should be performed in order to find the appropriate sonication settings. After sonication add 80 μl elution buffer (see
step 7 for recipe) to a 20-μl aliquot from the cleared cell lysate (step 4)
and incubate the tube overnight at 65°. Add 100 μl TE, pH 8.0, 40 μg proteinase K and incubate the tubes for 2–3 h at 55° as indicated earlier (step
8). Resuspend DNA pellet (Input DNA) in 20 μl TE, pH 8.0, after DNA
extraction and precipitation. Treat an aliquot (10 μl) with RNase A

(10 μg) 30 min at 37° and analyze the average DNA fragment size on a 1.5% agarose gel. Alter accordingly the sonication settings if DNA fragments are too large. Store the remaining RNase untreated DNA at −20° for later PCR amplification if needed (see later).

Double Crosslinking with Protein-Protein Crosslinking Agents and Formaldehyde

For certain proteins, such as the HDAC Rpd3, formaldehyde crosslinking alone is inadequate for efficient crosslinking of the enzyme to chromatin *in vivo*.[9] This may be due to the fact that Rpd3 is part of a large (~1 MDa) multiprotein complex and, unlike histones, may lie too far from DNA for efficient crosslinking by formaldehyde alone. The immunoprecipitation efficiency of Rpd3 is significantly improved when in addition to formaldehyde, a protein-protein crosslinking agent is also used.[9] In such double crosslinking scheme, the cells are sequentially treated first with a protein-protein crosslinking agent and then with formaldehyde. We have successfully used several protein-protein crosslinking agents such as dimethyl adipimidate (DMA), dimethyl pimelimidate (DMP), or 1-ethyl-3-(3-dimethylaminopropyl)carbodiimide (EDC) prior to formaldehyde treatment to ChrIP the Rpd3 deacetylase. This new crosslinking method has been described recently in more detail for various protein-protein crosslinkers[15] and is summarized as follows:

1. Grow cells as indicated earlier but do not add formaldehyde immediately when they have reached $A_{600} = 1$.
2. Pellet and wash the cells twice with 20 ml ice-cold 1× PBS in a 50-ml Falcon tube. Prepare a fresh 15 ml solution of 10 mM DMA (Pierce) in ice-cold 1× PBS containing 0.25% (v/v) dimethyl sulfoxide (DMSO). Since DMA reacts with amine groups, buffers such as Tris that contain primary amines should be avoided. Non–amine-containing buffers such as phosphate, carbonate, and HEPES are acceptable. Resuspend the cells in the DMA solution and incubate on a nutator at room temperature for 45 min. Pellet and wash the cells once again in 1× PBS and resuspend in 1× PBS containing 1% (w/v) formaldehyde. Incubate the cells on a nutator at room temperature for 10–11 h. This incubation time is necessary for maximal crosslinking of Rpd3 but needs to be empirically determined for the protein under study. Stop the reaction by adding 2.5 ml 2.5 M glycine and wash the cells again twice in ice-cold 1× PBS.

[15] S. K. Kurdistani and M. Grunstein, *Methods* **31,** 90 (2003).

3. Cell lysate is prepared as described earlier with the exception of a salt concentration of 150 mM NaCl in the lysis buffer (150-mM lysis buffer). DNA enrichment during chromatin immunoprecipitation may be further improved by partially purifying chromatin prior to sonication. To do this, pellet chromatin by centrifuging the crude extract for 3 min at 4° at 14,000 rpm. Discard the supernatant and carefully resuspend the chromatin pellet in 400 μl 150-mM lysis buffer.

4. Due to the longer formaldehyde crosslinking time, increase sonication time to two pulses of 20 s each (magnitude setting of 4) with a 60-s rest interval for more efficient shearing of chromatin.

5. ChrIP is then carried out essentially as described earlier with the following modifications. One hundred microliters of the lysate (WT and $rpd3\Delta$) was incubated with 5-μl of α-Rpd3 polyclonal antibody (Upstate) and incubated on a nutator at 4° for 2 h. After 1 h incubation with 50-μl a 50% v/v suspension of protein A beads, the beads are transferred to a 0.45-μm filter unit (Millipore Ultrafree-MC; cat#UFC30HV00) for the washing steps (2× in lysis buffer containing 500 mM NaCl and 2× in deoxycholate buffer, 15 min each). After each wash, the filter unit was spun at 3000 rpm for 30 s, and the flow through fraction was discarded. To elute chromatin off the beads, 100 μl elution buffer was added to the filter unit, incubated at 65° for 15 min, and spun at 3000 rpm for 30 s and repeated once more. The eluates were combined and incubated at 65° overnight. DNA is then recovered as described.

Probe Amplification by PCR

Low DNA yield after chromatin immunoprecipitation is not adequate for immediate labeling and hybridization onto DNA microarrays, thus requiring a DNA amplification step by PCR. This approach is adapted from Bohlander et al.[16] as described at http://www.microarrays.org/protocols.html. A similar method was recently covered by Horak and Snyder.[17] Since DNA sequences in the chromatin immunoprecipitation are heterogeneous, the first step in this approach will consist of the random incorporation of degenerated oligonucleotides (nanomer) attached to a linker sequence. The latter sequence will then be used for the probe amplification by PCR.

[16] S. K. Bohlander, R. Espinosa, III, M. M. Le Beau, J. D. Rowley, and M. O. Diaz, *Genomics* **13,** 1322 (1992).

[17] C. E. Horak and M. Snyder, *Methods Enzymol.* **350,** 469 (2002).

1. Use small 0.2-ml PCR tubes and perform all reactions in a PCR thermocycler. Add 2 μl 5× Sequenase buffer (200 mM Tris-HCl, pH 7.5, 100 mM MgCl2, and 250 mM NaCl) and 1 μl (40 pmol) oligonucleotide (5'-GTTTCCCAGTCACGATCNNNNNNNNN-3') to 7 μl immunoprecipitated DNA. Denature DNA for 2 min at 94°, cool the tubes down to 8°, and incubate at 8° for an additional 2 min. Pause the PCR machine and add 5 μl of reaction mix (1× Sequenase buffer, 0.9 mM dNTPs, 15 mM DTT, 0.75 μg BSA, and 4 U Sequenase 2.0). The oligonucleotide is annealed to DNA by slowly raising the temperature from 8° to 37° over a period of 8 min. DNA synthesis is allowed to proceed for an extra 8 min at 37°. Repeat once the whole denaturation-annealing-elongation process. However, since the DNA polymerase is heat sensitive, add 4 U Sequenase after the denaturation step at 94°. Finally, stop the reaction by diluting the reaction with 35 μl TE, pH 8.0. DNA (step 1 DNA) can be stored at −20° or used immediately for the PCR reaction described later.

2. The following step of the DNA amplification procedure is a regular PCR reaction using the fixed linker oligonucleotide sequence (5'-GTTTCCCAGTCACGATC-3') and DNA from step 1 as a template. Transfer an aliquot of 15 μl step 1 DNA into a fresh 0.5-ml PCR tube and carry the reaction in a final volume of 100 μl (20 mM Tris-HCl, pH 8.4, 50 mM KCl, 2 mM MgCl2, 1.25 nanomoles oligonucleotide, 0.25 mM dNTPs, and 5 U recombinant Taq polymerase [Invitrogen]). Denature DNA at 92° for 30 s, anneal with two consecutive short 30-s steps at 40° and 50°, respectively, and elongate at 72° during 90 s. Repeat this cycle 24 times and allow the final elongation to proceed for another 10 min at 72°.

3. Purify the PCR reaction through columns (QIAquick PCR purification kit, Qiagen) as indicated by the manufacturer and elute DNA with 50 μl 10 mM Tris-HCl, pH 8.5.

4. Estimate DNA yield under UV light by spotting serial dilutions of DNA in a Petri dish (0.5 μl DNA mixed with 0.5 μl EtBr (10 μg/ml) along with a standard of known concentration (2-fold serial dilution from 100 ng/μl to 6.25 ng/μl) and test the average size of DNA (5 μl aliquot) after PCR on a 1.5% (w/v) agarose gel. The probe average size is reduced from 500 bp (after sonication) down to 300 bp. This is due to the random incorporation of the oligonucleotide into DNA during step 1.

Troubleshooting

We routinely obtain between 3 and 5 μg DNA after amplification by PCR. It is not absolutely necessary to have such a DNA yield since the labeling reaction described later requires 500 ng DNA. However, if the amount of DNA is not satisfactory, try to start the amplification procedure with more material or scale-up the chromatin immunoprecipitation. We do not advise increasing the number of PCR cycles since it is kept relatively low in order to ensure amplification linearity.

The final PCR purification through columns (QIAquick PCR purification kit, Qiagen) is critical since unincorporated oligonucleotides and dNTPs may interfere with the subsequent labeling reaction efficiency.

For some microarray applications, it may be necessary to compare immunoprecipitated DNA with input DNA (i.e., study of a wild-type strain acetylation profile instead of analyzing the changes of acetylation between a WT and a HDAC mutant). The Input DNA is prepared from 20 μl cleared lysate as indicated earlier (see the Troubleshooting section for Chromatin Immunoprecipitation (ChrIP or ChIP)). Dilute 150-fold an aliquot of the Input DNA prior to starting the amplification procedure (step 1).

Klenow Labeling of the Probe and Hybridization

We favor labeling of the probe by Klenow random priming over a PCR-based method described elsewhere.[17] We have noticed that the fluorescent labels are not efficiently incorporated into DNA using a Taq DNA polymerase. Moreover, our first PCR amplification step described earlier yields enough DNA for direct labeling by Klenow random priming, thus bypassing the need for a second PCR reaction. It is advised to switch the fluorescent dye between the two probes when repeating the experiment to correct for the incorporation efficiency of each dye into DNA.

1. The labeling reaction makes use of reagents (random octamer and concentrated Klenow) from the Bioprime DNA labeling system (Gibco-BRL) as described at http://www.microarrays.org/protocols.html. Amplified DNA (500 ng) is mixed with 20 μl 2.5× random primer solution (125 mM Tris-HCl, pH 6.8, 12.5 mM MgCl2, 25 mM 2-mercaptoethanol, and 750 μg/ml random octamers). Complement the solution with water to a final reaction volume of 41 μl. Boil the samples for 5 min and transfer the tubes on ice immediately. Add 5 μl dNTPs mix (1.2 mM dATP, 1.2 mM dTTP, 1.2 mM dGTP, and 0.6 mM dCTP in TE, pH 8.0), 3 μl of either Cyanine 3-dCTP or Cyanine 5-dCTP (Renaissance, NEN),

and 1 μl of concentrated Klenow fragment (40 U/μl). Incubate the reaction in the dark for at least 2 h at 37°.

2. The probes are purified and concentrated through a microcon-30 filter (Amicon) as follows. Combine the Cy3- and Cy5-labeled probes and add 450 μl TE, pH 8.0, along with tRNA (10 μg) and salmon sperm DNA (10 μg). The addition of nucleic acid competitors (tRNA and salmon sperm DNA) at this stage presents the advantage of increasing the maximum probe volume (up to 4.6 μl) that can be subsequently mixed with the hybridization solution. Place the column on top of a collecting tube and centrifuge at room temperature for 7 min (Eppendorf tabletop centrifuge (model 5415 C) at 12,000 rpm). The labeled DNA (violet-purple color) should be clearly visible at the bottom of the column. Discard the flow-through, add another 450 μl TE, pH 8.0, in the column, and centrifuge again for 7 min. Check the remaining probe volume in the column and centrifuge by 1-min increments until it is concentrated down to 4.6 μl or less. Invert the column and recover the combined probes in a collecting tube by centrifugation (1 min).

3. Prepare the hybridization mix as follows: mix the concentrated probe with 5 μl 20× SSC (3 M NaCl, 0.3 M trisodium citrate, pH 7.0), 10 μl 100% formamide, and 0.4 μl 10% (v/v) SDS. Complement the solution with water up to a final volume of 20 μl if necessary (final concentration: 5× SSC, 50% formamide, 0.2% SDS, 0.5 mg/ml tRNA, and 0.5 mg/ml salmon sperm DNA). Denature the hybridization mix for 5 min at 95°. The probe can be cooled on ice briefly (few seconds) but should not remain there in order to prevent SDS precipitation. Pipet the hybridization mix on the glass cover slip and cover with the microarray slide. This is favored to the alternative (i.e., pipet the probe on the slide and cover with the cover slip), as it reduces the formation of bubbles. The slide is placed in a hybridization chamber (Corning) which in turn is arranged in a humid chamber consisting of a closed plastic box (Tupperware) containing paper soaked in water and a little stand. Hybridization is carried out overnight at 44° in a hybridization oven.

4. The cover slip is removed after a brief immersion in 400 ml 2× SSC at room temperature. The slides are then washed at room temperature for 5 min in 0.1× SSC/0.1% (v/v) SDS and twice in 0.1× SSC. All washing steps are done in staining jars with 400 ml solution and under mild shaking. Slides are dried by centrifugation in microtiter plate carrier for 5 min at room temperature at 500 rpm (57g in Beckman J2-HC, rotor JS-4.3). Drainage of liquid from the

slides is helped by placing a piece of folded paper towel between the slide holder and the microtiter plate carrier. Microarrays are scanned for both fluorescent dyes using a GMS 418 Array Scanner (Genetic Micro Systems).

Troubleshooting

In our hands, the labeling procedure is reliable and successful most of the time. The color of the purified probe is the best indicator for the procedure success. The combination of the two probes should give a dark purple color after concentration. Do not discard the flow-through if the probe is colorless after the first 7 min centrifugation step through the microcon-30 filter, as it is possible that the filter is defective. Simply reload the flow-through onto another filter and centrifuge again. If the problem persists, this is a clear indication that the labeling was not efficient and the probe should not be used for hybridization. Similarly, a pink (Cy3-labeling) or a blue (Cy5-labeling) color indicates that one of the probes was not labeled properly. Change the vial of dye if you suspect the lack of efficient labeling might stem from a bad batch. The dNTP mix is also an important factor for labeling. We have consistently observed a reduction in its efficiency when using the same dNTP mix stock several times. We strongly advise to prepare it fresh shortly before the labeling reaction.

We have observed that pre-hybridization of the microarray slide is not required when used for acetylation microarrays. Some antibodies used for immunoprecipitation of epitope-tagged protein (i.e., anti-hemagglutinin) appear to create unacceptable hybridization background levels. The following pre-hybridization conditions can be applied in order to alleviate or reduce the background problem. Incubate the slides in 3.5× SSC, 0.1% (v/v) SDS, and 10 mg/ml BSA in a staining jar for 20 min at 44°. Rinse the slides briefly first with water then with isopropanol. Dry the slides by centrifugation on a microtiter plate carrier as indicated earlier (step 4).

Data Quantitation, Normalization, and Analysis

Several commercial and freely available software packages can be used for data quantitation. Fluorescence intensities are quantified in our laboratory using Imagene software (version 4.1) from BioDiscovery (http://www.biodiscovery.com). The following site contains a non-exhaustive list of various data quantitation and analysis tools: http://genome-www5.stanford.edu/MicroArray/SMD/restech.html. Refer to their respective manuals to learn how to use them.

Normalization is probably one of the greatest challenges in data analysis. Indeed, the dye emission intensities are not equal and probes are not labeled similarly due mostly to the different DNA incorporation efficiency of Cy3 and Cy5. A broadly used normalization method makes the assumption that the sums of all intensities of both dyes are equal across the entire array. In other words, any acetylation enrichment should be compensated by a similar decrease in acetylation somewhere else in the genome. If this were true, one could use the ratio of the total intensities between both probes as a normalization factor. While this approach is used for expression arrays, it is not suitable for acetylation arrays since deletion of an HDAC will lead to an increase of general histone acetylation that will not be compensated by a global decrease. Therefore, we strongly suggest the use of internal controls for normalization of acetylation microarrays. Yeast telomeres are hypoacetylated and are not affected by most of the deacetylases (Rpd3, Hda1, Hos1, Hos2, and Hos3), with the notable exception of the telomeric Sir2, which deacetylates histone H4 lysine 16.[10,18] Telomeric sequences from chromosome 6R (coordinates 269505 to 270095, also labeled Tel6R in the intergenic array) are appropriate for a crude pre-normalization of the data. If possible, this DNA fragment should be spotted many times throughout the glass DNA microarray in order to be reliably used for normalization. Once the data are pre-normalized, identify other regions in the array (between 6 and 10 such regions) whose ratio of intensity between the two probes is very close to Tel6R (ratio of about 1). Confirm that the acetylation state of these newly identified regions is not affected in your experiment using a standard PCR analysis of the original ChrIP.[14] Use these regions to finalize normalization of the entire data set.

Different antibodies have inherently different affinities for their own epitope leading to variations in immunoprecipitation efficiency. For instance, a highly specific antibody raised against histone H3 lysine 18 may not be as efficient as another antibody recognizing histone H3 lysine 9. This is particularly important when comparing data obtained for different acetylation sites. Data from different experiments can be scaled by calculating the percentile ranking (using Microsoft Excel) of their respective enrichment ratio. This allowed us to compare acetylation with expression data (see Fig. 2).[10]

Basic data analysis does not require fancy and expensive software but can readily be performed using Microsoft Access, which is part of the Microsoft Office bundle. Access allows the creation and handling of databases. We have, for instance, created our own database containing gene annotation, published microarray data sets from other laboratories, and

---

[18] N. Suka, K. Luo, and M. Grunstein, *Nat. Genet.* **32**, 378 (2002).

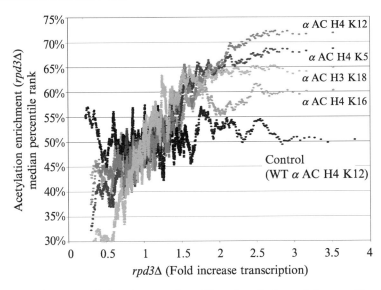

FIG. 2. Histone acetylation sites correlate differently with transcription resulting from *rpd3Δ*. This figure illustrates the scaling of different acetylation data sets using percentile ranking in order to compare them with a transcription data set. Due to inherent noise in microarray data, absolute correlation analysis between acetylation and transcription is not very informative. A moving average analysis that greatly reduces noise can be used to extract general trends. The moving average (window size, 100 data point; step, 1 data point) percentile rank of acetylation enrichment is plotted as a function of transcription increase resulting from *rpd3Δ* (B. E. Bernstein, J. K. Tong, and S. L. Schreiber, *Proc. Natl. Acad. Sci. USA* **97,** 13708 (2000)). Acetylation data are plotted for H4 K5 (dark blue), H4 K12 (red), H4 K16 (orange), and H3 K18 (light blue). Control corresponds to a comparison of two sets of probes amplified from the immunoprecipitation of acetylated H4 K12 in the WT strain and labeled separately with Cy3 and Cy5 prior to hybridization. Data show that increased acetylation at histone H4 K5 and K12 is associated most directly with increased gene transcription in *rpd3Δ*. H4 K16 show the poorest correlation with gene activity (*RPD3* disruption has no significant effects on the status of H4 K16 acetylation) (D. Robyr, Y. Suka, I. Xenarios, S. K. Kurdistani, A. Wang, N. Suka, and M. Grunstein, *Cell* **109,** 437 (2002)). Reprinted with permission from D. Robyr *et al., Cell* **109,** 437–446 (2002). (See color insert.)

gene functional categories as defined by MIPS (Munich Information Center for Protein Structure at: http://mips.gsf.de/proj/yeast/CYGD/db/pathway_index.html). Further data analysis (clustering, sequence analysis, etc.) will need software such as GeneXPress that is freely available for academics (http://genexpress.stanford.edu/).

Most of the data analysis software, however, was designed for ORF arrays and not for intergenic arrays. Promoter-containing intergenic fragments must first be assigned to their respective ORF, especially when comparing acetylation status at a promoter with the transcription level of the

ORF it regulates. Some intergenic regions are located between two divergent ORFs and are by default assigned to both ORFs. Other intergenic regions are located more than 1 kb away from any ORF. These are considered orphans and are not assigned to any particular ORF.

DNA Intergenic Microarray Preparation

Yeast intergenic microarrays include all sequences located between ORFs, including telomeric regions, rDNA, tRNA, centromeres, and transposable elements. A full description of the method to prepare microarray slides is not the scope of this review and was thoroughly discussed earlier.[19] Rather, we will briefly comment on the method we are currently using in our laboratory. We have amplified by PCR about 6700 intergenic regions from yeast genomic DNA using primer pairs available at ResGen (http://www.resgen.com/). The PCR amplification is performed in 96-well plates and is described at http://www.microarrays.org/protocols.html. The size of every single PCR product was checked by agarose gel electrophoresis. It is possible to coat glass slides in the laboratory with poly-L-lysine (http://cmgm.stanford.edu/pbrown/protocols/1_slides.html). However, we strongly recommend the use of amino-silane–coated slides from Corning (CMT-GAPS–coated slides) since they lead to better printing and reduced background during hybridization. PCR fragments (100 ng/μl in 3$\times$ SSC) were transferred onto 384-well plates for printing using 16 pins from Array-It (TeleChem International) and an arrayer built according to specifications provided at: http://cmgm.stanford.edu/pbrown/mguide/index.html. Printed slides are processed using succinic anhydride blocking according to a method provided by the slide manufacturer. Such a blocking procedure will dramatically reduce hybridization background. The preparation of ORF microarrays is identical. We suggest the printing of ORF and intergenic regions on the same slide if required for your application. Information concerning microarray arrayers and slide manufacturers is hosted at: http://www.biologie.ens.fr/en/genetiqu/puces/links.html.

Concluding Remarks

Acetylation microarrays have proved to be extremely useful in discovering new functions for the yeast HDACs, functions that would have been much more difficult to identify using classical genetic or molecular biology techniques.[10] Although the approach described here focuses on acetylation, it can be extended to other histone modifications as well. The

---

[19] M. B. Eisen and P. O. Brown, *Methods Enzymol.* **303,** 179 (1999).

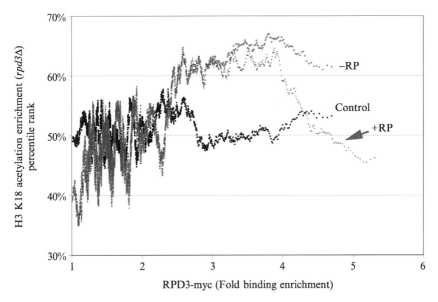

FIG. 3. Binding arrays are complementary to acetylation arrays. The moving average (window size, 100 data point; step, 1 data point) of Rpd3 enrichment (binding) over intergenic regions is plotted as a function of percentile rank of H3 K18 acetylation in *rpd3Δ*. Data sets with (+RP) and without (−RP) ribosomal protein genes are plotted as indicated. Rpd3 binds strongly at the promoter of ribosomal protein genes in logarithmically growing cells where these genes are highly active but under the same conditions, has little or no effect on acetylation of these promoters (S. K. Kurdistani, D. Robyr, S. Tavazoie, and M. Grunstein, *Nat. Genet.* **31,** 248 (2002); D. Robyr, Y. Suka, I. Xenarios, S. K. Kurdistani, A. Wang, N. Suka, and M. Grunstein, *Cell* **109,** 437 (2002)) or the expression of the RP genes (B. E. Bernstein, J. K. Tong, and S. L. Schreiber, *Proc. Natl. Acad. Sci. USA* **97,** 13708 (2000)). Thus, the ribosomal protein gene promoters as targets of Rpd3 are only detectable by binding arrays. This clearly illustrates the importance of combining different types of arrays (binding, acetylation, and expression) to fully comprehend HDAC function. Reprinted with permission from S. K. Kurdistani *et al., Nat. Gen.* **31,** 248–254 (2002). (See color insert.)

key to a successful analysis depends on the availability of highly specific antibodies and an appropriate crosslinker. We have used a similar approach to study genome-wide binding of Rpd3[9] using a double-crosslinking approach with a protein-protein crosslinking agent (DMA) and formaldehyde. The latter genome-wide binding study clearly showed that Rpd3 binds to many chromatin loci where it has no detectable effect on acetylation when deleted in logarithmically growing cells. These regions would be hidden in acetylation microarrays under the same condition (see Fig. 3), emphasizing the need for a combination of different types of arrays (acetylation, binding, and expression) for full understanding of Rpd3's function(s).

While acetylation microarrays are extremely powerful, they have some limitations. First, the relative low resolution of the microarrays (up to 1 kb large fragment spotted on the microarray) may not reveal regions where only one nucleosome is modified if nearby nucleosomes are not affected. High resolution is obtained by analyzing a region of interest by standard ChrIP[14] or by printing custom microarrays containing 100–200 bp DNA fragments along the regions of interest. Second, histone post-translational modifications create interacting surface for histone binding proteins (e.g., bromodomains of Gcn5 and Snf2[20]) that may potentially interfere with antibody recognition for the same site. Finally, aneuploidy between two strains[21] has to be corrected in order to ascertain that increase binding, acetylation, or expression is not the result of gene duplication in the analyzed strain versus the control. A simple hybridization of Input genomic DNA (obtained from the sonicated cell extract but not incubated with the antibody) from the experiment and the control strains can indicate potential aneuploidy problems.

The study of other histone modifications using similar approach will in the future provide important insights not only on their genome-wide respective patterns, but also on how they may relate and influence each other and correlate with transcription activity, unraveling the roles of histone modifications in chromosome functions.

[20] L. Zeng and M. M. Zhou, *FEBS Lett.* **513**, 124 (2002).

[21] T. R. Hughes, C. J. Roberts, H. Dai, A. R. Jones, M. R. Meyer, D. Slade, J. Buchard, S. Dow, T. R. Ward, M. J. Kidd, S. H. Friend, and M. J. Marton, *Nat. Genet.* **25**, 333 (2000).

[20] Use of Chromatin Immunoprecipitation Assays in Genome-Wide Location Analysis of Mammalian Transcription Factors

By BING REN and BRIAN D. DYNLACHT

One major challenge in the study of mammalian transcriptional regulation pertains to difficulties in studying direct targets of regulatory proteins in a physiological setting. This chapter describes methods for the use of chromatin immunoprecipitation (ChIP) in the genome-wide identification of binding sites in mammalian cells. This approach combines the adaptation of ChIP in the enrichment of mammalian transcription factor target

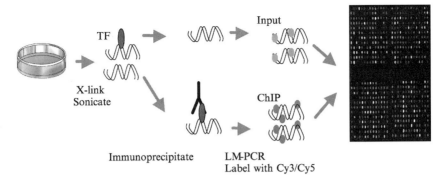

FIG. 1. A schematic diagram of genome-wide location analysis with mammalian transcription factors. The method can in principle be generalized for any DNA-binding regulatory protein. (See color insert.)

genes with DNA microarray technology. This method will enable researchers to overcome limitations in the study of eukaryotic transcriptional regulation by allowing the identification of direct, physiological targets of DNA-binding proteins in an unbiased, genome-wide manner.

Procedures

Overview

Genome-wide location analysis, described in detail in this chapter, is based on the rapid purification of specific genomic fragments associated with a particular protein followed by detection of enriched sequences with DNA microarray technology (see Fig. 1). Intact, living cells are treated with formaldehyde, a cross-linking agent that results in the covalent and reversible linkage between genomic DNA and associated proteins.[1] The cross-linked chromatin is extracted after disruption of cell and nuclear membranes and fragmented into an average length of 500 bp by sonication. The chromatin associated with a factor of interest is then purified through immunoprecipitation using a specific antibody against the protein. Subsequently, the cross-links are reversed, and the genomic DNA in the resulting sample is purified by successive steps to remove protein and RNA. DNA fragments are blunt-ended by T4 DNA polymerase to allow ligation of a universal oligonucleotide linker, and all DNA fragments in the sample are amplified through polymerase chain reaction (PCR) in an unbiased fashion. As a control, an aliquot of the input chromatin DNA is processed

[1] V. Orlando, *Trends Biochem. Sci.* **25,** 99 (2000).

using an identical procedure. The ChIP-enriched and control DNA are then labeled with different fluorescent dyes in a random priming reaction, and the labeled DNA samples are hybridized simultaneously to a single DNA microarray containing genomic DNA (promoter) sequences. An enriched genomic sequence is identified as a DNA spot in which there is significantly more fluorescence intensity with ChIP-enriched DNA than with the control input DNA.

This method has several important advantages over other binding site mapping methods, particularly, computational prediction and genome expression analysis. First, the method examines direct protein-DNA interactions throughout the genome, and it therefore reveals regulation by a transcription factor in a native setting under physiological conditions. Second, because the method does not require prior knowledge of a transcription factor's binding specificity, or its transcriptional activity, it is an unbiased approach and can uncover novel biological properties of a transcription factor. Third, in addition to revealing the location of promoter and enhancer binding proteins, the method can also be applied to examine *any* DNA-binding regulatory protein.

To adapt the method to mammalian transcription factors, the protocol described later has taken several factors into consideration. First, instead of using arrays that cover the entire genome, the method employs arrays that represent only gene promoters. This design is based on three considerations:

1. It is not yet feasible to produce DNA microarrays that cover the entire human genome in 1-kb segments. The effort to synthesize and print 3 million DNA fragments is not practical for individual laboratories.
2. Transcription factors most frequently target gene promoters. A survey of known transcription factor binding sites in the human genome showed that the vast majority occur within 1 kb from the transcription start site (TSS) of genes.[2]
3. To make a DNA microarray that represents all human gene promoters is both affordable and manageable by individual laboratories. The ease of production combined with high utility of the promoter array outweighs the potential lack of coverage for a small number of target sites in the genome.

[2] B. Ren, H. Cam, Y. Takahashi, T. Volkert, J. Terragni, R. A. Young, and B. D. Dynlacht, *Genes Dev.* **16**, 245 (2002).

The second factor to consider is the modification of the hybridization conditions. The human genome contains abundant repetitive DNA sequences. Certain repetitive DNA elements are represented by millions of copies and can lead to significant cross-hybridization problems. To reduce or eliminate such cross-hybridization, human Cot-1 DNA enriched for repetitive DNA sequences is included in the hybridization reaction. In addition, a stringent hybridization temperature is used to increase hybridization specificity.

Solutions

Cross-linking solution:	11% formaldehyde, 0.1 M NaCl, 1 mM Na-EDTA, 0.5 mM Na-EGTA, 50 mM HEPES, pH 8.0
Lysis buffer 1:	0.05 M HEPES-KOH, pH 7.5, 0.14 M NaCl, 1 mM EDTA, 10% glycerol, 0.5% NP-40, 0.25%, Triton X-100, with protease inhibitor cocktail (Roche Applied Science)
Lysis buffer 2:	0.2 M NaCl, 1 mM EDTA, 0.5 mM EGTA, 10 mM Tris, pH 8, protease inhibitor cocktail
Lysis buffer 3:	1 mM EDTA, 0.5 mM EGTA, 10 mM Tris-HCl, pH 8, protease inhibitor cocktail
Proteinase K stock solution:	20 mg/ml proteinase K (Sigma) 50 mM Tris-HCl, pH 8.0, 1.5 mM calcium acetate
RIPA buffer:	50 mM HEPES, pH 7.6, 1 mM EDTA, 0.7% DOC, 1% NP-40, 0.5 M LiCl, protease inhibitor cocktail
Elution buffer:	50 mM Tris, pH 8, 10 mM EDTA, 1% SDS
Hybridization buffer 1:	2.2 × SSC, 0.22% SDS
Hybridization buffer 2:	70% formamide, 3 × SSC, 14.3% dextran sulfate

Cross-Linking of Cells and Fragmentation and
  Immunoprecipitation of Chromatin

Formaldehyde Cross-Linking and Preparation of Chromatin

10^9 cells suspended in growth medium are transferred as 40-ml aliquots to 50 ml-tubes and placed on ice for 10 min. Then 1/10 volume of cross-linking solution is added directly to each tube. After a 10-min incubation on ice (or at room temperature, depending upon the cell type), 1/20 volume of 2.5 M glycine solution is added to each tube to stop the cross-linking reaction. The cells are harvested by centrifugation at 2000g for 10 min at 4°. The cell pellets are re-suspended in cold PBS (137 mM NaCl, 2.7 mM KCl, 10 mM Na$_2$HPO$_4$, 2 mM KH$_2$PO$_4$) and washed twice. The final cell pellet may be snap-frozen in liquid nitrogen. Adherent cells can also be fixed directly on plates, treated with glycine, washed with PBS, and harvested in cold PBS with a silicone scraper.

The cell pellet is re-suspended in 30 ml of Lysis Buffer 1 by pipetting and is mixed for 10 min at 4° on a rocking platform. After centrifugation at 2000g for 10 min at 4°, the cell pellet is re-suspended in 24 ml Lysis Buffer 2 by pipetting and mixed gently at room temperature for 10 min on a rocking platform. After centrifugation at 2000g for 10 min at 4°, the pellet is re-suspended in 10 ml of Lysis Buffer 3.

The mixture is divided into 5-ml aliquots in 15 ml-conical tubes. These tubes are then placed into 50-ml tubes containing ice. A sonicator (Branson Sonifier 450, with power setting at 5) is used to disrupt the cell and nuclear membranes and fragment the chromatin. To avoid foaming, the sonicator probe tip is first immersed in the mixture followed by a 25-s, continuous pulse of sonication. The tube is immediately placed on ice for at least 1 min to avoid overheating the sample. Sonication is repeated until the chromatin fragments are of the desired length. The fragment size can be examined by agarose gel electrophoresis of 10 μl of cell extract digested with Proteinase K for 1 h. The number of sonication cycles varies with cell type and cross-linking conditions, and pilot tests are recommended. Ten cycles of sonication are normally sufficient to achieve the desired (~500 bp) fragment size. Finally, the chromatin solution is adjusted to 0.5% Sarkosyl (sodium lauryl sarcosine) and gently mixed for 10 min at room temperature on a rocking platform. The chromatin solution is then transferred to a centrifuge tube and spun for 10 min at 10,000g to remove cell debris. The supernatant is collected for ChIP. At this stage, the concentration of DNA should be around 1–2 mg/ml. The solution can be stored at −80° in 1-ml aliquots. The amount of chromatin generated here should be sufficient for 10 immunoprecipitation reactions.

Immunoprecipitation of Chromatin

Magnetic beads (Dynal) pre-bound to polyclonal antibodies are used to immunoprecipitate the DNA associated with the protein of interest. To prepare the magnetic beads, 100 μl of sheep anti-rabbit IgG-conjugated Dynabeads (Dynal) are first washed three times with cold PBS containing 5 mg/ml bovine serum albumin (BSA) and then re-suspended in 5 ml of cold PBS. Ten micrograms of rabbit polyclonal antibody are added to the mixture and incubated overnight on a rotating platform at 4°. After collecting the magnetic beads by centrifugation and washing three times with cold PBS containing 5 mg/ml BSA, the beads are re-suspended in 100 μl of cold PBS with 5 mg/ml BSA and are then ready for immunoprecipitation.

In an Eppendorf tube, 2 mg of soluble chromatin are first adjusted to 0.1% Triton X-100, 0.1% sodium deoxycholate, and 1 mM PMSF, then mixed with 100 μl of magnetic beads pre-bound to the antibody. The mixture is incubated at 4° overnight on a rotating platform. The magnetic beads are then collected using a magnet (Dynal), and the supernatant is removed by aspiration. To remove material non-specifically bound to the beads, 1 ml of RIPA buffer is added to the tube, and the beads are gently re-suspended on a rotating platform in the cold room. The magnetic beads are again collected with the magnet and washed with RIPA buffer a total of five times. After washing once with 1 ml of TE, the beads are collected by centrifugation at 2000g for 3 min and re-suspended in 50 μl of elution buffer. To elute precipitated chromatin from the beads, the tubes are incubated at 65° for 10 min with constant agitation then centrifuged for 30 s at 2000g. Forty microliters of supernatant are removed and mixed with 120 μl of TE with 1% SDS. This solution is incubated at 65° overnight to reverse the cross-links. As a chromatin input control, 100 μg of chromatin are mixed with 120 μl of TE containing 1% SDS in a separate tube and incubated at 65° overnight.

Purification of DNA

After reversal of cross-links, proteins in the DNA sample are removed by incubation with 120 μl of Proteinase K solution (2% glycogen, 5% Proteinase K stock solution, and TE) for 2 h at 37°. The sample is then extracted twice with phenol and once with 24:1 chloroform/isoamyl alcohol. The sample is adjusted to 200 mM NaCl. After ethanol precipitation, the DNA is dissolved in 30 μl of TE containing 10 μg of DNase-free RNase A and incubated for 2 h at 37°. The DNA at this step can be further purified with a Qiagen PCR kit (Qiagen).

Blunting, Ligation of Linkers to DNA, and Amplification by PCR

The immunoprecipitated DNA is typically present in very small quantities (1–10 ng total), and it is therefore necessary to perform an amplification step prior to labeling and DNA microarray analysis. To achieve this, a ligation-mediated PCR (LM-PCR) procedure is used. Because the DNA at this stage contains recessed and heterogeneous ends as a result of the physical shearing process, it is first treated with T4 DNA polymerase to form blunt ends. Then a linker is ligated to the DNA fragments. The addition of this linker allows the DNA to be amplified by PCR using a universal oligonucleotide primer.

Blunting, Ligation, and PCR

In an Eppendorf tube, the immunoprecipitated DNA (or 20 ng of control input DNA) is combined with 11 μl of 10× T4 DNA polymerase buffer (New England Biolabs), 0.5 μl BSA (10 mg/ml) (New England Biolabs), 0.5 μl dNTP mix (20 mM each), 0.2 μl T4 DNA polymerase (3 U/μl) (New England Biolabs), and distilled H_2O in a total volume of 112 μl. After a 20-min incubation at 12°, 1/10 volume of 3 M sodium acetate (pH 5.2) and 10 μg of glycogen (Roche Applied Sciences) are added to the tube, and the DNA sample is extracted with phenol:chloroform:isoamyl alcohol (25:24:1) (Sigma) once. After ethanol precipitation, the DNA is dissolved in 25 μl of distilled H_2O.

The blunt-ended DNA is mixed with 8 μl of distilled H_2O, 10 μl of 5× ligase buffer (Invitrogen), 6.7 μl of annealed oligonucleotide linkers (oligo-1: GCGGTGACCCGGGAGATCTGAATTC, oligo-2: GAATT-CAGATC, annealed to make a 15 μM solution), 0.5 μl T4 DNA ligase (New England Biolabs), and distilled water in a total volume of 50 μl. The ligation reaction is allowed to proceed overnight at 16°. After ligation, the DNA is purified by ethanol precipitation and dissolved in 25 μl of distilled H_2O.

The ligated DNA is then mixed with 4 μl of 10× ThermoPol reaction buffer (New England Biolabs), 4.75 μl distilled H_2O, 5 μl of 10× dNTP mix (2.5 mM each), 1.25 μl oligo-1 (40 μM stock) in a final volume of 40 μl in a 500-μl thin-walled PCR tube. The tube is first incubated at 55° for 2 min in a thermal cycler to separate linker oligonucleotides not ligated to the DNA, then 10 μl of an enzyme mix [8 μl dH2O, 1 μl Taq DNA polymerase (5U/μl), 1 μl ThermalPol reaction buffer, and 0.025 units of Pfu polymerase (Stratagene)] is added. Subsequently, the following PCR protocol is performed:

Step 1:   72° for 5 min
Step 2:   95° for 2 min

Step 3:   95° for 1 min
Step 4:   60° for 1 min
Step 5:   72° for 1 min
Step 6:   go to step 3, 22 times
Step 7:   72° for 5 min
Step 8:   4° indefinitely

After PCR, the DNA is purified using the Qiaquick PCR purification kit (Qiagen) and eluted in 60 μl elution buffer provided with the kit.

Labeling of DNA and Microarray Hybridization

Labeling of Amplified DNA with Fluorophores

Two hundred nanograms of DNA from the previous step is mixed with 20 μl of 2.5× random primer solution (BioPrime kit, Invitrogen) and distilled H_2O in a final volume of 42.5 μl. The mixture is boiled for 5 min and then cooled on ice for 5 min. Subsequently, 5 μl of 10× low dCTP mixture (2.5 mM each for dATP, dTTP and dGTP, and 0.6 mM for dCTP), 1.5 μl of Cy5-dCTP (Amersham) or Cy3-dCTP (Amersham), and 40 units of Klenow DNA polymerase are added to tube. The tube is incubated at 37° for 2 h. After the reaction, the labeled DNA is purified using the Qiagen PCR kit (Qiagen).

Fabrication and Hybridization of Promoter DNA Microarrays

An array (1.5K array) containing approximately 1500 human promoter fragments was initially designed based on NCBI human genome assembly Build 24 (July 3, 2001 freeze). One thousand two hundred genes were selected from the collection of RefSeq entries based on three criteria:

1. Their expression varied during the mammalian cell cycle
2. The Transcription Start Site (TSS) for each of these genes was annotated based on full-length mRNA sequences and
3. Their biological function was well documented

In addition, approximately 200 genes that do not exhibit cell cycle periodicity were also included. Oligonucleotide primers were designed to amplify fragments with endpoints at −700 to +300 relative to the start site of these genes using PCR. After PCR, the amplified fragments were purified, verified by gel electrophoresis, and spotted onto GAPSII glass slides (Corning) using a Cartesian microarrayer (Genomic Solutions). After UV cross-linking, the glass slides were stored in a vacuum desiccator until use.

In a new Eppendorf tube, 2.5 μg of Cy5-labeled ChIP-enriched DNA is mixed with 2.5 μg of Cy3-labeled genomic (control input) DNA and 36 μg of human Cot-1 DNA (Invitrogen). DNA is precipitated with 1/10 volume of 3 M sodium acetate and 2 volumes of ethanol, and the DNA is dissolved in 22.4 μl of hybridization buffer 1. The mixture can be heated for 10 min at 37° to facilitate re-suspension of the DNA. Twenty microliters of hybridization buffer 2 are added to the mixture, and the tube is heated at 95° for 5 min to denature DNA, then incubated at 42° for 2 min. Four microliters of yeast tRNA (10 μg/μl; Sigma) and 3 μl of 2% BSA are then added to the mixture, which is then spotted onto a DNA microarray slide that has been incubated with pre-hybridization solution for 40 min at 42°. A 25 mm × 60 mm cover slip is then gently placed on top of the sample, and the hybridization is carried out in a hybridization chamber (Corning) at 60° overnight in a water bath.

After hybridization, the microarray slide is washed once with wash buffer 1 (2× SSC, 0.1% SDS) at 60° for 5 min in a glass slide staining dish. This is followed by a wash with buffer 2 (0.2× SSC, 0.1% SDS) for 10 min at room temperature and three times with buffer 3 (0.2× SSC) at room temperature. The slide is then dried by a brief spin at 1000g in a table top centrifuge.

Microarray Analysis and Identification of *In Vivo* DNA-Binding Sites

Microarray Scanning and Analysis

The microarray slides are scanned using a GenePix 4000B scanner from Axon Instruments, and each microarray image is first analyzed with the GenePix Pro 3.0 image analysis software to derive the Cy3 and Cy5 fluorescent intensity and background noise for all the spots on the array. The intensities for both Cy3 and Cy5 channels are first adjusted by subtracting the background intensity of each spot using the formula $I_{channel} - B_{channel}$, where channel represents either Cy5 or Cy3. The intensities are further adjusted by subtracting the median intensity of all the blank spots on the array. If any of these values are lower than 10, they are automatically raised to a value of 10.

A normalization factor is calculated based on the intensities of spots that are considered good (more than 65% of pixels have intensities higher than the background intensity plus 1 standard deviation). The median of the intensity ratios, I_{635}/I_{532}, is then used to adjust the Cy3 channel intensity to the same level as that of the Cy5 channel.

The quantitative amplification of small amounts of DNA generates some uncertainty in values for the low-intensity spots. In order to estimate that uncertainty and average repeated experiments with appropriate related weights, a single array error model[3] is used. The significance of a measured ratio at a spot is defined by a statistic X, which is formulated as

$$X = \frac{a_2 - a_1}{\sqrt[2]{\sigma_1^2 + \sigma_2^2 + f^2\left(a_1^2 + a_2^2\right)}},$$ (1)

where $a_{1,2}$ are the intensities measured in the two channels for each spot, $\sigma_{1,2}$ are the uncertainties due to background subtraction, and f is a fractional multiplicative error such as would come from hybridization non-uniformities, fluctuations in the dye incorporation efficiency, and scanner gain fluctuations. The distribution of X can be found to be close to a Gaussian distribution in experiments where Cy3 and Cy5 samples are identical. The significance of a change of magnitude X is then calculated using a one-sided probability model as follows:

$$P = 1 - \text{Erf}\left(\frac{x - \mu}{\sigma}\right),$$ (2)

where μ is the average of X and σ is the standard deviation of X. Since the intensities are normalized, μ should be near 0. The Erf(x) function is the standard normal accumulative distribution function corresponding to standard normal curve areas.

If the Cy3 and Cy5 samples are not identical, the Gaussian distribution can be skewed, because the ChIP can result in many DNA spots with significantly higher intensities in one channel than the other. Since the input DNA is always present, the intensity distribution of those non-enriched spots can be used to obtain the parameters of X distribution. First, the spots whose X is less than 0 are identified. These spots should be on the left half of the Gaussian distribution. Their mirror spots are then generated using $X^+ = -X^-$. The mean and standard distribution of this new X can be calculated.

When the genomic binding sites of a protein are investigated using the above-mentioned method, it is routine to perform several independent experiments so that the results are more reliable. To combine replicate data sets, each sample is first analyzed individually using the above-mentioned single-array error model. The average binding ratio and associated P values from these multiple experiments are then calculated using a weighted averaging analysis method.[3] For each spot, the uncertainty in the log(ratio) is defined as

$$\sigma = \log(a_2/a_1)/X_{norm},$$ (3)

[3] C. J. Roberts, B. Nelson, M. J. Marton, R. Stoughton, M. R. Meyer, H. A. Bennett, Y. D. He, H. Dai, W. L. Walker, T. R. Hughes et al., Science 287, 873 (2000).

where $a_{1,2}$ are the intensities measured in the two channels for each spot and X_{norm} is the normalized X for each spot. The weights for each spot are then defined as

$$w_i = 1/\sigma_i^2 \qquad (4)$$

The averaged log(ratio) is then calculated using the following formula

$$\bar{x} = \frac{\sum_{i=1,n} w_i x_i}{\sum_{i=1,n} w_i}, \qquad (5)$$

where n is the total number of experiments, x_i and w_i are the log(ratio) and weight, respectively, for a particular spot in each experiment; $\bar{x}$ is normally distributed and the averaged P value is then calculated using Eq. (2), where the variable for Erf( ) function is $\bar{x}$ instead. Target genes were selected on the basis of significant P values (e.g., less than 0.002) and binding ratios (e.g., greater than 2) in replicate experiments.

Conclusion

In summary, the method that we describe in this chapter is a direct, high-throughput and general approach to locate protein-DNA interactions in mammalian cells. Since most transcription factors function by binding to DNA and regulating gene expression, this method will allow for the rapid, specific identification of physiologically relevant target genes and provide key insights into the molecular mechanisms through which they function. We expect that this technique will have widespread applications in deciphering global gene regulatory networks in mammalian cells as it has in yeast.[4] In addition to revealing the location of promoter and enhancer binding proteins, the method can also be applied to examine regulatory elements bound by the DNA replication and recombination machinery and nucleosome modification complexes, and it should also allow a systematic identification of sites of histone modification and DNA methylation.[5–9]

[4] T. I. Lee, N. J. Rinaldi, F. Robert, D. T. Odom, Z. Bar-Joseph, G. K. Gerber, N. M. Hannett, C. T. Harbison, C. M. Thompson, I. Simon, J. Zeitlinger, E. G. Jennings, H. L. Murray, D. B. Gordon, B. Ren, J. J. Wyrick, J. B. Tagne, T. L. Volkert, E. Fraenkel, D. K. Gifford, and R. A. Young, *Science* **298**, 799 (2002).

[5] B. E. Bernstein, E. L. Humphrey, R. L. Erlich, R. Schneider, P. Bouman, J. S. Liu, T. Kouzarides, and S. L. Schreiber, *Proc. Natl. Acad. Sci. USA* **99**, 8695 (2002).

[6] Y. Blat and N. Kleckner, *Cell* **98**, 249 (1999).

[7] J. L. Gerton, J. DeRisi, R. Shroff, M. Lichten, P. O. Brown, and T. D. Petes, *Proc. Natl. Acad. Sci. USA* **97**, 11383 (2000).

Acknowledgments

We thank H. Cam, Y. Takahashi, T. Volkert, and members of Richard Young's laboratory for help with the development of techniques described in this chapter. Work in BDD's laboratory was supported by the American Cancer Society.

[8] D. Robyr, Y. Suka, I. Xenarios, S. K. Kurdistani, A. Wang, N. Suka, and M. Grunstein, *Cell* **109**, 437 (2002).

[9] J. J. Wyrick, J. G. Aparicio, T. Chen, J. D. Barnett, E. G. Jennings, R. A. Young, S. P. Bell, and O. M. Aparicio, *Science* **294**, 2357 (2001).

[21] High-Throughput Screening of Chromatin Immunoprecipitates Using CpG-Island Microarrays

By MATTHEW J. OBERLEY, JULISSA TSAO, PATRICK YAU, and PEGGY J. FARNHAM

Sequencing of the human genome has provided a wealth of information that is now allowing rapid advances in our understanding of gene regulation. For example, 5' exon prediction programs have provided estimates of the number of promoters in the human genome. Davuluri et al.[1] have developed a program called FirstEF and used it to predict 68,645 first exons. Such information now allows analysis of thousands of promoter sequences and provides a comprehensive data set for a bioinformatics approach to finding target promoters of human transcription factors. Using such an approach, programs have been developed to identify E2F sites in sets of mammalian promoters.[2,3] Similarly, scanning the human genome for Myc binding sites has identified a large set of promoters that contain Myc consensus sites.[4] Follow-up analyses of the predicted E2F and Myc target promoters using chromatin immunoprecipitation (ChIP) assays have indicated that many of the putative targets are in fact bound by E2F or Myc family members in living cells.[2,4] Therefore, such bioinformatics approaches are very instructive and have provided investigators with a large number of potential target genes. However, there are two major concerns

[1] R. V. Davuluri, I. Grosse, and M. Q. Zhang, *Nat. Genet.* **29**, 412 (2001).

[2] A. E. Kel, O. V. Kel-Margoulis, P. J. Farnham, S. M. Bartley, E. Wingender, and M. Q. Zhang, *J. Mol. Biol.* **309**, 99 (2001).

[3] S. Aerts, G. Thijs, B. Coessens, M. Staes, Y. Moreau, and B. De Moor, *Nucleic Acids Res.* **31**, 1753 (2003).

[4] P. C. Fernandez, S. R. Frank, L. Wang, M. Schroeder, S. Liu, J. Greene, A. Cocito, and B. Amati, *Genes Dev.* **17**, 1115 (2003).

that arise when one relies solely on a bioinformatics approach. First, many consensus sites may reside within inactive chromatin and may not be available for interaction with a factor due to inaccessibility. For example, computational identification of binding sites for the PIF/GMEB factor identified sites that were bound *in vitro* by the factor but which were extremely sensitive to the methylation status of the DNA.[5] Thus, these sites would not be occupied if the consensus site resided within hyper-methylated heterochromatin. Similarly, a binding site may reside within active (euchromatin) but be inaccessible due to specific nucleosomal positioning. We have shown that a consensus binding site for E2F that is located in exon 2 of the Myc gene is not occupied *in vivo*.[6] Also, overlap of a consensus binding site with the site for a more abundant and/or higher affinity DNA binding protein can prevent *in vivo* binding. For example, the consensus Ets site in the c-Myc promoter is occupied *in vivo* only if the overlapping E2F site is mutated.[7] Thus, methylation, nucleosomal positioning, and overlapping binding sites can lead to the identification of a set of false-positive target promoters using in silico methods. Conversely, many true positives may be overlooked if the factor is recruited to the DNA via a site that does not exactly match a consensus site. Recruitment via a non-consensus site could be accomplished via protein-protein interaction; the interaction between YY1 and E2F2 is believed to allow stable recruitment of E2F2 to the cdc6 promoter.[8] Such promoters would be overlooked by identification schema that rely solely on the presence of a high-affinity consensus site in a promoter region. Also, transcription factors such as steroid hormone receptors and POU domain proteins can recognize a variety of similar but not identical sites.[9,10] The lack of a strict consensus can make it difficult to identify true target promoters using only sequence inspection tools. Clearly, experimental methods that can allow global analyses and identification of binding sites for human transcription factors would complement the bioinformatics approaches.

Over the last several years, experimental methods have been developed to identify genomic targets of DNA binding factors. All of these approaches utilize as a first step the ChIP assay to selectively enrich for chromatin fragments bound by transcription factors. Investigators then use the

[5] E. Burnett, J. Christensen, and P. Tattersall, *J. Mol. Biol.* **314,** 1029 (2001).
[6] A. S. Weinmann, S. M. Bartley, T. Zhang, M. Q. Zhang, and P. J. Farnham, *Mol. Cell. Biol.* **21,** 6820 (2001).
[7] T. Albert, J. Wells, J. O. Funk, A. Pullner, E. E. Raschke, G. Stelzer, M. Meisterernst, P. J. Farnham, and D. Eick, *J. Biol. Chem.* **276,** 20482 (2001).
[8] S. Schlisio, T. Halperin, M. Vidal, and J. R. Nevins, *EMBO J.* **21,** 5775 (2002).
[9] S. Millevoi, L. Thion, G. Joseph, C. Vossen, L. Ghisolfi-Nieto, and M. Erard, *Eur. J. Biochem.* **268,** 781 (2001).
[10] M. A. Loven, V. S. Likhite, I. Choi, and A. M. Nardulli, *J. Biol. Chem.* **276,** 45282 (2001).

chromatin fragments to probe microarrays containing intergenic sequences. The list of factors that have been analyzed on a genome-wide basis using yeast as a model system is quite large.[11–17] In contrast, fewer studies have been performed using the human genome, mostly due to the lack of availability of appropriate microarrays. Because the human genome (3.2 Gb) is three orders of magnitude larger than the yeast genome (12 Mb), a set of chips containing all human intergenic regions has not yet been created. However, several groups have begun creating their own specific human intergenic microarrays. The largest effort to date consists of a set of oligonucleotide arrays that span the smaller chromosomes 21 and 22 at 35 bp intervals.[18] This array has not been used to identify sites for DNA binding factors; rather it has been used to assess the sum transcriptional activity of these chromosomes. In fact, no oligonucleotide arrays have been reported to be used in combination with chromatin IP, for either yeast or mammalian cell systems. It is not yet clear if this is due to technical problems or just due to a lack of availability of the arrays. A recent study utilized a DNA microarray comprising nearly all of the non-repetitive sequences of human chromosome 22.[19] In this case, PCR fragments (rather than oligonucleotide probes) were spotted onto the array, suggesting that this array could be useful for ChIP studies. However, the published use of this new array has been limited to analysis of mRNA expression profiles. Two different approaches, both using spotted PCR fragments, have successfully been used to identify target genes of human transcription factors. The microarrays used in these studies differ in whether they contain "investigator-selected" promoter regions or "randomly chosen clones."

[11] V. R. Iyer, C. E. Horak, C. S. Scafe, D. Botstein, M. Snyder, and P. O. Brown, *Nature* **409**, 533 (2001).

[12] J. D. Lieb, X. Liu, D. Botstein, and P. O. Brown, *Nat. Genet.* **28**, 327 (2001).

[13] B. Ren, F. Robert, J. J. Wyrick, O. Aparicio, E. G. Jennings, I. Simon, J. Zeitlinger, J. Schreiber, N. Hannett, E. Kanin, T. L. Volkert, C. J. Wilson, S. P. Bell, and R. A. Young, *Science* **290**, 2306 (2000).

[14] H. H. Ng, F. Robert, R. A. Young, and K. Struhl, *Genes Dev.* **16**, 806 (2002).

[15] M. Damelin, I. Simon, T. I. Moy, B. Wilson, S. Komili, P. Tempst, F. P. Roth, R. A. Young, B. R. Cairns, and P. A. Silver, *Mol. Cell.* **9**, 563 (2002).

[16] J. J. Wyrick, J. G. Aparicio, T. Chen, J. D. Barnett, E. G. Jennings, R. A. Young, S. P. Bell, and O. M. Aparicio, *Science* **294**, 2357 (2001).

[17] T. I. Lee, N. J. Rinaldi, F. Robert, D. T. Odom, Z. Bar-Joseph, G. K. Gerber, N. M. Hannett, C. T. Harbison, C. M. Thompson, I. Simon, J. Zeitlinger, E. G. Jennings, H. L. Murray, D. B. Gordon, B. Ren, J. J. Wyrick, J. Tagne, T. L. Volkert, E. Fraenkel, D. K. Gifford, and R. A. Young, *Science* **298**, 799 (2002).

[18] P. Kapranov, S. E. Cawley, J. Drenkow, S. Bekiranov, R. L. Strausberg, S. P. Fodor, and T. R. Gingeras, *Science* **296**, 916 (2002).

[19] J. L. Rinn, G. Euskirchen, P. Bertone, R. Martone, N. M. Luscombe, S. Hartman, P. M. Harrison, F. K. Nelson, P. Miller, M. Gerstein, S. Weissman, and M. Snyder, *Genes Dev.* **17**, 529 (2003).

A PCR fragment microarray including the sequence spanning from -700 to $+200$ of 1444 human genes has been utilized to find novel promoters bound by E2F family members.[20] Although a step in the right direction, such an array by necessity is biased in that the particular 1444 promoters analyzed were chosen by a set of criteria. Using this same approach, it would be feasible to prepare PCR fragments representing the sequences upstream of all predicted first exons in the human genome. However, this Herculean project has not been attempted to date. As an intermediate step, our laboratory has utilized a microarray that contains 7776 CpG island clones.[21] CpG islands tend to be found in intergenic regions and at the 5′ ends of genes. Thus, use of CpG arrays allows a relatively unbiased analysis of many thousands of human intergenic regions.

As described later, we have now developed protocols that allow these arrays to be utilized to perform a large-scale identification of genomic regions occupied by DNA-binding proteins under physiologically relevant, *in vivo* conditions. We have successfully used these arrays to identify genomic sites to which E2F and pRb family members are recruited,[21,22] whereas others have identified Myc binding sites using these same arrays.[23] As illustrated in Figs. 1 and 2. the assay subsumes six major areas:

1. ChIP
2. Generation of amplicons of the ChIPs to provide sufficient amounts of chromatin for microarray analysis
3. Amplicon labeling
4. Preparation of CpG microarrays
5. Hybridization to spotted CpG arrays
6. Data analysis and confirmation of functional binding

Details concerning each of these areas are provided (see also Figs. 1 and 2).

Chromatin Immunoprecipitation

We have recently published a detailed protocol that we have found will efficiently and specifically immunoprecipitate sequences associated

[20] B. Ren, H. Cam, Y. Takahashi, T. Volkert, J. Terragni, R. A. Young, and B. D. Dynlacht, *Genes Dev.* **16,** 245 (2002).

[21] A. S. Weinmann, P. S. Yan, M. J. Oberley, T. H. Huang, and P. J. Farnham, *Genes Dev.* **16,** 235 (2002).

[22] J. Wells, P. S. Yan, M. Cechvala, T. Huang, and P. J. Farnham, *Oncogene* **22,** 1445 (2003).

[23] D. Y. L. Mao, J. Watson, P. S. Yan, D. Barsyte-Lovejoy, F. Khosravi, W. W. Wong, P. Farnham, T. H. Huang, and L. Z. Penn, *Curr. Biol.* **13,** 882 (2003).

I. Chromatin immunoprecipitation

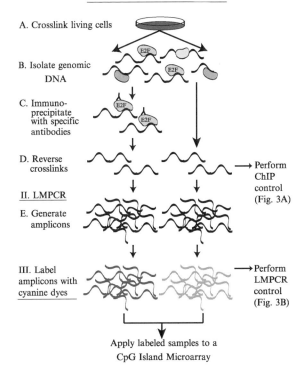

A. Crosslink living cells

B. Isolate genomic
   DNA

C. Immuno-
   precipitate
   with specific
   antibodies

D. Reverse
   crosslinks   → Perform
                                                   ChIP
                                                   control
   II. LMPCR   (Fig. 3A)

E. Generate
   amplicons

III. Label   → Perform
     amplicons with   LMPCR
     cyanine dyes   control
                                                    (Fig. 3B)

Apply labeled samples to a
CpG Island Microarray

FIG. 1. Preparation of samples for use in a ChIP-CpG assay (Steps I–III). Step I: cells are crosslinked in culture (A), which creates protein-protein and protein-DNA crosslinks. Nuclei are then isolated (B) and sonicated to create chromatin fragments of approximately 500–1000 bp. The chromatin is then immunoprecipitated with antibodies specific to the protein of interest; an IgG control sample can also be prepared (C). After extensive washing, the crosslinks are reversed and the DNA is purified, resulting in pools of sequences that are enriched by virtue of their binding to the protein of interest. Chromatin is also purified from crosslinked DNA that has not been immunoprecipitated; this serves as an input DNA control (D). At this point, PCR reactions should be performed using a positive control (see Fig. 3A). Step II: the pools of DNA are separately subjected to ligation-mediated PCR to generate amplicons (E). At this point, PCR reactions should be performed using the same set of primers as above (see Fig. 3B). Step III: the IP amplicon is labeled with Cy5, and an equal amount of the total input amplicon is labeled with Cy3; the two will be combined to hybridize to a CpG island array (see Fig. 2). The IgG amplicon can also be labeled with Cy5 and hybridized with the input amplicon labeled with Cy3 as an additional control (see text).

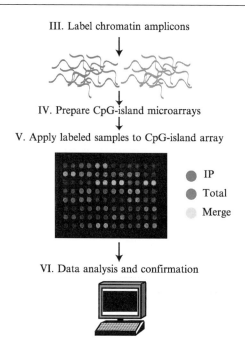

III. Label chromatin amplicons

IV. Prepare CpG-island microarrays

V. Apply labeled samples to CpG-island array

IP

Total

Merge

VI. Data analysis and confirmation

FIG. 2. Preparation and hybridization of CpG arrays (Steps IV–VI). Step IV: CpG islands are spotted onto glass slides and post-processed according to the manufacturer's protocols. Step V: the Cy5- and Cy3-labeled samples are hybridized to the CpG-island microarray in the presence of CoT-1 DNA overnight at 60°. The slides are then washed to remove unbound labeled sample and then dried. Step VI: the fluorescent intensity of each feature is read with an appropriate scanner and data are imported into analysis program for further data analysis. Features with intensities significantly higher in the ChIP-labeled samples relative to the input sample are regarded as putative binding targets. However, if a feature also is positive on the IgG versus input control array, it is removed from the list of targets. (See color insert.)

with a wide variety of different types of DNA binding factors.[24,25] We have utilized many different cell types (epithelial cells, fibroblasts, blood cells, etc.) and have also adapted them for use with mouse and human tissues. Although minor modifications may be required to develop the optimal set of conditions for a particular antibody and/or cell type, our published protocol will provide a good starting point for most transcription factors. This detailed protocol, as well as a discussion of how to prevent potential problems that may be encountered, is also available in our web site

[24] M. J. Oberley and P. J. Farnham, *Meth. Enzymol.* **371,** 579 (2003).
[25] A. S. Weinmann and P. J. Farnham, *Methods* **26,** 37 (2002).

(http://mcardle.oncology.wisc.edu/farnham/). Due to space constraints, we will not reiterate the entire ChIP protocol in this publication. Rather, we have briefly described several recent modifications to the protocol which we recommend be adopted when performing ChIPs (if a step in the protocol is preceded by an exclamation point, this indicates that the particular step has changed substantially from previous published protocols from this group).

Modifications to the Previously Published ChIP Protocol

! a. Although our original protocol describes the standard approach as proceeding to the cell lysis step directly after formaldehyde crosslinking, we now routinely freeze the crosslinked cells and store the pellets at -70°. Having a large number of cross-linked cells frozen in aliquots has proven to be quite useful when troubleshooting the entire assay.

! b. Due to multiple PCR reactions in the subsequent steps, it is critical to note that care must be taken to avoid cross-contamination of samples at all steps of the ChIP protocol. We recommend the use of barrier tips at all times.

! c. Another modification that we have adapted to the ChIP assay that differs from our published protocols is the use of Qiaquick PCR purification columns (Qiagen, 28104) in lieu of the phenol-chloroform extraction step following proteinase K digestion. We have found this shortens the time required to complete the assay, increases the yield of chromatin recovered, and eliminates the need for handling and disposal of the toxic phenol-chloroform mixture. Although we previously recommended using these columns only if the samples would subsequently be used to probe an array, we now suggest that investigators may wish to use this alternative to phenol-chloroform extraction for standard ChIP assays.

! d. The major difference between our most recently published protocol and how we currently design our experiments is concerned with sample size. We originally recommended beginning with approximately 15 standard ChIP reactions and pooling the resultant precipitated chromatin to probe an array. As described in detail later, we have now incorporated an amplification step such that a single ChIP assay will provide sufficient material for subsequent microarray analysis. However, as before, the most critical step of the ChIP assay is to confirm that the assay has been successful. After generating chromatin from the ChIP assay, do not proceed with the creation of amplicons unless the signal obtained in a test PCR experiment shows higher signal in the antibody samples than in the no antibody (or pre-immune) control (see Fig. 3A).

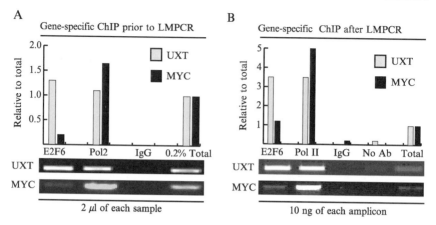

FIG. 3. ChIP controls performed prior to hybridization. (A) A traditional ChIP assay should be performed prior to subjecting the samples to LMPCR. In this example, ChIP was performed on 293 cells according to published protocols[25] using antibodies specific for E2F6 and RNA polymerase II (Pol2); a control was also performed using only a secondary antibody (IgG). After part D of Step I (see Fig. 1), PCR was performed with primers specific to the UXT or MYC promoters and the resulting signal was plotted as a fraction of the signal obtained using 0.2% of the total input chromatin. As expected, E2F6 is highly enriched on the UXT promoter but not the MYC promoter. Pol II occupancy is robustly detected on both promoters, in accordance with the fact that both genes are expressed in 293 cells (data not shown). Neither promoter is enriched by non-specific IgG alone. Therefore, these ChIP samples are appropriate for use in subsequent LMPCR reactions. (B) A PCR reaction should be performed after the samples are subjected to LMPCR. In this case, LMPCR was performed as described in section (Chromatin Amplicon Labeling) on the samples used in panel (A). Ten nanograms of each amplicon (E2F6, Pol II, IgG, and total) was then subjected to PCR with primers specific for the UXT and MYC promoters. The enrichment of UXT sequences seen in panel (A) by the E2F6 antibody is maintained after LMPCR, as the signal seen is at least 3-fold higher than the starting input. The E2F6 signal is not enriched relative to the input amplicon on the MYC promoter. These results indicate the LMPCR procedure did not introduce significant bias into the system and that the samples are appropriate for hybridization to an array.

Generating Chromatin Amplicons

In our previously published studies, we began by pooling between 28 and 50 individual ChIP assays (using 1×10^7 cells/assay) to obtain enough DNA to probe replicate arrays. Although tedious, it is generally possible to obtain this requisite number of cells using immortalized human cell lines. However, if multiple experimental points are to be analyzed or if primary human cells or tissues are to be used, collecting this large number of cells becomes onerous. Therefore, we have adapted a ligation-mediated PCR technique (LMPCR) that is based on the method of Ren et al.[13] to create amplicons of individual IPs. We have found that as little as one-half of one ChIP assay can be efficiently amplified to give amplicons that accurately represent the starting population (see Fig. 3B). The ligation-mediated PCR technique simply involves blunt ending the chromatin, ligating a unidirectional double-stranded oligonucleotide linker, and PCR amplifying the resultant DNA population. Other groups have utilized random priming of the starting chromatin material with comparable results.[12] Recently, a promising amplification method has been developed that utilizes *in vitro* transcription of the chromatin templates to create RNA amplicons.[26] Because this procedure results in a linear rather than exponential amplification, this method has the potential to reduce potential bias introduced by PCR-based amplification schemes. Indeed, this form of sample amplification is commonly used in microarray studies of gene expression where mRNA samples are limited.[27]

To ensure that the LMPCR reaction has not resulted in a loss of difference between the experimental and control samples, we amplify the IP samples as well as a small portion (10–100 ng) of the starting input DNA. This amplified input DNA will be used as a reference on the CpG microarray. We also commonly amplify a non-specific IgG ChIP control reaction to demonstrate that enrichment is due to the specific antibody-epitope interaction. Therefore, three different samples can be prepared that can be used in two different hybridizations comparisons (i.e., IP versus input and IgG versus input). Please note that this LMPCR amplification step is a new addition to our protocol; to call attention to this fact, both Day 1 and Day 2 have been identified with an exclamation point; all steps in these sections are new.

! *Day 1:*

1. The two unidirectional linkers oligoJW102 (5′ gcggtgacccgggagatct-gaattc 3′) and oligoJW103 (5′ gaattcagatc 3′) are annealed by combining 6.7 μl of 100 μM oligoJW102, 6.7 μl of 100 μM oligoJW103, and 86.6 μl H$_2$O. This mixture is boiled for 5 min in a water bath, then allowed to slowly cool to room temperature. The annealed linker can be stored indefinitely at $-20°$.

2. Because sonication of chromatin creates overhanging ends, the chromatin is blunt ended by mixing the chromatin sample derived from an experimental sample, an IgG sample, or 10 ng of input with the following: 11 μl T4 DNA polymerase buffer, 5 μl 10× BSA, 5 μl 2 mM dNTPs, 1 μl T4 DNA polymerase (New England Biolabs, M0203S), and H$_2$O to 110 μl.

[26] B. Bernstein and S. S. Schreiber, *Meth. Ezymol.* **376,** 350 (2004).
[27] C. C. Xiang, M. Chen, O. A. Kozhich, Q. N. Phan, J. M. Inman, Y. Chen, and M. J. Brownstein, *Biotechniques* **34,** 386 (2003).

This mixture is placed at 37° for 45–60 min, and then each sample is purified with a Qiaquick PCR purification kit, according to manufacturers protocol, eluting with 30 μl of elution buffer.

3. Next, double-stranded unidirectional linker from Step 1 is ligated to the blunted chromatin by adding 27 μl of the blunted chromatin, 10.3 μl H$_2$O, 5 μl 10× T4 DNA ligase buffer, 6.7 μl annealed JW102/JW103 linker, and 1.0 μl T4 DNA ligase (New England Biolabs, M0202S). This mixture is placed at 16° overnight.

! *Day 2:*

4. The following morning, each sample is purified with the Qiaquick PCR purification kit, as earlier, eluting with 30 μl of elution buffer.

5. The linker-ligated chromatin is then PCR amplified to create stocks of amplicons from which virtually unlimited amounts of DNA can be generated for use with the CpG-island microarrays. Each PCR reaction contains 5 μl 10× Taq polymerase buffer, 7 μl 2 mM dNTPs, 3 μl MgCl$_2$, 6.5 μl betaine, 2.5 μl oligoJW102 (20 μM), 1 μl Taq (Promega, M1861), and 25 μl of the linker-ligated chromatin.

6. PCR is performed using the following protocol:

$$\left.\begin{array}{l} 55° \text{ for 2 min} \\ 72° \text{ for 5 min} \\ 95° \text{ for 2 min} \end{array}\right\} 1 \text{ cycle}$$

$$\left.\begin{array}{l} 95° \text{ for 0.5 min} \\ 55° \text{ for 0.5 min} \\ 72° \text{ for 1 min} \end{array}\right\} 20 \text{ cycles}$$

72° for 4 min
 4° for ∞

7. Following the PCR, each reaction is purified using the Qiaquick PCR purification kit according to the manufacturer's instructions, eluting in 30 μl of elution buffer.

8. The PCR protocol (Steps 5–7) is repeated until enough amplicon is made for a master stock (1–10 μg), which can be stored at −20°. After the second 20 cycles, 5 μl of the amplicon should be visible on a 1% agarose gel, with fragments ranging from 200 to 600 bp, even if the starting chromatin was sheared to a much larger size. If sufficient chromatin is not produced after the second 20-cycle amplification, another round of PCR may be performed (Steps 5–7). However, always amplify all samples to be compared using the same number of overall cycles to avoid introducing bias between the IP and input sample. The number of times required to repeat the PCR protocols will depend on the quality and quantity of the

starting chromatin and the efficiency of linker ligation. After amplification, carefully quantitate the amplicon with UV spectrophotometry and store at $-20°$.

9. To prepare DNA for labeling, 10 ng of the master stock of the amplicons of the experimental IP, the IgG IP, and the input sample are amplified for 10 cycles, which should give 1–2 μg of DNA. DNA is purified using the Qiaquick PCR purification kit and then quantitated.

10. Using 10 ng of the experimental IP, the IgG IP, and the input amplicons, a standard PCR reaction is performed using primers corresponding to a known positive control (the same as used to verify that the original ChIP assays had been successful). Do not proceed if the positive control is not enriched in the experimental amplicon relative to the input amplicon (see Fig. 3B) or if the IgG amplicon shows non-specific enrichment.

Chromatin Amplicon Labeling

For our hybridization experiments we routinely compare the enrichment of the ChIP sample relative to the input chromatin. This provides several advantages: first, this comparison gives a direct (but non-linear) indication as to how robustly a given sequence is enriched during the ChIP procedure. Second, by comparing the IP labeled with Cy5 to the input chromatin labeled with Cy3, we generate a ratio that helps reduce bias introduced by microarray printing errors. Finally, using two-color fluorescence reduces the number of arrays required for each experiment.

There are two primary ways to label any given chromatin sample: direct versus indirect. Direct labeling involves denaturing the chromatin, random priming, and polymerizing new duplex DNA with a nucleotide analog that has the cyanine fluorescent dye conjugated to it. Because these fluorescent dyes are bulky, the polymerase will incorporate the conjugates at different rates than it would a natural nucleotide. This leads to labeling bias; for example, Cy5 tends to incorporate more readily than does Cy3. This has to be taken into account when co-hybridizing both fluorophores onto the same array, and the reciprocal experiment must be done to control for these differences where the test and control samples are each labeled in separate experiments with each dye. One can also directly label the amplicons by including the cyanine dye conjugated nucleotides directly in the PCR reaction; again because of labeling bias, this method also requires reciprocal experiments in which the samples are labeled with the other dye.

For these reasons, our laboratory primarily utilizes an indirect means of labeling the immunoprecipitated chromatin. Indirect means of labeling incorporates a non-fluorescent nucleotide analog such as aminoallyl dUTP or biotinylated dUTP, followed by conjugation of the fluorophore to the incorporated nucleotide analog. Because these small conjugates are incorporated into both the test and reference amplicons equally, this method helps to eliminate the incorporation biases that occur when directly labeling with Cy5-dUTP and Cy3-dUTP. The cyanine dyes can be purchased as an NHS-ester conjugate (Amersham, RPN5661), which will covalently bind to the aminoallyl nucleotide analogues, or as streptavidin conjugates which form a tight complex with the biotinylated chromatin. We routinely utilize aminoallyl labeling of our chromatin samples, followed by cyanine dye coupling via the protocol outlined later. In addition, the aminoallyl means of amplicon labeling has proved more cost-effective than using one of the direct labeling methods.

Day 1

! 1. Two micrograms of the experimental IP amplicon, 2 μg of the IgG IP amplicon, and 2 μg of the input amplicon are separately vacuum desiccated to complete dryness with heat and then each is resuspended in 33 μl molecular biologic-grade H_2O. Then, 30 μl of 2.5× random primer buffer (Invitrogen, 18094-011) is added to each.

! 2. The chromatin is denatured by holding at 95° for 5 min on a PCR block and then is immediately placed on ice for 3 min. The labeling reaction is initiated by adding 7.5 μl of the 10× dNTP mix (2 mM dATP, 2 mM dCTP, 2 mM dGTP, 0.35 mM dTTP), 1.8 μl of 10 mM aminoallyl dUTP (Sigma, A0410; make a 0.1-M stock in H_2O and store at −20°), and 2.5 μl of high-concentration Klenow (40 U/μl, Invitrogen, 18094-011) to each, and holding at 37° for 2 h.

! 3. The excess nucleotides are removed with the Qiaquick PCR purification kit, eluting with molecular biologic-grade H_2O, as the cyanine dyes that are subsequently coupled are sensitive to Tris base. It is very important at this stage to remove as much of the unincorporated aminoallyl dUTP as possible, because it will bind with the NHS-ester cyanine dyes and reduce the labeling efficiency of the chromatin. The samples are then dried with heat in a speed-vac.

! 4. To couple the NHS-ester cyanine dye to the chromatin, the aminoallyl-labeled chromatin is resuspended in 4.5 μl H_2O, and the cyanine dye is resuspended in 4.5 μl of 0.1 M NaHCO$_3$ (pH 9). The tubes are vortexed and spun several times to ensure that all the samples are dissolved. Next, the chromatin and Cy dye are combined and incubated at room temperature in the dark for 1.5 h, agitating and microfuging the samples every 15 min of the coupling. We typically couple the Cy5 dye to the experimental IP (or the IgG) amplicon, and Cy3 dye to the input reference amplicon, but these could be reversed as well. Dye swapping is not necessary because utilizing aminoallyl dUTP incorporation alleviates cyanine dye bias problems.

5. Thirty five microliters of 100 mM Na-acetate (pH 5.2) are added to lower the pH of the solution to allow the chromatin to bind to the Qiaquick columns, and then H_2O is added to a total of 100 μl for each sample.

6. We have found that removal of the unincorporated dye is quite efficient with Qiaquick PCR columns, and we follow the Qiagen protocol except that we again elute with 50 μl of water instead of the EB buffer provided with the kit. The elution is performed twice and the eluates are combined to give a total volume of 100 μl of H_2O.

7. Determination of the minimum amount of labeled DNA required for any given set of CpG microarrays needs to be empirically determined and this depends greatly on the quality of the microarrays that are produced. If one wants a measure of the success of the labeling reaction, the total dye incorporation can be determined by measuring the absorbance of the entire sample at 650 nm for Cy5 and 550 nm for Cy3. To determine the number of picomoles of incorporated Cy5 or Cy3, the A_{650} reading for Cy5 and A_{550} reading for Cy3 are taken on a spectrophotometer after dye conjugation to the aminoallyl-labeled amplicons. The picomoles incorporation of dye is calculated by using the following formula:

$$\frac{(A_{650})(\mu l \text{ of solution})(\text{dilution factor})}{(0.25 \text{ for Cy5, or } 0.15 \text{ for Cy3})} = \text{pmol of Cy dye incorporated}$$

By determining the picomoles incorporation necessary to give high quality data on the arrays, one can avoid wasting arrays due to poorly labeled samples. At this stage we also measure the amount of labeled DNA. After aminoallyl labeling of the amplicons, one can ascertain the success of the incorporation of the nucleotide analog if the measured amount of the DNA is amplified at least 2- to 5-fold over the starting amount. However, we have found that starting the labeling procedure with 2 μg of labeled chromatin is the least amount of chromatin one can use that will give a good signal/noise ratio on the array. If the arrays are not giving good median intensities per spot (i.e., >1000), one can start with two or three times as much amplicon. Because the cyanine dyes are easily photobleached, it is best to perform all steps under minimal light exposure conditions, both during and after the cyanine dyes have been conjugated. The cyanine dyes are also extremely susceptible to degradation by air humidity, so they should be stored dry with a desiccant at all times. Dry the labeled chromatin with heat in a speed-vac and store it dry at $-20°$ until ready for hybridization.

Preparation of the Arrays

CpG islands, found in or near approximately 50% of human promoters, are identified by three primary characteristics: they are more than 200 bp long, have over 50% GC composition, and retain an observed/expected ratio of CpG dinucleotides greater than 0.6.[28] The CpG library used on the microarrays was originally prepared in the laboratory of Bird[29] and consists of approximately 60,000 independent clones that represent CpG islands that were not hypermethylated in human male genomic DNA. Approximately 8000 clones representing hypomethylated CpG islands were arrayed by the Huang lab[30] onto slides for high-throughput study of DNA methylation changes in human cancer. We have successfully used these arrays to identify genomic binding sites for E2F4, E2F1, and Rb.[21,22] However, these clones were not sequence verified and therefore cultures of each putative positive target clone had to be grown and DNA prepared and sequenced to determine the location on the genome of the identified CpG island. To circumvent this limiting aspect of the CpG arrays, a new version of the arrays is now in production. For these new arrays, the clones are sequenced prior to spotting. This prior knowledge of the identity of each clone will greatly speed the data analysis and target gene identification steps.

The CGI library[29] was obtained from the United Kingdom Human Genome Mapping Project (http://hgmp.mrc.ac.uk/geneservice/reagents/) as frozen bacterial cultures (methylation tolerant *Escherichia coli* strain XL1-Blue MRF′) harboring individual clones in the pGEM-5Zf($-$) vector. The clone set consists of approximately 12,200 independent clones stored in 96-well plates. The entire clone set was amplified using a 96-well format. The cultures were stabbed and transferred into a total volume of 100 μl containing 4 U *Taq* DNA polymerase, 10 mM Tris, pH 8.4, 50 mM KCl, 1.5 mM MgCl$_2$, 800 μM dNTPs, and 0.2 pM each primer. The PCR mixture was held at 94° for 2 min and then cycled 40 times at 94° for 30 s, 53.5° for 30 s, and 72° for 1 min, followed by holding at 4° after the final cycle. From each reaction 2.5 μl were electrophoresed on a 2% agarose gel to check the yield and quality of PCR products. The remaining reaction volume was filtered through Unifilter 800 filter plates (Whatman, 7700-2803) and the purified PCR products were quantified by spectrophotometric measurements at 260 and 280 nm. Then approximately 2 μg of DNA from each clone was transferred into 384-well polyfiltronic plates

---

[28] I. P. Ioshikhes and M. Q. Zhang, *Nat. Genet.* **26,** 61 (2000).
[29] S. H. Cross, J. A. Charlton, X. Nan, and A. P. Bird, *Nat. Genet.* **6,** 236 (1994).
[30] P. S. Yan, C. M. Chen, H. Shi, F. Rahmatpanah, S. H. Wei, and T. H. Huang, *J. Nutr.* **132,** 2430S (2002).

(Whatman, 7701-5101) using Evolution P3 liquid-handling robotics (Perkin Elmer). The PCR products were subsequently dried by overnight centrifugation in vacuum dessicator and resuspended in 10 μl of 3× SSC for array printing. The CpG microarrays were printed from 32 384-well plates containing around 0.2 μg/ul of DNA in 3× SSC in the VersArray ChipWriter Pro System (Bio-Rad, 169-0006) using a 48-pin (SMP-3 pins, ArrayIT, product #SMP3) configuration. Each clone was spotted once on GAPS II–coated slides (Corning, 40003). The printed arrays were processed following manufacturer's protocol (Corning) and kept in a dessicator for later use. (Contact the Microarray Centre, University Health Network, Toronto, ON M5G2C4 at orders@microarrays.ca for information concerning the distribution of the human CpG arrays.)

Hybridization of the Arrays

Day 1:

! 1. The CpG-island microarray is placed in the hybridization chamber probe side up. A 24 × 60 mm Lifterslip (Erie Scientific, 25 × 60I-2-4789) is carefully placed over the area containing the DNA probes to create a uniform space between the array and coverslip.

! 2. The Cy5- and Cy3-labeled chromatin are each resuspended in 15 μl of 1.0 μg/μl CoT-1 DNA (Invitrogen, 15279-011), and the two are then combined. CoT-1 DNA is included to bind to repeat elements in the chromatin which helps to prevent non-specific binding to the array. The tube should be vortexed several times to completely dissolve the chromatin and then spun in a microfuge. Seventy microliters of the hybridization solution (Genisphere Buffer 6, cat#100V600) are added without vortexing to avoid bubble formation. The mixture is denatured at 95° for 2 min and then held at 37° for 30 min to allow the CoT-1 DNA to hybridize with repeats. The mixture is then applied to the microarray by carefully adding the solution to one end of the array and allowing it to wick underneath the Lifterslip. Note that the size of the Lifterslip and the final volume of the hybridization solution may need to be adjusted to match the printed area of the microarray.

! 3. We utilize a dual hybridization chamber (Genemachines, HYB-03) and a water bath for the hybridization to maintain a constant temperature. It is important that the hybridization chamber not directly contact the bottom of the water bath because it will cause inconsistent heat transfer. We typically use a small piece of foam to separate the chamber from the metal bottom of the water bath. The arrays are hybridized overnight at 60° for up to 18 h.

Day 2:

11. To remove the coverslip, the microarray(s) are inverted in a glass dish filled with 1× SSC and 0.1% SDS preheated to 50°. If no drying has occurred overnight, the Lifterslip will immediately slide off.

12. The arrays are then agitated in 1× SSC, 0.1% SDS at 50° for 5 min, followed by agitation in 1× SSC, 0.1% SDS at room temperature for 5 min. Finally, the arrays are washed with 0.2× SSC for 5 min at room temperature to remove residual SDS (which autofluoresces), and then dried by centrifugation at 600 rpm for 5 min in a 50-ml conical tube.

13. The arrays are scanned immediately using an Axon 4000B scanner (Axon Instruments).

Analysis

The identification of positive clones is a multi-step procedure, first requiring normalization of the Cy5 and Cy3 channels on the array because of potential labeling efficiency differences, followed by clone selection, confirmation, and ultimately, functional analysis.

1. The hybridized microarrays are analyzed using the Genepix Pro 4.0 (Axon Instruments) software package. This provides a set of raw values for each feature on each array. To identify clones that are selectively enriched during IP relative to the starting population, one must first normalize the Cy5 and Cy3 channels on the array. To do so, we make the assumption that the vast majority of loci will not be enriched in the IP relative to total; thus, we can normalize across the entire array, by taking the ratio of the medians for all features and normalizing them to unity.

2. After normalization, a ratio is generated that is the intensity in the Cy5 channel minus background divided by the intensity in the Cy3 channel minus background. Features in areas of the array that are obviously blemished are manually flagged and excluded from the data set, as are features with signal that have low intensities (<500). The ratio is generated for all features that meet this criterion, and features with ratios above 2 are isolated for further analysis.

3. Because of the complexity of the mammalian genome and the relatively large number of steps in this procedure, false positives are inevitable. Great care must be taken to reduce these errors to prevent further wasting of time and resources following up on potential artifacts. It is imperative to reduce the error rate by repeating the entire procedure at least three times, starting with independent chromatin IPs from a different batch of crosslinked cells each time (see Fig. 4B).

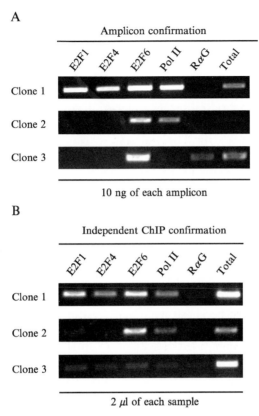

FIG. 4. Confirmation of putative targets. (A) Gene-specific PCR should be performed to verify that hybridization signals resulting from the microarray accurately represent the relative amounts of sequences present in the amplicons. In this case, 10 ng of each amplicon was subjected to PCR with promoter-specific primers from genes that were positive on the CpG-island array analysis. In our hands >90% of the genes that are positive on the arrays demonstrate at least 2-fold enrichment in the experimental IP amplicon relative to input amplicon. (B) Reproducible identification of the putative clones should be performed using traditional ChIP analysis in an independent assay.[25] In this example, 2 μl of each IP from a second ChIP experiment was PCR amplified using the same gene-specific primers indicated in panel (A) and the signal was compared to 0.2% of the input chromatin. Clones 1 and 2 show the same pattern in the original amplicon and the subsequent, independent ChIP assay. However, clone 3 is an example of a clone that appears to be robustly enriched in panel (A) but does not appear to be bound by E2F6 in subsequent ChIP assays. This highlights the fact that multiple, independent analyses are required due to inherent noise generated during ChIP. The independent validation rate varies from antibody to antibody but can be increased by utilizing amplicons from three independent ChIPs, performing three independent hybridizations, and discarding any promoters from the data set that are not repeatedly robustly enriched relative to the input amplicon.

4. To ensure that positive signals are not simply a consequence of non-specific enrichment of certain loci by the ChIP procedure, a comparison of the IgG ChIP sample to input DNA must also be performed. CpG islands that are non-specifically enriched as a result of the procedure are then removed from the putative target list.

5. For an initial test to demonstrate that the positive signals derived from microarray hybridization accurately reflect the relative abundance of sequences present in the labeled chromatin, the amplicons should be subjected to gene-specific PCR analysis starting with 10 ng of LMPCR experimental IP amplicon, 10 ng of the LMPCR IgG amplicon, and 10 ng of the LMPCR input amplicon. This should be done for a selected number (10–20) of the CpG islands that appear enriched after data analysis. This gene-specific PCR analysis will indicate which CpG islands are enriched significantly over input (see Fig. 4A), giving an empirical measure of the hybridization accuracy. In our initial experiments, we employed arrays in which the clones corresponding to the CpG islands were not sequenced prior to spotting the arrays. Therefore, to perform the test PCRs, the clones had to be grown as overnight cultures, plasmid DNA prepared, and then sequenced using vector primers. Gene-specific primers for the putative positives could then be constructed and used in the confirmation assays. However, future experiments will be much easier due to the fact that the new arrays will be composed of sequence-verified clones.

6. Reproducible binding is demonstrated using an independent standard ChIP assay in which 2 μl of an individual IP is examined via PCR with gene-specific primers and compared to an IgG IP and a fraction (0.2%) of the input chromatin (see Fig. 4B). This will identify false-positive clones that may have been enriched independently of the antibody used; these clones should be discarded from the putative target list. Ultimately, an ideal experimental validation can be derived from cells with specific gene deletions and their wild-type counterparts to show the PCR signals are the result of the desired specific antibody-epitope interaction. For example, Wells et al.[31] identified several E2F-1–specific genes and validated them using ChIPs in wild-type and E2F-1–knockout mouse embryo fibroblasts.

7. The final step is to determine the biological significance of the experimentally verified binding of the factor of interest to the identified genomic site. For example, RNAi represents an experimental system where one can knock-down the expression of a particular gene in a transient way.[32]

[31] J. Wells, C. R. Graveel, S. M. Bartley, S. J. Madore, and P. J. Farnham, *Proc. Natl. Acad. Sci. USA* **99,** 3890 (2002).
[32] R. Cao, L. Wang, H. Wang, L. Xia, H. Erdjument-Bromage, P. Tempst, R. S. Jones, and Y. Zhang, *Science* **298,** 1039 (2002).

This system allows one to perturb a normal biological system without some of the concerns with traditional gene-ablation technology where the knock-out of a specific gene may be compensated for during development of the animal by upregulation of other family members. One could use RNAi to specifically knock-down the expression of the DNA binding factor of interest and then use RNase-protection, Northern, or traditional microarray gene expression analysis to determine whether the lack of binding to a putative target has a functional consequence. Shi *et al.*[33] have developed protocols that allow the CpG-island microarray to be probed with mRNA that utilizes RNA ligase-mediated full-length cDNA synthesis. With this protocol, it will be possible to examine the effect of DNA binding factor knock-down on the resultant mRNA expression of the identified target genes using the same CpG-island microarrays as used for the target gene identification.

Conclusions

Studies to date employing the ChIP-chip approach using human cells have been limited by the need to refine the methodology used for similar studies of less complex eukaryotes. However, great strides have been made recently in overcoming issues related to sample size and genomic complexity. A major problem in studying human transcription factors is the fact that no appropriate microarrays have been widely available that can be used to study intergenic DNA; that is, all commercially available microarrays have allowed only the detection of transcribed sequences. Although individual investigators have begun spotting promoter-specific[13] or promoter-enriched[30] arrays, individual labs cannot serve as a source for arrays for all the investigators who wish to employ the ChIP-chip approach. Unfortunately, commercial microarray companies have not yet entered into the production of human promoter arrays. However, now that the feasibility of using such arrays has been demonstrated, greater efforts to provide investigators with a comprehensive set of human intergenic DNA chips must be made. The Encode (Encyclopedia of DNA Elements) project, which has the long-term goal of identifying all functional elements in the human genome, has chosen 30 Mb (approximately 1% of the human genome) for preliminary analysis (see http://www.genome.gov/ Pages/ Research/ENCODE/). It can be hoped that the technologies and products arising from such projects will allow much greater access of all investigators to the required reagents needed for a thorough analysis of the regulatory elements encoded in the human genome.

[33] H. Shi, P. S. Yan, C. M. Chen, F. Rahmatpanah, C. Lofton-Day, C. W. Caldwell, and T. H. Huang, *Cancer Res.* **62**, 3214 (2002).

Acknowledgments

We would like to thank P. S. Yan and T. Huang for their generous gift of the CpG arrays. M.J.O would like to thank A. Kirmizis for help adapting the LMPCR protocol and for stimulating discussion. The Farnham lab would also like to thank A. Kirmizis and A. S. Weinmann for critical readings of this manuscript. The Microarray Centre, which would like to thank J. Woodgett and Neil Winegardin for technical input, was supported by ORDCF and Genome Canada. This work was supported in part by Public Health Service grants CA22484, CA09135, and CA45240.

[22] Chromatin Immunoprecipitation in the Analysis of Large Chromatin Domains Across Murine Antigen Receptor Loci

By David N. Ciccone, Katrina B. Morshead, and Marjorie A. Oettinger

In the nucleus of eukaryotic cells, DNA is wrapped around nucleosomes forming a regulatable polymer called chromatin. Chromatin structure has been shown to influence many processes, including transcription, replication, recombination, DNA repair, and chromosome segregation. The intimate association between DNA and histone proteins of the nucleosome can be regulated by histone-modifying enzymes and chromatin remodeling machines to permit or restrict the occupancy of specific regulatory sequences by DNA binding proteins. The chromatin fiber is a very dynamic structure that must be opened and subsequently closed in a strict temporal and spatial manner to ensure the proper gene expression pattern and the developmentally appropriate accessibility of certain chromatin domains.[1] Therefore, mapping the distribution of covalently modified histones, regulatory proteins, or DNA binding factors relative to specific sequences within a genomic locus will ultimately provide valuable insight into the molecular mechanisms underlying localized chromatin alterations as well as developmentally regulated, large-scale reorganizations of chromatin structure into active and inactive domains.

Chromatin immunoprecipitation (ChIP) is a powerful technique used to analyze protein:DNA interactions in a native chromatin context. This versatile method has been used with cell types from a wide range of organisms to determine the localization of specific histone modifications with respect to promoter regions and large chromosomal domains.[2,3] In

---

[1] W. Fischle, Y. Wang, and C. D. Allis, *Curr. Opin. Cell Biol.* **15**, 172 (2003).
[2] M. H. Kuo and C. D. Allis, *Methods* **19**, 425 (1999).
[3] V. Orlando, *Trends Biochem. Sci.* **25**, 99 (2000).

conjunction with previously favored procedures such as nuclease sensitivity mapping, ChIP also provides an additional tool to search for specific DNA regulatory elements. For example, at the 47-kb silent mating-type locus of the fission yeast *Saccharomyces pombe,* the euchromatin region was found to be associated with high levels of histone H3–lysine 4 (Lys4) methylation. Alternatively, the heterochromatin interval, located completely within the euchromatic domain and defined by two flanking boundary elements, is associated with histone H3 (Lys9) methylation and Swi6 (yeast HP1 homologue) localization.[4] In a separate study, an inverse correlation between acetylation and methylation of histone H3 (Lys9) was observed during the developmental changes associated with erythropoiesis at the 53-kb chicken β-globin locus.[5]

The murine immunoglobulin heavy chain (IgH) locus represents another example of a large, highly regulated, chromosomal domain (see Fig. 3A for a depiction of the IgH locus). The IgH locus consists of many variable *(V)* region gene segments located upstream of diversity *(D)* and joining *(J)* gene segments spread across a genetic domain that exceeds a megabase of DNA. Coordination of the series of DNA rearrangement events required to assemble an Ig receptor from component *V, D,* and *J* segments is a process called *V(D)J* recombination.[6] This recombination process generates a diverse repertoire of Ig receptors through multiple combinations of possible joining events between various germline gene segments. A vast amount of genetic and biochemical evidence suggests that *V(D)J* joining is developmentally regulated by specific alterations in chromatin structure that render a particular locus, or certain regions within a locus, accessible or inaccessible to the recombinase machinery.

Here we present a ChIP protocol used to identify the histone modification patterns at this locus. Data generated from ChIP experiments have provided valuable insight into how specific gene segments within a locus are made accessible to the recombination machinery while simultaneously restricting access to other gene segments during lymphocyte development. The ChIP method involves three basic steps:

a. *In vivo* formaldehyde crosslinking of intact cells
b. Micrococcal nuclease (MNase) digestion and sonication of the chromatin to generate a pool of small chromatin fragments
c. Selective immunoprecipitation (IP) of protein:DNA complexes from the chromatin pool with specific antibodies

[4] K. Noma, C. D. Allis, and S. I. Grewal, *Science* **293,** 1150 (2001).
[5] M. D. Litt, M. Simpson, M. Gaszner, C. D. Allis, and G. Felsenfeld, *Science* **293,** 2453 (2001).
[6] C. H. Bassing, W. Swat, and F. W. Alt, *Cell* **109** (Suppl.), S45 (2002).

This chapter is devoted to a full, detailed description of this ChIP method, including the subsequent analysis of the immunoprecipitated DNA.

Chromatin Immunoprecipitation

The basic procedures used for ChIP have been developed by and represent a fusion of contributions made by a large number of researchers. A handful of these pioneering studies are outlined here: In 1981, whole cell formaldehyde fixation was shown to crosslink and preserve chromatin structure[7,8]; a few years later, antisera specific for RNA polymerase[9] and topoisomerase I[10] were used to precipitate UV crosslinked protein:DNA complexes; in 1988, histone antibodies were used in ChIP experiments[11] and a correlation between core histone acetylation and transcriptionally poised chromatin was observed.[12] Recently, with the development and subsequent commercialization of highly specific antibodies recognizing various histone modifications, ChIP has become a very popular method for analyzing chromatin structure *in vivo*. Although many variations of the basic ChIP method are currently used, the backbone of our protocol relies largely on the method reported by Kuras and Struhl[13] and is described in detail.

Procedure

1. The number of cells harvested for a single ChIP can vary tremendously depending on the total number of DNA sites being analyzed. The DNA we recovered from a single ChIP experiment was used to examine the association of various histone modifications at approximately 30 loci across a large genomic region. However, the total number of cells used per ChIP can be reduced if the analysis involves only a few DNA sites. A transformed recombinase-deficient Pro B cell line was chosen as the most practical line for examining the chromatin structure of the murine IgH locus. However, this ChIP method has also been successfully applied to additional suspension cell lines as well as two non-adherent fibroblast cell lines.

---

[7] V. Jackson and R. Chalkley, *Proc. Natl. Acad. Sci. USA* **78,** 6081 (1981).

[8] V. Jackson and R. Chalkley, *Cell* **23,** 121 (1981).

[9] D. S. Gilmour and J. T. Lis, *Proc. Natl. Acad. Sci. USA* **81,** 4275 (1984).

[10] D. S. Gilmour, G. Pflugfelder, J. C. Wang, and J. T. Lis, *Cell* **44,** 401 (1986).

[11] M. J. Solomon, P. L. Larsen, and A. Varshavsky, *Cell* **53,** 937 (1988).

[12] T. R. Hebbes, A. W. Thorne, and C. Crane-Robinson, *EMBO J.* **7,** 1395 (1988).

[13] L. Kuras and K. Struhl, *Nature* **399,** 609 (1999).

Use approximately 2×10^8 cells for a single IP. To assess the efficiency of the ChIP experiment, a Control IP lacking an antibody is performed in parallel with one containing the specific antibody of interest. Therefore, for a properly controlled experiment, approximately 4×10^8 cells are needed. From a larger cell preparation, such as 8×10^8 cells, three independent ChIP experiments can be performed with a single Control IP. Harvest the cell population (collect adherent cells by trypsinization and wash the cell pellet twice in ice-cold 1× PBS) and pellet at 640g for 4 min. Resuspend the cell pellet thoroughly in 20 ml of culture media.

2. Fix the cells by adding formaldehyde to a final concentration of 1% directly to the cell suspension. Allow the fixation reaction to proceed for 10 min in a fume hood at room temperature. Invert the suspension every few minutes to mix the cells. Add 1/20 the volume of 2.5 M glycine to stop the fixation reaction, and incubate for an additional 5 min at room temperature. Transfer the cell suspension to an ice bucket.

3. Pellet the cells at 640g for 4 min (all centrifugations should be performed at 4° until otherwise noted). Discard the supernatant into an appropriate receptacle designated for formaldehyde waste. Resuspend the cell pellet in 10 ml of 1× PBS and collect the cells by centrifugation. Discard the supernatant and repeat the 1× PBS wash to remove all traces of formaldehyde from the cell suspension

4. Isolate nuclei from fixed cells by washing the cell pellet three times with 10 ml of cold lysis buffer (10 mM Tris-HCl, pH 7.5, 10 mM NaCl, 3 mM MgCl$_2$, and 0.5% NP-40). After each wash, pellet the cells at 300g for 5 min.

5. Resuspend the pelleted nuclei in 3 ml of MNase reaction buffer (10 mM Tris-HCl, pH 7.5, 10 mM NaCl, 3 mM MgCl$_2$, 1 mM CaCl$_2$, 4% NP-40, and 1 mM PMSF added just prior to use). Add 50 units of MNase (Sigma) to the nuclei suspension, mix, and incubate for 10 min at 37°. The amount of enzyme and the length of incubation should be specifically tailored for optimum digestion when using different cell types or cell numbers. Add EGTA to a final concentration of 3 mM to stop the digestion reaction. At this point, save a 10-μl aliquot and store at −20°. This aliquot should be decrosslinked (described later) at the end of this procedure and then run out on an agarose gel, providing a clear indication of the efficiency of MNase digestion.

6. Add 60 μl of 50 mM PMSF, 3 μl of 2 mg/ml aprotinin, 3 μl of 2 mg/ml leupeptin, 3 μl of 2 mg/ml pepstatin, 300 μl of 10% SDS, and 120 μl of 5 M NaCl to the MNase-digested nuclei. Sonicate the nuclei to lyse the nuclear membrane and further shear the chromatin into a sample containing primarily 0.2–2 kb DNA fragments. The settings and conditions used during sonication must be determined empirically and tend to vary

significantly between different sonicators. (*Note:* the sonication conditions outlined here have been used successfully with different murine cell types.) We use a Heat Systems-Ultrasonics sonicator (model W-375) set at power level 4 and 50% duty cycle. Place the nuclei on ice and sonicate the sample for 1 min, followed by a 1-min incubation on ice. Repeat the sonication/ incubation step. Finally, sonicate the sample one last time for 30 s. Split the sonicated sample equally into microcentrifuge tubes and spin at maximum speed for 10 min at $4°$ to remove all traces of cellular and nuclear debris prior to IP. This spin can be repeated if some cellular remnants remain in the supernatant. Again, save a 10-μl aliquot and store at $-20°$. This aliquot should also be decrosslinked (described later) and run out on an agarose gel to determine the sonication efficiency and ultimately the size range of the DNA fragments in the chromatin sample used for IP. At this point the chromatin can be stored at $-80°$ and the IP can be performed at a later date.

7. If the sample was stored at $-80°$, spin down any additional precipitates at maximum speed for 5 min at $4°$. Transfer the supernatant to a 50-ml conical tube. Add IP dilution buffer (20 mM Tris-HCl, pH 8.0, 2 mM EDTA, 1% Triton X-100, 150 mM NaCl, 1 mM PMSF, 2 μg/ml aprotinin, 2 μg/ml leupeptin, and 2 μg/ml pepstatin) to the tube bringing the volume up to 15 ml (remember to always add protease inhibitors to all appropriate buffers just before use). The chromatin sample is now pre-cleared with Protein A Sepharose 4B resin; however, alternate resins may be needed depending on antibody isotype. Since the resin will be used to pull down antibody-bound chromatin fragments from the IP solution, it is helpful to pre-clear the sample with resin, removing any chromatin fragments that non-specifically bind to the resin prior to IP. Equilibrate 50 μl resin in IP dilution buffer and add to the diluted chromatin sample. Pre-clear the chromatin at room temperature for 15 min while rocking. Pellet the resin at 2000g for 2 min. Transfer the supernatant to a new 50-ml conical tube and discard the resin. Save a 50-μl Input aliquot and store at $-20°$. This aliquot represents a fraction of the total chromatin sample used in the IP.

8. Split the chromatin sample into an appropriate number of 50-ml conical tubes and bring the final volume of each tube up to 10 ml with IP dilution buffer. One of these tubes will be used for the Control IP; therefore, no antibody should be added. To the other tube or tubes add the appropriate amount of specific antibody, which must also be determined empirically by comparing the amount of DNA recovered (see DNA Quantitation later) from a Control IP versus a specific IP and adjusting the antibody concentration accordingly (for antibodies specific to covalently modified histones, we typically use dilutions ranging from 1:500 to 1:2000). The optimum antibody dilution used for ChIP experiments can vary

greatly depending on the specificity and target of the antibody. It is generally accepted that polyclonal antibodies are better suited for ChIPs than monoclonal antibodies, presumably due to the wide range of epitopes, some of which may be masked in a native chromatin context, recognized by the polyclonal antibody pool. Incubate at room temperature for 3 h while rocking. Add 50 μl of pre-equilibrated resin to all tubes and continue rocking at room temperature for an additional 3 h.

9. Collect the resin by spinning at 2000g for 2 min (these conditions should be used for all subsequent spins unless indicated otherwise). Discard the supernatant. Resuspend the resin in 1 ml of IP dilution buffer and transfer the solution to a microcentrifuge tube (or a dolphin-nosed tube, Corning Cat. No. 3213). Spin down the resin and discard the supernatant. The subsequent steps are designed to disrupt non-specific binding that may have occurred during the IP by washing the resin in a series of buffers with increasing salt concentrations. Add 1 ml of ChIP wash buffer I (20 mM Tris-HCl, pH 8.0, 2 mM EDTA, 1% Triton X-100, 0.1% SDS, 150 mM NaCl, 1 mM PMSF), collect the resin by centrifugation and discard the supernatant. Add 1.5 ml of ChIP wash buffer I to the resin and rock at room temperature for 5 min. Spin down the resin and discard the supernatant. Add 1.5 ml of ChIP wash buffer II (20 mM Tris-HCl, pH 8.0, 2 mM EDTA, 1% Triton X-100, 0.1% SDS, 500 mM NaCl, 1 mM PMSF) to the resin and rock at room temperature for 5 min. Collect the resin by spinning and discard the supernatant. Add 1.5 ml of ChIP wash buffer III (10 mM Tris-HCl, pH 8.0, 1 mM EDTA, 0.25 M LiCl, 1% NP-40, 1% deoxycholic acid (sodium salt)) to the resin and rock at room temperature for 5 min. Spin down the resin and discard the supernatant. Add 1 ml of TE buffer (10 mM Tris-HCl, pH 8.0, 1 mM EDTA) to the resin and rock at room temperature for 1 min. Collect the resin by spinning and discard the supernatant. Repeat the last wash with TE buffer, spin, and discard the supernatant.

10. Add 200 μl of elution buffer (25 mM Tris-HCl, pH 7.5, 10 mM EDTA, 0.5% SDS) to the extensively washed resins from each sample, and incubate for 15 min at 65° to dissociate the resin from the antibody-chromatin complexes. Collect the resin by spinning at 5000g for 2 min and transfer the supernatant from each sample to a new tube. Resuspend the resin in 100 μl of elution buffer, pellet the resin at 5000g for 2 min, and again transfer the supernatants to the new tubes.

11. Remove the formaldehyde crosslinks by adjusting the volume of the post-MNase, post-sonication, and Input aliquots to 300 μl with elution buffer and add SDS, as necessary to each sample for a final concentration of 0.5%. Add Pronase (Roche Applied Science) to all samples to a final concentration of 1.5 μg/μl. Incubate the samples at 42° for 1 h and then at

$65°$ for 5 h. Add LiCl to the samples to a final concentration of 0.8 M. To remove all traces of protein, extract the samples in an equal volume of phenol/chloroform/isoamyl alcohol, followed by an extraction in an equal volume of chloroform/isoamyl alcohol to remove any residual phenol. Precipitate the DNA by adding 20 μg of glycogen and 1 ml of 100% ethanol to each sample, vortex and incubate at $-80°$ for at least 1 h. Spin at maximum speed for 15–30 min in a microcentrifuge. Wash the DNA pellet in 1 ml of 70% ethanol. Allow the pellet to air-dry completely, resuspend the DNA in an appropriate volume of TE buffer, and store at $-20°$.

12. At this point the DNA precipitated from the post-MNase and post-sonication aliquots should be run out on an agarose gel to assess the efficiencies of both treatments. To demonstrate the role of MNase in the outlined procedure, Fig. 1 shows the DNA collected after sonication from a ChIP experiment utilizing MNase (+MNase, lane 1) and ChIP in which the enzyme was omitted, relying solely on sonication (−MNase, lane 2). In lane 1 there is a ladder consisting of approximately 200-bp increments characteristic of efficient MNase digestion at nucleosome boundaries. Since the length of the DNA fragments seen in the post-sonication aliquot is a direct indication of the length of the chromatin fragments used in the IP step, it is extremely important to the overall resolution of the ChIP that the size of the DNA fragments reside between 200 bp and 2 kb (see Fig. 1, lane 1). If this range is greatly extended beyond 2 kb, then the sonicator settings and conditions should be adjusted. In the absence of MNase (see Fig. 1, lane 2), some smaller DNA fragments are generated through sonication, but a large percentage of the DNA remains as high molecular weight fragments.

DNA Quantitation

A clear indication of the overall efficiency of the ChIP is assessed by comparing the amount of DNA present in the IP sample to the amount present in the Control IP. However, the amount of DNA recovered in both of these samples is much too low for accurate quantitation using standard procedures such as UV absorbance. The PicoGreen dsDNA Quantitation Assay from Molecular Probes provides a rapid and sensitive technique for quantitating DNA ranging from 25 pg/ml to 1 μg/ml. Picogreen is a cyanine dye that is essentially non-fluorescent when free in solution, but exhibits a greater than 1000-fold enrichment in fluorescence upon binding to DNA. The amount of DNA typically recovered in the Control IP is approximately 20 ng. This amount should not vary drastically between independent ChIP experiments; however, the amount of DNA recovered

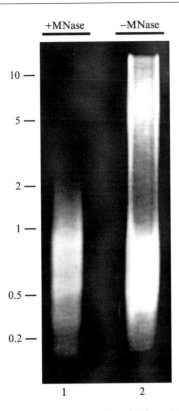

Fig. 1. The efficiency of MNase digestion and sonication. A representative example of post-sonication aliquots taken during ChIP experiments performed in the presence of MNase (+MNase) and without using MNase (−MNase). The DNA collected from these aliquots were electrophoresed into a 1% agarose gel, stained with ethidium bromide, and photographed under UV radiation. Lane 1 contains 15 μl of DNA from a 50-μl +MNase sample and lane 2 contains the same amount of DNA from a 50-μl −MNase sample. DNA size markers in kilobases are indicated to the left of the gel.

in the IP sample, although enriched by at least a factor of 2 as compared to the Control IP, can vary greatly depending on the antibody. For example, the anti-acetyl histone H3 (Lys9 and 14) antibody (Upstate Biotechnology) immunoprecipitates approximately 90 ng of DNA, while the anti-methyl histone H3 (Lys9) antibody (Upstate Biotechnology) yields approximately 150 ng of DNA from a typical ChIP experiment with approximately 200 million cells. Indeed, some antibodies that are used at low concentrations immunoprecipitate much more DNA as compared to other antibodies used at higher overall concentrations.

Real-Time PCR Primer Design

A majority of the antigen receptor loci of the mouse are sequenced and available on-line. To design primers suitable for real-time PCR analysis from these large chromosomal regions, we utilized the computer program Oligo 4.0 (National Biosciences). Oligo 4.0 is designed to search for and analyze oligonucleotides for PCR, sequencing, and hybridization applications. The software is very good at analyzing sequences for primer design, calculating melting temperatures, and identifying problems, such as hairpins, primer-dimers, temperature-mismatched pairs, etc. We follow four simple rules when searching for primers to be used in real-time PCR analysis.

1. *Avoid large discrepancies in melting temperatures between template and primers.* When the difference between template and primer melting temperatures exceeds 35°, then the PCR yield will be less than optimal.

2. *Do not let the difference in melting temperatures between each primer exceed 2°.* Highly disparate melting temperatures between the primers may cause the less stable primer to decrease the annealing temperature, thus increasing the likelihood of non-specific priming.

3. *Avoid primer dimers with highly negative free energies (ΔG) less than −5.0 kcal/mol.* A DNA duplex is stable when the ΔG is negative. If the free energy of a primer dimer exceeds −5.0 kcal/mol, the resulting PCR reaction is less than optimal.

4. *Avoid primers with stable hairpin loops.* Priming is extremely inefficient if the hairpin loop melting temperature approaches the annealing temperature.

Once primers have been designed to meet all aspects of the rules stated earlier, the amplification potential of the primers should subsequently be tested in PCR to ensure the accumulation of a single product of the appropriate size. Due to the highly repetitive nature of the murine IgH locus, the accumulation of a specific product is not guaranteed even though all of the specifications listed earlier for proper primer design have been fulfilled. In a PCR tube add 10 pmol of each primer, 50 ng of sheared genomic DNA (template), 100 μM dNTPs, 1× PCR buffer, and 0.5 units HotStarTaq DNA polymerase (Qiagen). Each reaction is subsequently spiked with 1 μCi α^{32}P-dNTP, which dramatically increases the sensitivity of detecting non-specific products and misprimings as compared to traditional ethidium bromide staining. The PCR reactions can be run out on an acrylamide gel and subjected to autoradiography. The ability of the primers to amplify a single product of the appropriate size is exceedingly important for SYBR Green real-time PCR as is discussed later. Another important feature of the reaction is the intensity of the product as observed by autoradiography. Through the course of evaluating hundreds of primer pairs in the linear range of amplification we have noticed that even though a particular primer pair amplifies a single product, the intensity may be exceedingly low when compared to other products in the same linear range, directly correlating with a low rate of amplification (discussed later), and thus unable to generate useful real-time PCR data.

Real-Time Quantitative PCR Analysis

Real-time quantitative PCR analysis is a reliable method used to measure the accumulation of products generated during each cycle of the PCR reaction.[14,15] These products are directly proportional to the amount of template DNA present in each reaction prior to the onset of amplification. Real-time PCR has become very widely used in recent years, partially due to the technological advances made in instrumentation and fluorescent reagents, which have caused market prices to drop, increasing the number of researchers with access to this technology. This method of PCR analysis has many applications including, but not limited to, measuring DNA copy number and mRNA expression levels. Real-time PCR is appealing because it is much more rapid and reliable than endpoint quantitation of PCR products as bands on a gel. Inhibitors of the polymerase chain reaction, inherent to the reaction itself, include template DNA concentration, limiting PCR reagents, and the accumulation of pyrophosphate molecules. All of these conditions act in concert to slow the amplification process and eventually cause the reaction to enter the plateau phase, in which the generation of the template no longer proceeds at an exponential rate. Real-time PCR is our method of choice for analyzing the DNA recovered from a ChIP experiment because of its ability to measure PCR product formation in "real-time," at a point in which amplification is considered exponential. Only during the exponential phase of the PCR reaction it is possible to extrapolate back to the starting material to determine the original amount of specific DNA template added. In addition, it is during the exponential accumulation of product that a fluorescence signal is detected (a description of the most common fluorescent reagents used in real-time PCR analysis is provided later). The signal corresponds to the accumulation of specific PCR products and is immediately detected and recorded in real time by the

[14] A. Giulietti, L. Overbergh, D. Valckx, B. Decallonne, R. Bouillon, and C. Mathieu, *Methods* **25**, 386 (2001).
[15] D. G. Ginzinger, *Exp. Hematol.* **30**, 503 (2002).

same instrument that regulates the steps involved in thermocycling (e.g., iCycler iQ, Bio-Rad; ABI7900, Applied Biosystems).

The three most commonly used fluorescent reagents for detecting PCR amplification in real time are TaqMan probes, molecular beacons, and SYBR Green I dye. A TaqMan probe is designed to hybridize to the template amplified between two standard PCR primers. The probe is conjugated to a fluorescent dye at its 5' end and a quencher at the 3' end. The fluorescent tag is quenched by fluorescence resonance energy transfer (FRET) and does not emit a signal while the probe is intact. However, during amplification the 5' fluorescent dye is cleaved and released by the 5' to 3' exonuclease activity of Taq polymerase. Once released from the quenching ability of the intact probe, the fluorogenic dye is able to emit its signal, which is recorded during the exponential phase of the reaction. Molecular beacons are very similar to TaqMan probes in that they utilize the ability of a 3' quencher to suppress the fluorescent signal normally emitted by a 5' dye by FRET. However, the quenching effect is achieved by introducing a hairpin loop in the probe. When the beacon hybridizes to a complementary nucleotide sequence, the template, the conformational change resulting from the hairpin loop structure being converted to a linear DNA molecule provides enough distance between the dye and quencher to allow a fluorescent signal to be emitted and recorded. The third reagent used in real-time PCR is SYBR Green I dye, which differs from the above-mentioned reagents in that there is no need for a modified oligo or fluorogenic hybridization probe. The SYBR Green I intercalating dye has an undetectable fluorescence when free in solution but emits a signal on binding to dsDNA. We have chosen this dye because it can be used in combination with any pair of primers, specific to any desired target sequence, thus making it extremely versatile and the least expensive choice. However, this dye cannot discriminate between true DNA template and non-specific products such as primer dimers and misprimings, unlike TaqMan and molecular beacons, which recognize and emit a fluorogenic signal only when hybridized to the template DNA. This is why primer design for SYBR Green–mediated real-time PCR is so important.

In our studies, differences in the amount of specific DNA sequences enriched in the IP sample relative to the Input sample are determined by SYBR Green real-time PCR analysis using the iCycler iQ from Bio-Rad.

Procedure and Calculations

1. Each PCR reaction for both the IP sample and the Input sample is to be done in triplicate. The PCR cocktail consists of 2 ng of Input or IP

DNA, 10 pmol of each primer, and SYBR Green PCR Master Mix (Applied Biosystems). Although a 50-μl minimum reaction volume is suggested for the Bio-Rad iCycler, we have consistently obtained reproducible results with a 20-μl reaction.

2. During amplification, the fluorescence signal intensity is plotted as a function of the cycle number (see Fig. 2). After the reaction has finished, a fluorescence signal cycle threshold (Ct) level must be set. The threshold level should reside in the exponential phase of the real-time PCR reaction and represents the point at which the Input sample can now be compared to the IP sample. The number of cycles of each PCR reaction required to reach the set threshold level is known as the Ct. Cycle threshold values are directly proportional to the amount of DNA template present in each reaction prior to amplification and provide the basis for calculating relative fold enrichment values. The Ct values for each reaction within the triplicate are averaged according to the following rule; if a Ct value within the triplicate fluctuates greater than or less than 0.5 cycles in either direction from the other two values in the triplicate (likely due to pippetting errors), then this data point is not considered when calculating the overall cycle threshold for that particular target sequence. Next, the rate of amplification (R) should be calculated for each PCR reaction. Due to the exponential nature of the PCR the theoretical optimum rate of amplification is 2.0, but to make accurate and meaningful comparisons between primer pairs, the differences in amplification rates should be calculated. PCR primers that have been designed according to the rules stated earlier typically have R values ranging between 1.85 and 2.0. These values remain relatively constant for a particular primer pair, although on rare occasions, variability in the rate is seen, which is likely due to evaporation of a particular reaction from the 96-well PCR plate. The R values for each reaction within a triplicate are averaged, except for those individual R values that are less than 1.75 or greater than 2.25. If an R value falls outside this bracket, then the data point is not used in the calculation of overall primer amplification rate. The fold enrichment (see mock calculation illustrated in Fig. 2) of a specific target sequence in the IP sample relative to the Input sample is calculated using the following equation:

$$\text{Fold enrichment} = R^{(Ct_{Input} - Ct_{IP})}.$$

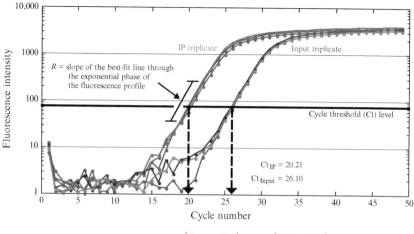

$$\text{Fold enrichment} = R^{\left(\text{Ct}_{\text{Input}} - \text{Ct}_{\text{IP}}\right)} = 1.91^{\left(26.10 - 20.21\right)} = 45.68$$

FIG. 2. A graphic demonstration of the data generated by real-time PCR and the calculations used in determining the enrichment of specific DNA sequences in an IP sample relative to an Input sample. Fluorescence data generated in real time were plotted as a function of the cycle number during PCR amplification of a particular target sequence from representative IP and Input triplicate reactions. The cycle number is listed along the x-axis and the fluorescence intensity is plotted on the y-axis. For this example, the Ct level was set to a fluorescence of 80, which resides within the exponential phase of amplification. Extrapolation of the fluorescence data from the set Ct level down to the x-axis provides Ct values for both the IP and Input samples. Taking into account the rate of amplification (R), the fold enrichment of a particular DNA sequence in the IP sample relative to the Input sample is 45.68. (See color insert.)

Results and Conclusions

This ChIP protocol has an important key difference when compared to the more widely used ChIP techniques reported in the literature. In addition to sonication, or mechanical shearing, this procedure has implemented the use of MNase prior to the conventional sonication step. This additional step increases the resolution of the ChIP method significantly by reducing the number of large DNA fragments that may remain with sonication alone. This additional layer of resolution is exceedingly important in determining the fold enrichment levels of certain histone modifications. For example, due to the relatively high overall levels of methylated histone H3 (Lys9) in mammalian genomes (it has been reported that approximately 20% of chicken histone H3 lysine 9 residues are methylated[16]),

---

[16] J. R. Davie and V. A. Spencer, *J. Cell Biochem.* **32/33** (Suppl.), 141 (1999).

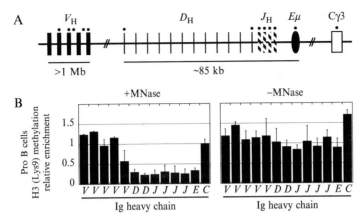

FIG. 3. The addition of MNase prior to sonication in the outlined ChIP method provides the resolution needed to establish an enrichment pattern of the association of certain histone modifications with genomic loci. (A) A schematic of the immunoglobulin heavy chain (IgH) locus. Solid, black rectangles represent families of *V* segments, black vertical lines depict *D* regions, and *J* gene segments are illustrated as hatched bars. The regulatory enhancer, *Eμ* (*E*), and the constant region, *Cγ3* (*C*), are depicted as a solid, black oval and a white rectangle, respectively. Black dots located immediately above the schematic mark the location of each PCR primer pair used in (B) and Fig. 4. (B) The enrichment of histone H3 (Lys9) methylation across the IgH locus in Pro B cells as determined from multiple independent ChIP experiments with and without the addition of MNase. Along the *x*-axis is the type of gene segment amplified by each primer pair (*V*, *D*, *J*, *E*, and *C*). These segments are arranged from left to right in the 5′ to 3′ orientation as they are located in the IgH locus. The fold enrichment of each DNA sequence in an IP sample relative to an Input sample is shown.

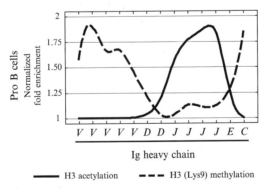

FIG. 4. The enrichment of histone H3 (Lys9) methylation and H3 acetylation at lysines 9 and 14 across the IgH locus in Pro B cells. The particular amplified gene segment is provided along the *x*-axis and the normalized fold enrichment values for both modifications are shown on the *y*-axis. The data for each modification were normalized such that, among the gene segments shown, the lowest fold enrichment value was set to 1 and the highest fold enrichment value was set to 2.

the level of enrichment of this modification seen at specific DNA sequences as calculated by real-time PCR analysis may appear slightly reduced. This is exactly the case at the β-globin locus in chicken cells, where enrichments in H3 (Lys9) methylation rarely exceed twofold.[5] Indeed, Fig. 3B emphasizes this point by illustrating that the conventional ChIP method, relying only on sonication (−MNase), generates an ambiguous enrichment pattern of methylated histone H3 (Lys9) at the IgH locus in Pro B cells. However, by combining sonication with MNase digestion (+MNase) a clear and consistently reproducible pattern of enrichment is observed. Furthermore, in the presence of MNase, the low level of H3 (Lys9) methylation seen across the *DJ* region as well as the high level of methylation observed at the upstream *V* segments correlates closely with the active and inactive domains of the IgH locus in Pro B cells, respectively (discussed later). In the absence of MNase, the measured levels of methylation across the IgH locus are not coincident with any known aspects of the developmental progression of Pro B cells, providing experimental support for the inclusion of MNase in our ChIP method.

Due to the lack of the recombinase machinery, recombinase-deficient Pro B cells are arrested just prior to the first stage of rearrangement, *D* to *J* recombination. Thus, *D* and *J* segments of the IgH locus reside in an active, accessible domain, while the *V* segments in these arrested cells should remain repressed and inaccessible. Using the protocol outlined here, we have defined these active and repressed long-range chromatin domains by histone H3 acetylation and H3 (Lys9) methylation, respectively (see Fig. 4). In conclusion, the ChIP method described in this chapter provides another versatile tool, in addition to DNaseI hypersensitivity mapping, which can be used in a number of unique cell types to identify regulatory elements, such as promoters, enhancers, or boundary elements as well as define the histone modification pattern across chromatin sub-domains within a large, complex genomic locus.

Acknowledgments

We would like to acknowledge Joe Geisberg and other members of Kevin Struhl's laboratory for their help with the design of the ChIP protocol, PCR primers, and real-time PCR analyses. We would also like to thank Laura Corey for her help with implementing MNase into our protocol. We thank Mary Donohoe for her helpful discussions and critical reading of the manuscript. This work was supported by NIH grant GM48026 to M.A.O.

[23]   The Use of Chromatin Immunoprecipitation
        Assays in Genome-Wide Analyses of
        Histone Modifications

By BRADLEY E. BERNSTEIN, EMILY L. HUMPHREY,
CHIH LONG LIU, and STUART L. SCHREIBER

Introduction

Systematic studies made possible by a variety of technological advances are impacting many areas of biology. Of special relevance to the field of chromatin are approaches that combine chromatin immunoprecipitation (chromatin IP) with microarrays to analyze histone modifications genome-wide in yeast.[1,2] The resulting genomic maps provide a unique global perspective on the functions of, and inter-relationships between, different modifications.

Chromatin IP experiments use antibodies to immunoprecipitate a protein of interest and associated DNA from a solubilized chromatin preparation.[3–7] Specialized antibodies can be used to enrich for DNA associated with histones that exhibit specific post-translational modifications. A considerable number of antibodies that recognize histones acetylated, methylated, or phosphorylated at specific residues have been developed, and many are commercially available. Chromatin IP experiments typically assess whether a particular gene, gene promoter, or genomic region is enriched in the IP sample relative to a whole cell extract (WCE) control. For example, quantitative polymerase chain reaction (PCR) using pre-selected primer pairs can be used to compare the representation of specific DNA species in the IP and WCE. Although several regions can be queried simultaneously, these conventional studies are limited to a subset of genes or regions chosen by the investigator.

Global studies have the potential to overcome selection bias by analyzing comprehensively many or all elements in a system. Microarray

[1] B. E. Bernstein, E. L. Humphrey, R. L. Erlich, R. Schneider, P. Bouman, J. S. Liu, T. Kouzarides, and S. L. Schreiber, *Proc. Natl. Acad. Sci. USA* **99**, 8695 (2002).

[2] D. Robyr, Y. Suka, I. Xenarios, S. K. Kurdistani, A. Wang, N. Suka, and M. Grunstein, *Cell* **109**, 437 (2002).

[3] M. J. Solomon, P. L. Larsen, and A. Varshavsky, *Cell* **53**, 937 (1988).

[4] M. Braunstein, A. B. Rose, S. G. Holmes, C. D. Allis, and J. R. Broach, *Genes Dev.* **7**, 592 (1993).

[5] V. Orlando, H. Strutt, and R. Paro, *Methods* **11**, 205 (1997).

[6] M. H. Kuo and C. D. Allis, *Methods* **19**, 425 (1999).

[7] A. Hecht and M. Grunstein, *Methods Enzymol.* **304**, 399 (1999).

technology, typically used for mRNA analysis,[8,9] can be adapted for the comprehensive analysis of chromatin IP samples.[10,11] These studies rely on several innovations, including:

1. The development of molecular biology protocols to amplify and label chromatin IP DNA samples in a sequence independent manner and
2. The production of DNA microarrays that contain the open reading frames (ORFs) as well as the regulatory regions

Recently this approach has been used to map relative levels of histone acetylation and methylation genome-wide in yeast, and to identify regions deacetylated by the various histone deacetylase enzymes in yeast (see Fig. 1).[1,2]

Protocols

The protocol for analyzing histone modifications genome-wide in yeast is presented in four sections:

1. Chromatin IP to isolate DNA associated *in vivo* with modified histones
2. DNA amplification and labeling
3. Microarray hybridization and data acquisition
4. Data analysis and interpretation

Chromatin IP to Isolate DNA Associated *In Vivo* with
Modified Histones

Solutions

     YPD: Yeast extract/peptone/dextrose
     PBS: Phosphate-buffered saline pH 7.4
     Buffer L: 50 mM HEPES-KOH, pH 7.5, 140 mM NaCl, 1 mM EDTA, 1% Triton X-100, 0.1% sodium deoxycholate. For 5 ml of buffer L, 20 μl of protease inhibitor cocktail (Sigma), and one mini complete protease inhibitor tablet (Boeringer Mannheim) are added just before use

[8] J. L. DeRisi, V. R. Iyer, and P. O. Brown, *Science* **278**, 680 (1997).
[9] L. Wodicka, H. Dong, M. Mittmann, M. H. Ho, and D. J. Lockhart, *Nat. Biotechnol.* **15**, 1359 (1997).
[10] V. R. Iyer, C. E. Horak, C. S. Scafe, D. Botstein, M. Snyder, and P. O. Brown, *Nature* **409**, 533 (2001).
[11] B. Ren, F. Robert, J. J. Wyrick, O. Aparicio, E. G. Jennings, I. Simon, J. Zeitlinger, J. Schreiber, N. Hannett, E. Kanin *et al.*, *Science* **290**, 2306 (2000).

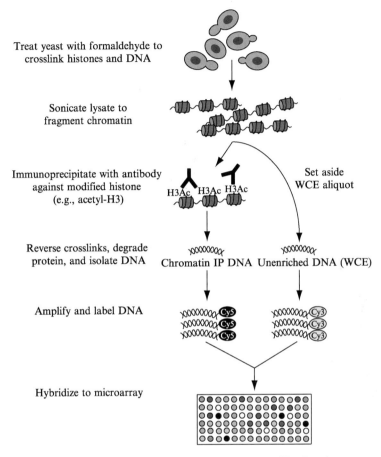

Treat yeast with formaldehyde to
crosslink histones and DNA

Sonicate lysate to
fragment chromatin

Immunoprecipitate with antibody
against modified histone
(e.g., acetyl-H3)

Set aside
WCE aliquot

H3Ac   H3Ac   H3Ac

Reverse crosslinks, degrade
protein, and isolate DNA

Chromatin IP DNA   Unenriched DNA (WCE)

Amplify and label DNA

Hybridize to microarray

Fig. 1. Steps in the genome-wide analysis of histone modifications in yeast.

Buffer W1:	Buffer L with 500 mM NaCl
Buffer W2:	10 mM Tris-HCl, pH 8.0, 250 mM LiCl, 0.5% NP-40, 0.5% sodium deoxycholate, 1 mM EDTA
TE:	10 mM Tris-HCl, pH 8.0, 1 mM EDTA
Proteinase K solution:	TE with 0.4 mg/ml glycogen, 1 mg/ml proteinase K
Elution buffer:	TE, pH 8.0, with 1% SDS, 150 mM NaCl, and 5 mM DTT

DNA is isolated using the following chromatin IP protocol, adapted from published protocols by Kuo and Allis,[6] and Hecht and Grunstein.[7]

A 180-ml culture of yeast *(Saccharomyces cerevisiae)* is grown in YPD at 30° to an OD_{600} of 1.0. To cross-link proteins and DNA, 4.9 ml of 37% formaldehyde is added and the culture incubated at room temperature for 15 min with occasional shaking. The formaldehyde is quenched by the addition of 9 ml of 2.5 M glycine, and the cells are left at room temperature for an additional 5 min. The yeast are pelleted by centrifugation at 2000g for 5 min at 4°, and washed twice with 180 ml ice-cold PBS. Washed cells are divided into four aliquots, pelleted, frozen in liquid nitrogen, and stored at −80°.

To lyse cells, one aliquot (representing 45 ml of the original culture) is resuspended in 400 μl of buffer L. Five hundred microliters of acid-washed glass beads are added and the mixture is vortexed on high for 45 min at 4°. With the lid of the microfuge tube open, the bottom of the tube is punctured with a hot 25-gauge needle and the cell lysate is spun to a new tube (5 s full speed, cut off the top of the punctured tube before spinning). The total volume of lysate is then adjusted to 700 μl by addition of buffer L. Chromatin is fragmented by sonication using a Branson 250 Sonifier (4 × 20 s at setting 3 and 70% duty, with 20-s rests), and the insoluble fraction precipitated by centrifugation at 14,000g for 15 min at 4°. The supernatant contains the solubilized chromatin and is referred to as the WCE. The size distribution of the chromatin fragments can be examined by gel electrophoresis (see Fig. 2). A desired distribution is achieved by varying sonication time.[7]

Prior to immunoprecipitation, 15 μl of the WCE are set aside for a control. The remaining WCE is added to 50 μl of packed protein A beads (pre-equilibrated in buffer L) and rotated at 4° for 1 h to remove proteins that bind nonspecifically to the beads. The supernatant is then transferred to another tube that contains 20–30 μl of anti-modified histone antibody and rotated at 4° for 4 h. Antibody-histone complexes are precipitated by the addition of 50 μl of pre-equilibrated protein A beads and rotation at 4° for 1 h. The beads are collected by centrifugation at 3000g for 30 s and washed successively with 1 ml of buffers L (twice), W1 (twice), W2 (twice), and TE (twice) for 5 min each. The beads are then incubated in 125 μl of elution buffer at 65° for 10 min with frequent mixing. The beads are pelleted by centrifugation at 10,000g for 2 min and the supernatant is retained. After a second elution, the eluted samples are combined, overlaid with mineral oil, and left at 65° overnight to reverse cross-links. In parallel, 15 μl of WCE is combined with 235 μl of elution buffer, overlaid, and incubated at 65° overnight.

To isolate DNA, the IP sample and WCE control are incubated with an equal volume of proteinase K solution at 37° for 2 h, extracted with 1 volume phenol (twice), 1 volume chloroform/isoamyl alcohol (25:1) (once), and ethanol precipitated. The samples are centrifuged at 15,000g for

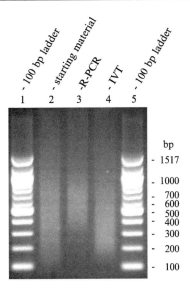

Fɪɢ. 2. WCE starting material and products amplified by R-PCR or IVT. Lanes 2–4 of a 2% non-denaturing agarose gel contain 200 ng nucleic acid. Lanes 1 and 5: 500 ng 100 bp ladder (NEB). Lane 2: Yeast WCE. Lane 3: DNA amplified by R-PCR from WCE. Lane 4: RNA amplified by IVT from WCE.

15 min at 4°, washed with cold 70% ethanol, dried, resuspended in 20 μl of TE with 10 μg of RNase A, and incubated at 37° for 1 h. Prior to amplification, the DNA is further purified with a MinElute PCR Purification Kit (Qiagen).

DNA Amplification and Labeling

Chromatin IP experiments typically yield low nanogram quantities of DNA. An amplification step is required to obtain the 2–4 μg of material necessary for DNA microarray analysis. Amplification of a heterogeneous population of fragmented genomic DNA poses unique challenges. The PCR requires a known flanking sequence for priming. Two PCR-based protocols have been developed for the amplification of genomic DNA. In "ligation-mediated PCR,"[11] the genomic DNA is blunt-ended, and ligated to double-stranded linkers. Ligated DNA is then amplified by PCR using the linker sequence for priming. In random-primer PCR (R-PCR),[10,12,13]

[12] S. K. Bohlander, R. Espinosa III, M. M. Le Beau, J. D. Rowley, and M. O. Diaz, *Genomics* **13,** 1322 (1992).
[13] J. D. Lieb, X. Liu, D. Botstein, and P. O. Brown, *Nat. Genet.* **28,** 327 (2001).

the genomic DNA is copied twice using a primer adaptor that has a conserved 5' sequence and a degenerate 3' sequence that anneals to the ends of the genomic DNA. The copied DNA is amplified by PCR using the 5' conserved sequence for priming. Both methods have been used to amplify DNA obtained by chromatin IP.

Due to its exponential kinetics, PCR is particularly susceptible to sequence- and length-dependent biases. Recently, a linear amplification strategy has been applied to the amplification of genomic DNA.[14] In this procedure, a T7 promoter sequence is added to the ends of the genomic DNA, creating a template suitable for *in vitro* transcription (IVT). Although this IVT protocol has the added complication of working with RNA, it preserves the species representation of the chromatin IP DNA more effectively than the PCR methods. Here we present the R-PCR and IVT amplification protocols as we have applied them in studies of histone modifications.

R-PCR Amplification Protocol

Solutions

5× Sequenase buffer (Amersham):	200 mM Tris, pH 7.5, 100 mM MgCl₂, 250 mM NaCl
dNTP mix:	3 mM each of dATP, dCTP, dGTP, and dTTP
Sequenase reaction mix (for 2 reactions):	2.5 μl 5× Sequenase buffer, 3.75 μl dNTP mix 1.875 μl 0.1 M DTT, 3.75 μl 0.5 mg/ml BSA, 0.75 μl Sequenase 2.0 (Amersham)
Sequenase Dilution buffer (Amersham):	10 mM Tris, pH 7.5, 5 mM DTT, 0.1 mM EDTA
50× aa-dNTP mix:	25 mM dATP, 25 mM dCTP, 25 mM dGTP, 10 mM dTTP, 15 mM aminoallyl-dUTP
PCR buffer:	500 mM KCl, 100 mM Tris, pH 8.3
Primer A:	40 pmol/μl 5'-GTTTCCCAGTCACGAT-CNNNNNNN- NN-3'
Primer B:	100 pmol/μl 5'-GTTTCCCAGTCA-CGATC-3'

DNA from the chromatin IP sample and the WCE control are amplified in parallel using a variant of the Round A/Round B DNA amplification protocol used by Iyer *et al.*[10] and described at www.microarrays.org. This variant protocol minimizes PCR rounds in an attempt to reduce amplification bias. In Round A, 7 μl sample, 2 μl 5× Sequenase buffer, and 1 μl

[14] C. L. Liu, S. L. Schreiber, and B. E. Bernstein, *BMC Genomics* **4**, 19 (2003).

Primer A are combined in a thin-walled PCR tube. In a thermal cycler, the mixture is heated at 94° for 2 min, cooled to 10°, and held for 5 min while 5.05 μl of Sequenase reaction mix is added. The reaction is ramped from 10° to 37° at a rate of 0.1° per second, held at 37° for 8 min, and then heated at 94° for 2 min. It is then cooled to 10° and held for 5 min while 1.2 μl of Sequenase 2.0 diluted 1:4 with Sequenase Dilution Buffer is added. Finally, the reaction is ramped from 10° to 37° at a rate of 0.1° per second and held at 37° for 8 min. The samples are diluted with water to a final volume of 60 μl.

In Round B, 15 μl diluted Round A product, 8 μl 25 mM MgCl$_2$, 10 μl 10× PCR buffer, 2 μl 50× aa-dNTP solution, 1 μl Primer B, 1 μl Taq (5 U/ μl), and 63 μl nuclease-free water are combined in a thin-walled PCR tube. The following PCR program is carried out for 30 cycles: 92° for 30 s, 40° for 30 s, 50° for 30 s, 72° for 1 min. The cycles are followed by a 5-min extension at 72°. PCR product is purified using a QIAquick PCR Purification Kit (Qiagen). Amplification is verified by spectrophotometry and gel electrophoresis. The size distribution should approximate that of the WCE, though we note that lower molecular weight species are inefficiently amplified by this method (see Fig. 2). Products from 2–3 Round B PCR reactions are combined to obtain the 8–10 μg of DNA required for microarray analysis. Equal quantities of DNA amplified from IP and WCE are used for subsequent labeling and analysis.

IVT Amplification Protocol

Solutions

NEB buffer 3 (New England Biolabs):	100 mM NaCl, 50 mM Tris, pH 7.9, 10 mM MgCl$_2$, 1 mM DTT
TdT buffer (Roche; *note* *different source than enzyme*):	1 M potassium cacodylate, 125 mM Tris, pH 6.6, 1.25 mg/ml BSA
8% nucleotide tailing solution:	4.6 μM dTTP, 0.4 μM ddCTP
Eco Pol Buffer (New England Biolabs):	10 mM Tris-HCl, pH 7.5, 5 mM MgCl$_2$, 7.5 mM DTT
50× aa-dNTP mix:	25 mM dATP, 25 mM dCTP, 25 mM dGTP, 10 mM dTTP, 15 mM aminoallyl-dUTP
Priming mix:	5 μg/μl 18–22-mer oligo dT (optional), 5 μg/μl random primer pd N6 (Invitrogen)
T7-A$_{18}$B primer:	25 pmol/μl 5'-GCATTAGCGGCC GCGAAATTAATACGACT- CACT ATAGGGAG(A)$_{18}$[B]-3' ([B] stands for any base other than A)

DNA from the chromatin IP sample and the WCE control are treated with calf intestinal alkaline phosphatase (CIP) to remove $3'$ phosphate groups prior to IVT. DNA (up to 50 ng), 1 μl NEB Buffer 3, and 0.25 μl of 10 U/μl CIP (NEB) are combined and H_2O is added to a total volume of 10 μl. The reaction is incubated at $37°$ for 1 h, and cleaned up with a MinElute Reaction Cleanup Kit (Qiagen). The kit is used after several steps in this protocol, always according to manufacturer's instructions except with the elution volume increased to 20 μl. Terminal transferase (TdT) is used to add $3'$ tails of relatively uniform length to the ends of the genomic DNA. Ten microliters of CIP-treated DNA, 4 μl TdT Buffer, 3 μl 5 mM CoCl$_2$, 1 μl 8% nucleotide solution, and 2 μl TdT enzyme (10 U/μl) are combined, overlaid with mineral oil, and incubated at $37°$ for 20 min. The reaction is halted by the addition of 2 μl 0.5 M EDTA, pH 8.0, and subjected to the MinElute Reaction Cleanup.

T7 promoters are incorporated at the ends of the DNA fragments as follows: 20 μl tailing reaction product, 0.6 μl T7-A$_{18}$B primer, 5 μl 10× EcoPol buffer, 2 μl 5.0 mM dNTPs, and 20.4 μl nuclease-free water are combined. In a thermal cycler, the samples are incubated at $94°$ for 2 min, ramped down at $-1°$ per second to $35°$, held at $35°$ for 2 min, ramped down at $-0.5°$ per second to $25°$ and held while 2 μl 5 U/μl Klenow (NEB) is added. The reaction is then incubated at $37°$ for 90 min, halted by the addition of 5 μl 0.5 M EDTA, pH 8, and subjected to a MinElute Reaction Cleanup. (*Note:* mineral oil is not used at this stage.)

Prior to the IVT, the volume is reduced to 8 μl in a vacuum centrifuge at medium heat. IVT is carried out using the T7 Megascript Kit (Ambion). The samples are purified using an RNeasy Mini Kit (Qiagen) per manufacturer's RNA cleanup protocol, except with an additional 500-μl wash with buffer RPE. RNA is quantified by absorbance at 260 nm and visualized on a 2% native TAE agarose gel. Yields range from 10 to 60 μg. The size distribution should approximate that of the WCE (see Fig. 2).

Equal quantities of RNA from the IP sample and WCE control ($\sim$2–4 μg) are reverse-transcribed to generate DNA probes. In a 14.5-μl volume, the RNA is primed by addition of 1 μl priming mix, incubation at $70°$ for 10 min, and incubation on ice for 10 min. Then, 6 μl 5× RT buffer, 3 μl 0.1 M DTT, 3 μl nuclease-free H_2O, 0.8 μl 50× aa-dNTP mix, and 2.5 μl 200 U/μl Superscript II (Invitrogen) are added, and the mix is incubated at $42°$ for 2 h. The reaction is halted by the addition of 10 μl 0.5 M EDTA, pH 8.0, and the RNA is hydrolyzed by the addition of 10 μl 1 N NaOH and incubated at $65°$ for 15 min. The solution is neutralized with 25 μl 1 M HEPES, pH 7.5. The entire samples are used for labeling and hybridization.

Probe Labeling

Amplified samples are fluorescently labeled using monofunctional Cy5 and Cy3 fluorescent dyes as described by Carroll *et al.*,[15] and at: www.microarrays.org. These derivatized dyes covalently couple to the aminoallyl groups incorporated in Round B of R-PCR, or in the RT step following IVT. DNA samples are diluted with 400 μl of nuclease-free water, concentrated to $\sim$70 μl in a YM-30 microcon (Millipore) by centrifugation at 11,000g, re-diluted and re-concentrated. After a final dilution, the samples are concentrated to a volume of 18–20 μl (microcons are pre-weighed, and final volume determined by weight). Samples are collected by inverting the microcon into a new microfuge tube and spinning at 4000g for 5 min. One microliter of 1 M NaHCO$_3$, pH 9, is added and the samples are transferred to a microfuge tube containing a dried aliquot of NHS cyanine dye (Cy5 for IP sample; Cy3 for WCE control) and incubated at room temperature for 75 min in the dark. [The dye aliquots are prepared by dissolving one vial of monofunctional Cy5 or Cy3 dye (Amersham) in 20 μl DMSO, distributing the solution into 10 microfuge tubes, and drying in a vacuum centrifuge under low heat.] Reactions are diluted with 70 μl H$_2$O and purified using a QIAquick PCR Purification Kit.

Microarray Hybridization and Data Acquisition

Microarray production, hybridization, and data acquisition are carried out using protocols developed by Patrick Brown and colleagues, available on-line at www.microarrays.org and recently reviewed by Gasch.[16] Comprehensive coverage of the yeast genome is achieved by hybridizing amplified DNA to microarrays that contain all ORFs and all promoter-containing intergenic regions (INTs). The INTs are designated in such a way that each gene promoter can be assigned to a specific INT. DNA fragments representing all ORFs and INTs can be amplified from genomic DNA using primers from Research Genetics. The ORFs and INTs are printed on separate poly-lysine–coated slides by pin transfer using a robotic arrayer as described at www.microarrays.org, or using alternative protocols described elsewhere.[11,17] The inclusion of INT and ORF arrays in a global chromatin analysis can enable an experimenter to discriminate between promoters and coding regions of genes.[1,13]

[15] A. S. Carroll, A. C. Bishop, J. L. DeRisi, K. M. Shokat, and E. K. O'Shea, *Proc. Natl. Acad. Sci. USA* **98,** 12578 (2001).
[16] A. P. Gasch, *Methods Enzymol.* **350,** 393 (2002).
[17] C. E. Horak and M. Snyder, *Methods Enzymol.* **350,** 469 (2002).

Hybridization

The competitive hybridization used in DNA microarray analyses allows flexibility in determining which samples to compare. In this presentation of the protocol, the relative enrichment of a particular modification is analyzed genome-wide by comparing a chromatin IP sample to its WCE control. However, it is also possible to query changes in modification patterns resulting from gene deletion or small-molecule treatment by comparing two chromatin IP samples obtained from different sources using the same antibody.

Reagents and solutions:

WS1:   387 ml Milli-Q water, 12 ml 20× SSC, 1 ml 10% SDS
WS2:   1 ml of 20× SSC in 399 ml of Milli-Q water

The following is adapted from the hybridization procedure described at www.microarrays.org. Cy5-labeled chromatin IP probe and Cy3-labeled WCE probe are combined, and concentrated in a microcon YM-30 filter until 50 μl of solution remain on the filter. The sample is collected by inverting the microcon into a new microfuge tube and spinning at 4000g for 5 min. Six microliters of 20× SSC, 3 μl of 10 μg/μl poly(A), and 0.96 μl of 1 M HEPES, pH 7.0, are added and the solution is filtered with a pre-wet Millipore 0.45-μ filter spun at 12,000 rpm for 2 min. Ten percent SDS, 0.9 μl, is added, the mixed probe is placed in a 100° heat block for 2 min, placed at room temperature for 10 min, and then applied to the microarray. Slides printed with ORFs and INTs are placed in Telechem hybridization chambers with the arrays facing up. One clean lifterslip (Erie Scientific) is placed on each array so that the white strips face down. Half of the mixed probe (about 20 μl) is applied to the ORF array and the other half to the INT array. Each well of the hybridization chambers is filled with 40 μl of 3× SSC, the chambers are then sealed and placed in a 60° water bath for 12–15 h.

Four Wheaton glass tanks containing WS1 or WS2 are prepared. A metal slide rack is placed in the first tank of each solution. The microarray slides are removed from the hybridization chambers and turned upside down in the first tank of WS1. Slides are tilted to allow the lifterslips to fall off, placed in the slide rack, and plunged up and down 20 times. This dunking is repeated in the second tank of WS1 and in the two WS2 tanks. The slides are dried by centrifugation at 1000 rpm in a Beckman tabletop centrifuge for 2 min.

Data Acquisition and Processing

Data acquisition and analysis are detailed in a variety of sources.[16–19] Briefly, data are acquired by scanning the microarrays with a scanning laser device (Axon Instruments) that simultaneously excites and measures fluorescence emission for the Cy5 and Cy3 dyes. Image analysis software (Axon) is used to quantify fluorescent features and associate them with the appropriate ORF or INT. The data output consists of Cy5 and Cy3 intensities and deviations for each ORF or INT. The Cy5/Cy3 ratios reflect the amount of DNA in the chromatin IP sample relative to the WCE control. The data are normalized so that the relative fluorescence signal is the same for most spots.[19] Alternatively, control samples spiked into the probes can be used for normalization.[16] To eliminate questionable measurements, array features having a signal-to-noise ratio of less than a cutoff value (e.g., 2.5) are removed from further analysis. Replicate data sets can be merged using a weighted-averaging procedure.[18]

Data Analysis and Interpretation

Microarray experiments generate vast amounts of data, and it can be challenging to convert these data into informative observations. Each histone modification data set contains Cy5/Cy3 ratios for each microarray feature. These ratios reflect the degree to which histones associated with the corresponding ORF or INT exhibit a particular modification. A high ratio indicates a high level of modification. A ratio of around 1 indicates a modification level not significantly different from the genome average. A low ratio suggests a low level of modification, with the caveat that some genomic regions may be less available to IP. (To control for this potential caveat, antibody against an invariant region of a histone can be used.) It is possible to assign p-values that reflect the probability that the modification level of histones associated with a particular ORF or INT is significantly higher (or lower) than the genome average.

The data can be used as a screen to identify genes or gene promoters associated with highly modified histones. These can be verified by conventional chromatin IP and followed up by other biological means. Alternatively, since the genomic location of each ORF and INT is known, modification status can be visualized across the yeast genome using chromosome mapping programs or plotted with respect to distance from telomere ends. The continuity of a modification along adjacent genomic

[18] C. J. Roberts, B. Nelson, M. J. Marton, R. Stoughton, M. R. Meyer, H. A. Bennett, Y. D. He, H. Dai, W. L. Walker, T. R. Hughes *et al.*, *Science* **287,** 873 (2000).

[19] J. Gollub, C. A. Ball, G. Binkley, J. Demeter, D. B. Finkelstein, J. M. Hebert, T. Hernandez-Boussard, H. Jin, M. Kaloper, J. C. Matese *et al.*, *Nucleic Acids Res.* **31,** 94 (2003).

regions can be assessed using statistical methods, though such analyses must account for DNA fragments in the chromatin IP experiment that span adjacent ORFs and INTs.

Data sets that assess different modifications can be compared in several ways. The Pearson correlation provides a single measure of similarity between the global patterns of two modifications (this can be visualized by plotting pairwise relationships between the modifications). Alternatively, lists of outliers (e.g., features with high ratios) can be collated and compared. The significance of an observed overlap can be calculated using a hypergeometric probability model.[20] Two-dimensional hierarchical clustering, commonly used in expression profiling, can be performed and visualized using Cluster and Treeview.[21] Clustering can be used to identify sets of genes that exhibit similar modification patterns. These comparative analyses can also be extended to other microarray-based data sets to identify functional relationships. This kind of cross-comparison was used to identify associations between H3 Lys4 methylation and transcription,[1] and between histone deacetylation and repression.[1,2,22] Although these analyses have revealed valuable insights into roles of histone modifications, a realization of the full potential of the histone modification data sets awaits implementation of more sophisticated analytical approaches.

Conclusions and Future Directions

Methods for analysis of histone modifications genome-wide in yeast have been developed and applied toward a limited subset of site-specific acetylation[2] and methylation[1] marks. Future studies will assess the potential of this systematic approach to provide more comprehensive insight into the functions of, and inter-relationships between, the many other modifications that occur on histones in yeast. Furthermore, the development of microarrays able to query regulatory regions of the human genome[23,24] should allow for global analyses of histone modifications in higher organisms. The recent development of a method for linear amplification of chromatin IP samples should facilitate this transition.

[20] S. Tavazoie, J. D. Hughes, M. J. Campbell, R. J. Cho, and G. M. Church, *Nat. Genet.* **22**, 281 (1999).
[21] M. B. Eisen, P. T. Spellman, P. O. Brown, and D. Botstein, *Proc. Natl. Acad. Sci. USA* **95**, 14863 (1998).
[22] B. E. Bernstein, J. K. Tong, and S. L. Schreiber, *Proc. Natl. Acad. Sci. USA* **97**, 13708 (2000).
[23] A. S. Weinmann, P. S. Yan, M. J. Oberley, T. H. Huang, and P. J. Farnham, *Genes Dev.* **16**, 235 (2002).
[24] B. Ren, H. Cam, Y. Takahashi, T. Volkert, J. Terragni, R. A. Young, and B. D. Dynlacht, *Genes Dev.* **16**, 245 (2002).

[24] Sequential Chromatin Immunoprecipitation from Animal Tissues

By DINA CHAYA and KENNETH S. ZARET

The chromatin immunoprecipitation (ChIP) technique for identifying transcription factors and histones bound to specific DNA sites in cells has revolutionized the analysis of gene regulation. In this procedure, cells or nuclei are exposed to formaldehyde, which crosslinks proteins bound to each other and to DNA in native chromatin. The crosslinked chromatin is isolated and fractionated, the proteins are immunoprecipitated and assessed by a quantitative PCR reaction to determine the relative extent to which the antigen is associated with the DNA. While the majority of ChIP studies have employed populations of freely growing cells, such as cultures of yeast or animal cell lines, many important questions in gene regulation must be addressed with native tissues. First, it is not yet clear that biologically relevant chromatin states, which for this context consists of bound factors and modified histones, are accurately represented by cell lines, although of course the latter provide excellent model systems for study. Second, there are well-documented cases where the transcriptional regulation of all known cell lines is markedly different from that of the native tissue, such as for liver-specific genes in hepatic cell lines versus native liver or cultured liver slices.[1] Finally, native tissues provide the opportunity to investigate gene-regulatory transitions during changes in organismal physiology that are simply absent in cell lines, such as during organogenesis or diabetes. This is particularly important now that advances in the genetics of flies, zebra fish, frogs, and mice are such that the effects of virtually any genetic perturbation can be tested directly in the animal.

Despite the technical obstacles regarding the handling of animal tissues for ChIP assays, a growing number of laboratories have made advances in the area.[2,3] This chapter will present two different protocols for formaldehyde crosslinking of mouse tissue chromatin, followed by methods for chromatin solubilization, purification, fragmentation, and immunoprecipitation; some of these steps require special adaptations for animal tissues. Although the methods presented here were developed for liver chromatin,

[1] D. F. Clayton, A. L. Harrelson, and J. E. Darnell, Jr., *Mol. Cell Biol.* **5,** 2623 (1985).

[2] M. Parrizas *et al., Mol. Cell Biol.* **21,** 3234 (2001).

[3] D. Chaya, T. Hayamizu, M. Bustin, and K. S. Zaret, *J. Biol. Chem.* **24** (2001).

experience in our laboratory and others has shown that they represent a good starting point for working with other tissue types.

Note that our method for chromatin fragmentation involves a partial sonication step followed by a complete restriction digestion step. The restriction digest, with an enzyme that recognizes a frequently occurring four-base sequence, allows the target DNA to be definitively trimmed to a known length and therefore limits the final PCR analysis solely to the defined sequence. This eliminates a concern arising when only sonication is used, where the immunoprecipitated protein is crosslinked to target sequences that are statistically but not precisely sized by the extent of DNA fragmentation, and hence may not be bound at the DNA site of interest.

Another novel aspect of our approach is that we developed a method to release antigen from the initial immunoprecipitation to allow a second, sequential immunoprecipitation of the same sample. This allows the investigator to evaluate whether two distinct proteins were bound to the same DNA fragment within the starting population of molecules. The resolution of this approach is enhanced by the aforementioned restriction digestion of chromatin prior to the first immunoprecipitation.

Preparing Formaldehyde-Crosslinked Chromatin from Tissues

Here we present two methods for delivering a formaldehyde crosslinking agent to nuclei within tissues. In the first, formaldehyde is delivered to cells in the native tissue *in situ;* in the second, nuclei are isolated from tissues and then treated with formaldehyde. Since it is well established that even the more careful nuclear isolation procedures lead to loss of proteins bound to target sequences within chromatin,[4] the *in situ* formaldehyde delivery method is preferred.

Formaldehyde crosslinking can alter protein epitope, and therefore it is essential to empirically determine whether crosslinking treatment is sufficient or excessive. We have found it useful to use a constant 1% formaldehyde concentration and vary the temperature and time of the crosslinking reaction.

Liver Retrograde Perfusion and Chromatin Crosslinking

This procedure utilizes the existing vascular system within a tissue to deliver formaldehyde to cells *in situ*. Although the procedure was adapted from well-established perfusion protocols for intact livers, our laboratory

---

[4] G. Pfeifer and A. D. Riggs, *Genes Dev.* **5,** 1102 (1991).

has further adapted it for pancreas, using pancreatic ducts, and it should be adaptable to other organ systems containing vessels or ducts. In this procedure, the animal is sacrificed and the liver is perfused with phosphate-buffered saline (PBS) and heparin to flush out blood cells. The chromatin is then crosslinked *in vivo* by perfusing formaldehyde into the liver. In order to avoid the body temperature of the animal to drop too quickly during these steps and thereby inhibit the effect of formaldehyde, the perfusion is carried out on a heat block at $30°$ with $30°$ solutions. Liver nuclei are then crudely purified.

The day before the experiment, prepare buffers A and B (see Table I) but without 2-mercaptoethanol, which is added just prior to use. Store buffers at $4°$ overnight, along with the following which are kept in the cold room: pipettes, beakers, 15-ml Corex tubes, rotor and adaptors, slides and cover slips, tissue homogenization device (dounce), cheesecloth, and funnel. For tissue dissection, prepare: ethanol, 4 pins, big and small sharp scissors, big and small tweezers, string, cotton-tipped applications, catheter

TABLE I
SOLUTIONS FOR TISSUE PERFUSION PROTOCOL

		Final concentrations	
	Prepare from stock (M)	Buffer A	Buffer B
HEPES pH 7.6	1	15 mM	15 mM
KCl	3	60 mM	60 mM
NaCl	4	15 mM	15 mM
EDTA	0.5	0.2 mM	0.1 mM
EGTA	0.5	0.5 mM	0.1 mM
Sucrose	2.5	0.34 M	2.1 M
2-Mercaptoethanol	2	0.15 mM	0.15 mM
Glycine	2.5	125 mM	—

Formaldehyde	[20% in water, stored at $-20°$ (Fischer Scientific, ref. F79–500)]	
Sonication buffer		
	Stock	Final concentration
Tris, pH 8.1	1 M	50 mM
EDTA	0.5 M	2 mM
N-Lauroylsarcosine	10%	0.5% (filtered through 0.2 μm)
PMSF	50 mM	0.5 mM
Protease inhibitors	5 mg/ml	Each 5 μg/ml (leupeptin, trypsin inhibitor, antipain)

(Becton Dickinson, Angiocath, ref. 381137, 20 GA, 1.88 IN, 1.1 × 48 mm), two 10-ml syringes, plastic wrap, paper towels, small beaker, heat block, aluminum-wrapped cardboard, hybridization oven at 30°, and a timer.

Prepare a beaker with 6 ml solution A, keep on ice. Place the following solutions used for perfusion and pre-warm in the 30° oven: PBS with 1% formaldehyde, PBS with 50 ng/ml heparin. Fill respective syringes with 10 ml of each solution. Prepare the dissecting area by placing heat block in the center of the lab bench with the aluminum-wrapped cardboard cover on top. Layer plastic wrap and then a paper towel above the cardboard. Have all the tools ready on the side, including 10-cm long pieces of string.

Sacrifice one mouse at a time and place it on its back on the heat block. Pin down the four legs and cut open the abdomen posterior to anterior. Cut through the diaphram, damaging the least possible tissue to avoid bleeding. Gently push the intestines on the side of the animal with a cotton-tipped applicator in order to have better access to the liver. Seal the superior vena cava, anterior to the liver, with a double knot using small tweezers and string. Insert the catheter in the inferior vena cava, just above the kidney junction. Remove the needle partway and blood should flow back into the catheter. Cut the portal vein with small scissors. Screw the syringe containing the PBS/heparin onto the needle and start very slowly perfusing the liver (10 ml total). All the lobes of the organ should rapidly blanch white. You should also see the buffer and blood flow out of the liver through the portal vein. If the animal is sick or too old (over 3 months), the perfusion will be difficult. If the lobes do not fully blanch, massage gently the liver with a cotton-tipped applicator and keep perfusing very slowly. Switch to the formaldehyde-containing syringe that is sitting in the 30° oven and retrograde perfuse 10 ml slowly. The perfusions should take about 5–10 min. Let the formaldehyde sit in the perfused liver for 5 min in the animal while on the heat block at 30°. Dissect out the liver and place in a beaker in the 30° oven for another 5 min. Place the liver in 6 ml buffer A on ice. The cold as well as the glycine will stop the crosslinking reaction. Mince the liver with small scissors and let sit on ice while perfusing other livers. Repeat this retrograde perfusion procedure on one or two more animals.

In the cold room, pour minced livers into a cold douncing apparatus on ice. Rinse beaker with 5 ml buffer A and add to dounce. Aspirate excess buffer from dounce. Dounce mechanically 12 times. Check a sample under a phase-contrast microscope for quantitative release of nuclei and cells and very few lumps of tissue. Filter through 8 layers of cheesecloth prewet with buffer A. Layer gently onto a 1.5-ml cushion of a 1:1 mixture of buffers A and B in 15-ml Corex tubes (if the tissue lysate volume is 4–6 ml, use one tube; if 6–12 ml, use 2 tubes). Centrifuge in fixed-angle rotor (e.g., Sorvall SS-34) at 10,000 rpm for 10 min at 4°. Gently decant the supernatant.

Resuspend the pellets gently in 5–10 ml of solution A. Layer onto a 1.5-ml cushion of 1:1 buffers A and B and centrifuge at 5000 rpm for 10 min at 4°. Gently decant supernatant, resuspend pellet in 2–5 ml sonication buffer, and follow the protocol for nuclear sonication.

Isolation of Nuclei from Liver and Chromatin Crosslinking

Nuclei are first crudely prepared from a mouse liver and then treated with formaldehyde. The nuclear isolation protocol works for many other tissues; the main parameter to adjust is the number of strokes in the tissue homogenizer.

The day before the experiment, prepare buffers A, B, and C (see Table II) but without 2-mercaptoethanol, spermine, or spermidine, which are added just prior to use. Store at 4° overnight, along with the same materials described in the section earlier except for those required for perfusion. The best way to obtain high-quality nuclei is to keep all tissues and materials ready and cold before you begin, and to work fast. Prepare a beaker with 12 ml of buffer A, another with a 1:1 mix of solutions A and B, and another with 8 ml of solution A. Keep on ice.

Sacrifice one mouse 2–3 months old. Carefully dissect out the liver, rinse in the 12 ml of buffer A, and place in the beaker with 8 ml buffer A on ice. Use small sharp scissors to mince the liver into pieces of a few millimeters. Perform same operation on the next mouse. In the cold room, pour

TABLE II
SOLUTIONS FOR ISOLATING NUCLEI

		Final concentrations		
	Prepare from stock (M)	Buffer A	Buffer B	Buffer C
HEPES, pH 7.6	1	15 mM	15 mM	15 mM
KCl	3	60 mM	60 mM	60 mM
NaCl	4	15 mM	15 mM	15 mM
EDTA	0.5	0.2 mM	—	—
EGTA	0.5	0.5 mM	—	—
Sucrose	2.5	0.34 M	2.1 M	0.34 M
2-Mercaptoethanol	2	0.15 mM	0.15 mM	0.15 mM
Spermine	0.5	0.15 mM	—	—
Spermidine	0.5	0.5 mM	—	—
MgCl$_2$	1	—	—	2 mM
Glycine	2.5 (kept at room temperature)			
Formaldehyde	20% in water, stored at −20°			

minced livers into the cold dounce on ice. Rinse the beaker with 5 ml solution A and add to dounce, to get all tissue fragments. After the tissue settles, aspirate excess supernatant from the dounce. Dounce 10 times with a motor-driven pestle. Check under the microscope that you have nuclei and cells and very few lumps of tissue. Dounce further accordingly. Filter through 8 layers of cheesecloth prewet with solution A. Layer gently onto a 1.5-ml cushion of a 1:1 mixture of buffers A and B in 15-ml Corex tubes (if the tissue lysate volume is 4–6 ml, use one tube; if 6–12 ml, use 2 tubes). Centrifuge in fixed-angle rotor (e.g., Sorvall SS-34) at 10,000 rpm for 20 min at 4°. Gently decant the supernatant. Resuspend the pellets gently in 5–10 ml of solution C. Measure the volume. Add formaldehyde (20%) to a final concentration of 1% and incubate for 10 min at 30°. Mix gently every 2 or 3 min. Stop the crosslinking reaction by adding room temperature glycine to a final concentration of 125 mM and let incubate on ice for 5 min. Layer onto a 1.5-ml cushion of a 1:1 mixture of solutions A and B. Centrifuge in fixed-angle rotor at 5000 rpm for 10 min at 4°. Gently decant the supernatant, resuspend the pellet into 2–5 ml sonication buffer, and follow the protocol for nuclear sonication.

Sonication and Purification of Crosslinked Chromatin from Tissues

The crosslinked nuclei are sonicated partially in order to solubilize the chromatin and facilitate the later step of restriction enzyme digestion. The sonication product is then loaded on a CsCl step gradient and ultracentrifuged to separate the crosslinked chromatin from the cell debris. This purification step is crucial for the restriction digest and for obtaining a good signal to noise ratio in the immunoprecipitation (IP). After the spin, fractions are collected from the gradient and each is assayed for the presence of DNA. DNA containing fractions are pooled and the chromatin is dialyzed to TE to eliminate the CsCl and allow a treatment with RNase, which we found necessary for quantitative restriction digestion.

Nuclear Sonication

These conditions are optimized for chromatin that has been crosslinked within tissue and should be reduced for chromatin that has been crosslinked within isolated nuclei. Nuclear pellets that were resuspended into 2–5 ml of sonication buffer should be incubated for 5 min at room temperature, then 5 min on ice. Sonicate with a Branson Sonifier 250, mounted with a microtip, using duty cycle constant, output #2. Pulse alternatively for 30 and 15 s, incubating 30 s on ice between pulses. Avoid bubbles and be careful not to heat the sample during the sonication, as the heat will reverse the crosslinking in presence of Sarkosine.

CsCl Step Gradient

Prepare CsCl step gradient solutions of 1.35 g/ml, 1.5 g/ml, and 1.75 g/ml CsCl in 0.5% Sarcosine, 1 mM EDTA. If CsCl does not solubilize, heat the solutions. Keep all solutions and materials at room temperature. Prepare the step gradient in ultraclear centrifugation tubes for a SW50.1 rotor. Gently layer 1 ml of 1.75 g/ml, then 1 ml of 1.5 g/ml, and finally the 1.35 g/ml CsCl solution to near the top of the tube. Gently layer the sonication product on top of the step gradient. Do not overload the gradient; use at least 2 tubes for each mouse liver, hence loading no more than about 1 mg of DNA per tube. Ultracentrifuge in SW50.1 rotor at 31,000 rpm at 20° for 20 h. To collect fractions, puncture a hole in the bottom of the centrifuge tube with a 25G needle and let the CsCl gradient drip into 1.5-ml tubes. Collect 5–10 drops per fraction, ~500 μl each, for about 10 fractions. Crosslinked chromatin has a density equal to 1.39 g/ml CsCl and will hence be closer to the top of the gradient. Repeat for each gradient tube.

To determine which fractions contain the crosslinked chromatin, remove 10-μl samples from each, incubate at 68° for 2 h to reverse the crosslinking, then add 1 μl of 100 ng/μl RNase and incubate 37° for 30 min. min. Add 3 μl of 1% SDS (final 0.25%) to denature the protein, add electrophoretic loading buffer, and analyze the DNA by agarose gel electrophoreses. The samples will still contain RNA at this stage; however, the desired fractions will yield DNA as a faint smear of higher molecular weight, ethidium bromide–staining material. Pool the gradient fractions containing crosslinked chromatin and measure the volume. Generally, we pool fractions 5–8; although fractions 9 and 10 contain DNA, they are typically less clean as seen by adherence of staining material in the wells of the analytical gel.

Tissue Chromatin Dialysis and RNase Treatment

At this stage, the crosslinked chromatin has been purified from the cell debris. For further analysis of the crosslinked chromatin, RNA and CsCl should be eliminated. Measure the volume of the purified crosslinked chromatin and place the material in a dialysis bag (3500 M.W. cutoff) that has been rinsed with water. Dialyze against 500 ml of 10 mM Tris (pH 7.5), 1 mM EDTA (TE buffer) 5 min at 37°. Add RNase A (10 mg/ml stock) to a final concentration of 25 ng/μl directly into the dialysis bag, incubate 37° for 20 min. Change dialysis buffer to fresh TE and dialyze overnight at 4°. Collect chromatin, dispense into 0.5–1 ml aliquots, and store at −80°.

For subsequent quantitative analysis, it is useful to remove an aliquot of 200 μl or so at this stage for DNA purification and quantification. Knowing the concentration of DNA will help in establishing conditions for restriction enzyme digestion and immunoprecipitation. The concentration of DNA of the purified crosslinked chromatin obtained by perfusion is usually about 0.5 mg/ml. For each immunoprecipitation condition, 50–100 μg of chromatin DNA are used.

Tissue Chromatin Fragmentation by Restriction Enzyme Digestion

Restriction enzyme digestion of the chromatin will be highly sensitive to the extent of crosslinking. A control digest reaction can be performed on DNA samples where the crosslinks have been reversed first, in cases where it is suspected that protein crosslinked to the DNA is inhibiting digestion, rather than the digestion reaction conditions. Typically, digestion conditions on crosslinked, partially sonicated chromatin are as indicated by the enzyme manufacturer. For 500 ng/μl chromatin, 2 ml total volume, add 1.5 ml H_2O, 400 μl of 10× digestion buffer buffer, 500 units of enzyme, and incubate at 37° for 3 h. The concentration of the purified crosslinked chromatin after the restriction enzyme digestion is about 250 ng/μl and can be stored in aliquots at −20°.

To ascertain complete digestion, reverse the crosslinks on aliquots of control and enzyme-treated sonicated, crosslinked chromatin. Run the products on an agarose minigel and perform Southern blot analysis with a 32-P–labeled, PCR-generated probe of the region of interest. The hybridization signal should appear as a smear in the undigested control and as a unique, appropriately sized band in the enzyme treated sample (see Fig. 2C of Chaya et al. (2001)[3]).

Sequential Immunoprecipitation of Purified, Restriction-Digested Chromatin from Tissues

Sequential immunoprecipitation requires that the primary antibody be immobilized, so that antigen complexes can be released and subjected to a secondary immunoprecipitation. Methods are described later. Excellent references exist for working out conditions for immunoprecipitation.[5] Conditions described here were optimized for rabbit antisera against the transcription factor HNF3α.

[5] E. Harlow and D. Lane, in "Antibodies: A Laboratory Manual." Cold Spring Harbor Laboratory, NY, 1988.

Bead Preparation and Preclearing

This protocol describes the preparation of 1.5 g of Sepharose CL-4B beads containing protein A (Sigma). Make all solutions fresh and keep solutions and beads at 4°. Add 10 ml PAS (see Table III) to 1.5 g protein A-Sepharose beads in a 50-ml plastic tube. Rinse bottle containing original beads twice with 10 ml PAS and add to 50-ml tube. Incubate on ice 30 min, occassionally inverting the tube gently. Centrifuge at 1500 rpm for 5 min at 4° in a swinging bucket rotor. Aspirate supernatant and resuspend in 30 ml cold BSB by inverting the tube gently a few times. Centrifuge at 1500 rpm for 5 min at 4° in a swinging bucket rotor and then aspirate supernatant. Repeat this wash for a total of 5 times, and then resuspend the beads in 1 volume of BSB containing protease inhibitors. Aliquot and store at 4°.

Before using the protein A beads for immunoprecipitation, nonspecific DNA binding sites need to be blocked. Centrifuge an aliquot of beads at 2500 rpm for 1 min in a microcentrifuge. Discard the supernatant, add 1 volume RIPA buffer and sonicated salmon sperm DNA to a final concentration of 100 μg/ml, and incubate overnight at 4° while rotating. Centrifuge the beads at 2500 rpm for 1 min in a microcentrifuge, discard the supernatant, and add 1 volume RIPA buffer. Beads at 1:1 ratio in RIPA buffer are ready for immunoprecipitation.

Primary Antibody Crosslinking to Protein A-Sepharose beads

Incubate 500 μl of protein A-Sepharose beads in 1:1 RIPA buffer with 75 μl of specific antiserum or preimmune serum (120–150 mg/ml) in a 12-ml polypropylene centrifuge tube. Incubate 1 h at room temperature while rotating. Add 10 volumes (5 ml) of 0.2 M sodium borate, pH 9. Incubate 10 min at room temperature while rotating. Centrifuge at 3000 rpm for 1 min in swinging bucket rotor. Decant and repeat wash. Resuspend beads in 5 ml of 0.2 M sodium borate, pH 9. Remove a 20-μl aliquot for subsequent SDS-PAGE analysis (see later). Add fresh dimethyl pimelimidate (solid) to a final concentration of 20 mM and incubate at room temperature for 30 min while rotating. Remove a 20-μl aliquot for subsequent SDS-PAGE analysis. Stop the reaction by washing with 5 ml of 0.2 M ethanolamine, pH 8. Centrifuge at 3000 rpm for 1 min in swinging bucket rotor. Discard supernatant, add 5 ml of 0.2 M ethanolamine, pH 8, and incubate 2 h at room temperature while rotating. Centrifuge at 3000 rpm for 1 min in swinging bucket rotor. Resuspend beads in 250 μl RIPA buffer. Store at 4°. Analyze the pre- and post-crosslinked samples on a 10% SDS-PAGE gel and stain with Coomassie. The crosslinking reaction should have prevented SDS from solubilizing antibody protein from the beads.

TABLE III
Solutions for Preparing Antibody Covalently Bound to Protein A Beads,
Chromatin Immunoprecipitation, and DNA Purification

	Final concentration	Amounts
PAS (30 ml)		
BSA	0.3 mg/ml	900 μl of 10 mg/ml stock
Methionine	0.1 M	0.45 g
(Sigma, L-methionine)		
SDS	0.01%	15 μl to 30 ml of 20% stock H$_2$O
BSB (200 ml)		
NaCl	150 mM	7.5 ml of 4 M stock
Borate adjusted	10 mM	20 ml to 122.5 ml of 0.1 M stock
to pH 8 (boric acid)		H$_2$O
Protease Inhibitors, PI		
(added to 20 ml of BSB)		
Leupeptin, trypsin inhibitor,	5 μg/ml	20 μl stock 5 mg/ml
antipain mix		
Benzamidine	1 mM	20 μl stock 1 M
PMSF	0.5 mM	200 μl stock 50 mM

RIPA buffer	Stock	Final concentration
Tris, pH 8	1 M	50 mM
NaCl	4 M	150 mM (or 500 mM for I.P. washes)
IGEPAL CA-630	100%	1% (ICN)
Deoxycholic acid (Na salt)	10%	0.5% (Sigma)
SDS	20%	0.1%
RIPA buffer 2× without Tris		
NaCl	4 M	300 mM
IGEPAL CA-630	100%	2% (ICN)
Deoxycholic acid (Na salt)	10%	1% (Sigma)
SDS	20%	0.2%
TNESK 5X		
Tris, pH 7.5	1 M	50 mM
NaCl	4 M	500 mM
EDTA	0.5 M	5 mM
SDS	20%	5%
Proteinase K	20 mg/ml	0.5 mg/ml (added just before use)

Chromatin Immunoprecipitation

For each immunoprecipitation, use about 50–100 µg of chromatin DNA. Dialyze the total amount of purified crosslinked chromatin overnight at 4° in 3500 cut-off dialysis bags against 500 ml RIPA buffer (without IGEPAL). The next morning, dialyze a few hours against RIPA buffer containing IGEPAL, then transfer to a polypropylene tube. To block nonspecific sites on the chromatin, add 2 µl preimmune serum per milligram chromatin and 200 µl of 1:1 precleared protein A:RIPA. Incubate rotating for 3 h at 4°. Centrifuge at 2500 rpm for 1 min at 4° and save supernatant in fresh tube. Remove an aliquot of chromatin as an "input" sample.

Dispense 50–100 µg aliquots of supernatant chromatin and add relevant antibodies. Examples of experimental conditions include: no antibodies, 2 µl nonimmune serum (stock 130 mg/ml), 2 µl specific antiserum, 40 µl preimmune serum coupled to protein A, 40 µl antiserum coupled to protein A. Incubate on rotator overnight at 4°. Add 40 µl 1:1 precleared protein A:RIPA for samples that are not coupled to protein A and incubate rotating for 2 h at 4°. Centrifuge all samples at 2500 rpm for 1 min at 4°. Save supernatant for DNA analysis. Wash beads 4 times with 20 volumes RIPA buffer containing 500 mM NaCl; for each wash, add buffer, let sit on ice for 10 min, mixing occasionally, centrifuge, discard supernatant. For final wash, transfer suspended beads to a fresh tube to eliminate chromatin bound nonspecifically to the plastic. For a single immunoprecipitation experiment, beads are finally resuspended in 1 volume of TE. For sequential immunoprecipitation, chromatin will be eluted from the beads, as described in the following.

Releasing Primary Chromatin Antigens from Protein A-Antibody Beads

Add to beads 5 volumes of phosphoric acid and incubate at room temperature for 5 min, gently mixing occasionally. Centrifuge at 2500 rpm rpm for 1 min. Transfer supernatant to fresh tube containing 1 M Tris, pH 8 to neutralize. Add 1 volume RIPA buffer 2× without Tris. Centrifuge at 2500 rpm for 1 min and carefully collect supernatants without beads into fresh tubes. Divide chromatin samples for the second, sequential immunoprecipitations. Follow protocol "D.3" for immunoprecipitation.

DNA Purification from Chromatin Immunoprecipitates

Reverse crosslinking for all samples to be analyzed by incubating for at 68° for 3 h. Digest proteins by adding 5× TNESK to a final of 1× and incubating overnight at 37°. Extract proteins with one phenol extraction followed by one chloroform extraction. For maximal recovery of DNA, back-extract the phenol phase with TE buffer. Add 5 μg of glycogen (stock 20 mg/ml) as a carrier to facilitate the precipitation of small amounts of DNA. Add 0.1 volumes of 3 M Na acetate and 2.5 volumes of ethanol to each sample and incubate at $-20°$ overnight. Centrifuge for 5 min, decant supernatant, gently dry and resuspend pellets in TE buffer. The DNA is ready for analysis by PCR.

Acknowledgment

Work on this project was supported by a grant from the NIH (GM47903).

[25]   Immunofluorescent Staining of Polytene Chromosomes: Exploiting Genetic Tools

By Gena E. Stephens, Carolyn A. Craig, Yuhong Li, Lori L. Wallrath, and Sarah C. R. Elgin

Immunostaining of *Drosophila* polytene chromosomes is a powerful tool for investigating the components of chromatin on a genome-wide scale. Techniques for immunostaining were developed in the 1970s and have evolved since their introduction.[1–4]

Many of the nuclei of *Drosophila* undergo multiple rounds of DNA replication without cell division during the larval stages of development, a process known as endoreduplication. While the euchromatic regions are copied ca. 10 times, the pericentric heterochromatin undergoes only a few rounds of replication, and centromeric satellite DNA and the Y chromosome are not amplified due to their heterochromatic nature. The replicated chromosome arms remain tightly aligned and the heterochromatic regions fuse in a common chromocenter. The amplification of the DNA provides additional substrate for binding proteins and allows one to obtain a map of protein distribution on polytene chromosomes with the use of an antibody specific for a given protein. The largest polytene chromosomes, the product of 10 rounds of replication, are found in the salivary glands of

[1] L. M. Silver and S. C. R. Elgin, *Proc. Natl. Acad. Sci. USA* **73,** 423 (1976).

[2] C. Rodriguez-Alfageme, G. T. Rudkin, and L. H. Cohen, *Proc. Natl. Acad. Sci. USA* **73,** 2038 (1976).

[3] M. Jamrich, A. L. Greenleaf, and E. K. F. Bautz, *Proc. Natl. Acad. Sci. USA* **74,** 2079 (1977).

[4] B. E. Schwartz, J. K. Werner, and J. T. Lis, *Meth. Enzymol.* **376,** 393 (2004).

the third instar larvae (see also Schwartz *et al.*[4] for more information on endoreduplication).

Differential packaging of the DNA along each chromosome arm results in a distinct banding pattern. Some of the constrictions represent sites of underreplication, but in most cases, the bands and the interbands differ only in their compaction of the DNA and not in their level of polyteny.[5] In 1935, Bridges drew a detailed map of each chromosome arm; more recently photographic representations have been made.[6] A physical map has been established through the technique of in situ hybridization. This technique has allowed a specific band or region within a band to be correlated with the location of a specific gene or repetitive sequence at a resolution of ca. 30 kb.[7]

Immunological methods for studying the association of proteins with polytene chromosomes have been used to address a variety of biological questions. In wild-type flies, immunostaining has been used to determine the global distribution of one or more proteins; colocalization studies have been done to determine if a protein of interest might be in close association or part of a multiprotein complex with other proteins.[8] Immunological staining has also been performed on a variety of *Drosophila* species to monitor conservation of a chromosomal protein and detect changes in its distribution through evolution. Much has also been gained by taking advantage of various genetic tools. Fly lines with chromosome rearrangements have been used to show when the localization of a protein reflects local as opposed to global structural features, for example, proximity to the chromocenter.[9] Mutations in chromosomal proteins can be assessed by immunofluorescent staining to monitor both the extent of DNA binding and distribution of the mutant protein, and the impact on chromosome organization and distribution of other proteins.[10] Finally, *Drosophila melanogaster* can be transformed using a transposable element (usually a *P*-element) carrying a cloned DNA fragment, allowing experiments to examine the protein complexes that associate with a given construct.

[5] A. Sperier and P. Sperier, *Nature (London)* **307**, 176 (1984).
[6] G. Lefevre, *in* "The Genetics and Biology of *Drosophila*" (M. Ashburner and E. Novitski, eds.), **1a**, p. 31. Academic Press, London, 1976.
[7] J. Gall and M. L. Pardue, *Proc. Natl. Acad. Sci. USA* **63**, 378 (1969).
[8] C. D. Shaffer, G. E. Stephens, B. A. Thompson, L. Funches, J. A. Bernat, C. A. Craig, and S. C. R. Elgin, *Proc. Natl. Acad. Sci. USA* **99**, 14332 (2002).
[9] T. C. James, J. C. Eissenberg, C. Craig, V. Dietrich, A. Hobson, and S. C. R. Elgin, *Eur. J. Cell Biol.* **50**, 170 (1989).
[10] G. Schotta, A. Ebert, V. Krauss, A. Fischer, J. Hoffman, S. Rea, T. Jenuwein, R. Dorn, and G. Reuter, *EMBO J.* **21**, 1121 (2002).

Such assemblies can be inherited through mitosis,[11] and sometimes through meiosis, an event termed "cellular memory."

Squashing and Staining Protocols

The technique presented here is used for determining the in situ distribution of chromosomal proteins in polytene chromosomes using specific antibodies. We will focus on a procedure using formaldehyde fixation to crosslink the proteins to the DNA with subsequent squashing in acetic acid.[1,12] Other labs have published similar techniques.[2–4] The procedure to be described produces polytene chromosome spreads of good morphology and preserves most bound proteins in an immunologically reactive state. Antibodies specific for a chromosomal protein are applied to polytene chromosomes, and a fluorescent or enzyme-linked secondary antibody that is directed against the primary antibody is then incubated on the chromosomes. The chromosomes are viewed and photographed with a fluorescence microscope and digital camera. This technique allows for reproducible results in determining the localization of proteins on polytene chromosomes to the resolution of individual bands (ca. 30 kb).

Fixation of Polytene Chromosomes

Third instar larvae that have been grown with minimal crowding are the best source of polytene chromosomes. Choose the fattest larvae that have just crawled out of the food and up the container wall. A visual montage of the dissection of salivary glands from third instar larvae is shown in Fig. 1. Carefully remove larvae from the bottle with dissection forceps and wash in Cohen buffer[13] in microwells. Place the larva in a microwell with fresh Cohen buffer and using two pairs of dissection forceps, pull apart the larva in one motion, grasping the mouth hooks at the anterior of the larva with one pair of forceps and the posterior of the larva with the other pair of forceps. The salivary glands will most likely be removed along with the brain, eye-antennal imaginal discs, and fat tissue as one mass. Then, dissect the salivary glands away from the fat body and any other extraneous tissue, while allowing the glands to remain attached to each other through their common duct. This duct is a convenient place to grasp the glands while removing any extraneous tissue and while transferring the glands to new solutions. Be careful not to damage the glands while removing any tissues. If

---

[11] G. Cavalli and R. Paro, *Science* **286**, 955 (1999).

[12] L. M. Silver, C. E. C. Wu, and S. C. R. Elgin, *in* "Methods in Chromosomal Protein Research" (G. Stein, J. Stein and L. Kleinsmith, eds.), p. 151. Academic Press, New York, 1977.

[13] L. H. Cohen and B. V. Gotchel, *J. Biol. Chem.* **246**, 1841 (1971).

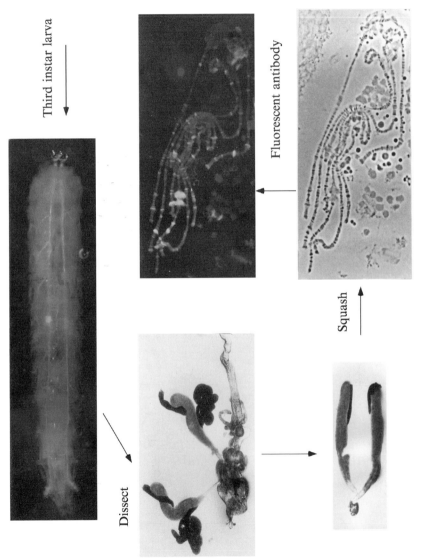

FIG. 1. Diagram of isolating, squashing, and staining salivary gland polytene chromosomes. Reprinted from: *Methods Cell Biol.* **35**, 203 (1991), with permission from Academic Press.

the glands are damaged, the morphology of the chromosomes will suffer. If the adhering fat body cannot be removed at this step without damaging the glands, it can be removed later during acetic acid fixation. The paired salivary glands will either appear slightly milky or clear and are somewhat wishbone-shaped. Incubate the excised glands in Cohen buffer for 8–10 min. Incubation of glands in this solution with detergent allows for the dissolution of cytoplasmic membrane structures. Poor morphology of the chromosomes will result if an 8- to 10-min incubation is exceeded. Next, the glands are incubated for 3–25 min in a formaldehyde fixative. Shorter incubations can be used but may result in incomplete fixation and hence loss of some chromosomal proteins, while longer incubations result in chromosomes that are difficult to spread out well. The glands are then transferred to 45% acetic acid and incubated for 3–60 min. During this incubation, any adhering fat body may be removed. This acetic acid fixation is necessary to attain good phase morphology of the spread polytene chromosomes. The phase morphology will suffer if the glands are left in the acetic acid for more than 1 h. This formaldehyde fixation technique minimizes the extraction of chromosomal proteins while maintaining good antigenicity.[1]

Reagents. All stock solutions should be kept at 4° unless otherwise indicated (see Table I).

TABLE I

Final concentration	Volume/5 ml	Stock solution
Cohen buffer		
10 mM MgCl₂	50 μl	1 M
25 mM sodium glycerol 3-phosphate	125 μl	1 M, pH 7 (−20°)
3 mM CaCl₂	150 μl	0.1 M
10 mM KH₂PO₄	250 μl	0.2 M
0.5% NP40	250 μl	10%
30 mM KCl	750 μl	0.2 M
160 mM sucrose	1.00 ml	0.8 M (−20°)
H₂O	2.425 ml	
(Can be kept at 4° for 2–3 days.)		
Formaldehyde fixative		
0.1 M NaCl	100 μl	5 M
2 mM KCl	50 μl	0.2 M
10 mM NaH₂PO₄	50 μl	1 M, pH 7
2% NP40	1.00 ml	10%
2% formaldehyde	270 μl	37%
H₂O	3.53 ml	
(Must be made fresh daily. Use 37% formaldehyde stock within 6 months of purchase.)		

TABLE II

Final concentration	Volume/5 ml (ml)	Stock solution (%)
Formaldehyde fixative squashing solution		
45% acetic acid	2.25	100
H_2O	2.75	
(Must be made fresh daily.)		
Acetic acid/formaldehyde squashing solution		
45% acetic acid	2.25	100
3.7% formaldehyde	0.25	37
H_2O	2.25	
(Must be made fresh daily.)		

Squashing without prior formaldehyde fixation may result in loss of most of the histones and approximately 10% of the nonhistone chromosomal proteins. In particular, lysine-rich proteins will be extracted.[1] However, depending on the protein, it may be necessary to skip the formaldehyde fixation prior to squashing. In some instances, formaldehyde can hide antigenic determinants and/or lead to denaturation of the protein. Forty five percent acetic acid does not extract or perturb most nonhistone chromosomal proteins. If omitting the 1 h formaldehyde fixation, the salivary glands are fixed less harshly by excising in acetic acid/formaldehyde squashing solution (45% acetic acid, 3.7% formaldehyde, see Table II) and incubating for 3–40 min. This solution usually gives better preservation of morphology and antigenicity than 45% acetic acid alone.[14] Glands that are fixed in this solution are also squashed in this solution. This "conventional fixation technique" results in polytene chromosomes with good band morphology and good antigenicity of many proteins, but a significant number of proteins are extracted.

Squashing of Polytene Chromosomes

After the glands are fixed, they are now ready to be squashed. Place a drop of 45% acetic acid into the center of a clean siliconized coverslip[15] and transfer the glands to the drop with forceps. Clean a glass slide with 95% ethanol, dry it, and then position the middle of the slide so that it is on top of the coverslip. Break open the salivary gland nuclei by picking up the slide and moving the coverslip back and forth by gently tapping

[14] G. Holmquist, *Chromosoma* **36**, 413 (1972).
[15] J. Sambrook and D. W. Russell, "Molecular Cloning: A Laboratory Manual," Cold Spring Harbor Laboratory Press, Cold Spring Harbor, NY, 2001.

the edges of the slide between the thumb and forefinger. Once the salivary gland cells and nuclei break open, the arms of the chromosomes will spread out. Turn the slide over so that the coverslip is now on top of the slide and monitor the spreading of the chromosome arms by looking at them under a phase-contrast microscope at 400×. Mark the location of the squash with a marker so that it may be easily found when looking under the fluorescent microscope. If the chromosome arms have not spread sufficiently, place the slide on the bench, coverslip up, and tap the coverslip with the eraser end of a pencil. When the spreading of the chromosome arms is satisfactory, flatten the chromosomes by firmly applying thumb pressure to the coverslip. Place a folded tissue between the thumb and the coverslip so that grease from the hands does not get on the slide. Be sure that the coverslip does not move under pressure. If movement occurs, stretching and breaking of the chromosome arms will result. If too much pressure is applied, the result will be fragmented chromosomes. Also, be sure that the squash does not dry out. If the squash is allowed to dry, subsequent staining will be poor. Once the spread is satisfactory, quickly submerge the slide in liquid nitrogen using hemostatic forceps. Once bubbling has ceased, remove the slide from the liquid nitrogen and flick off the coverslip with a razor blade. Be sure not to scrape the slide or the chromosomes may be scraped off. Before the specimen is allowed to thaw, immerse the slide in Tris-buffered saline (TBS-Tween, see Table III). If the slides will not be used within 2–4 h, you may store them in slide storage medium at −20° for up to a month. If the squash is allowed to dry out at any time during the procedure, the result will be poor staining of the chromosomes; the stain will outline the chromosomes rather than staining specific bands (see Fig. 2). The squashes are

TABLE III

Final concentration	Volume	Stock solution
10× TBS-Tween		
0.2 *M* Tris-HCl	200 ml	1 *M*, pH 8
17% NaCl	170 g	
Tween-20 (Sigma)	10 ml	100%
H$_2$O	Bring to 1 L	
(Dilute 10-fold in H$_2$O before use.)		
Slide storage medium		
67% glycerol	335 ml	100%
33% PBS	165 ml	[a]

[a] One liter of phosphate-buffered saline (PBS) is made with 20 g of NaCl, 0.5 g of KCl, 0.5 g of KH$_2$PO$_4$, and 1.45 g of Na$_2$HPO$_4$.

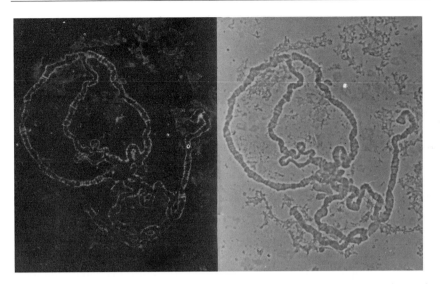

FIG. 2. Fluorescent outlining of chromosomes. *Left panel:* phase-contrast. *Right panel:* immunofluorescence. Adapted, with permission, from: L. M. Silver, "Methods for Analyzing the *In situ* Distributions of Chromosomal Proteins by Immunofluorescence." Ph.D. thesis, Harvard University, 1977.

most vulnerable to drying out after they are flattened. After flattening, the liquid layer is very thin. If the acetic acid is seen to recede from the edges of the coverslip, particularly at the corners, the specimen may be too dry to be stained well. Therefore, during the squashing procedure, perform all steps as quickly as possible without sacrificing the quality of your specimen. A humidifier on the bench may help if the air is dry. Results will improve with practice. A troubleshooting guide is provided at the end of this section to help solve any problems that are encountered.

Antibody Staining Procedure

To determine the localization pattern of chromosomal proteins on the polytene chromosomes, the spreads are first incubated with a primary antibody against the protein(s) of interest and then incubated with a fluorescent or enzyme-linked secondary antibody against the primary antibody. Immunofluorescent labeling will be described in detail here. Other secondary labeling techniques may also be used. The use of a fluorescently labeled secondary antibody to localize to primary antibodies was first achieved by Coons in the early 1940s.[16,17] In this technique, 5–10 secondary antibodies can bind to each primary antibody, greatly increasing the original signal.[18–20] The antibody staining procedure described here is essentially as performed in Weller and Coons,[21] with some minor modifications as described by Silver and Elgin.[1,22]

Several criteria must be followed when using antibodies for immunofluorescent staining of polytene chromosomes. First, the titer of both the primary and the secondary antibodies must be determined empirically. If too high a concentration of either antibody is used, extensive nonspecific staining or the loss of specific signal might result. Too high of a dilution will result in a loss of signal. Typical dilutions used in our lab range from 1:50 to 1:5000. Second, appropriate controls must be carried out to be sure that the antiserum is specific for the protein of interest. Preimmune serum from many animals will carry antibodies that react generally with the chromosomes and some will carry antibodies that will give a very specific banding pattern. The background signal will often be lost upon dilution since nonspecific antibodies are usually of low titer, but it is best to screen animals by checking the preimmune serum prior to selecting animals for immunization. Antibodies used for staining should be tested for specific binding to *Drosophila* chromosomal proteins by western blotting or immunoprecipitation to be sure that the interaction is specific. Unfortunately, the antibody may react well on western blots but not on polytene chromosomes or vice versa. Chromosome staining with only the secondary antibody should be done with each new batch of antibody obtained to be sure that the secondary antibody by itself does not label the chromosomes.

To begin the staining procedure, wash the slides with the polytene chromosome squashes three times with gentle agitation for 5 min each in cold TBS-Tween. The antiserum or monoclonal antibody should be diluted in TBS-Tween as determined empirically. Blocking agents may be added to the solution to help reduce nonspecific labeling. Commonly used blocking agents are fetal calf or goat serum (10%), 5 mg/ml γ-globulin, and/or 5% nonfat milk (all final concentrations). Use of these blocking agents will not alter antibody labeling patterns. With a tissue, dry all regions of the slide except for the squash area. Place the slide horizontally in a humidity chamber. The humidity chambers in our lab consist of a plastic box with a

[16] A. H. Coons, H. J. Creech, and R. N. Jones, *Proc. Soc. Exp. Biol. Med.* **47**, 200 (1941).

[17] A. H. Coons, H. J. Creech, R. N. Jones, and E. Berliner, *J. Immunol.* **45**, 159 (1942).

[18] M. Goldman, *in* "Fluorescent Antibody Methods." Academic Press, New York, 1968.

[19] L. A. Sternberger, *in* "Immunocytochemistry." Prentice-Hall, Englewood Cliffs, NJ, 1974.

[20] C. A. Williams and M. W. Chase, *in* "Methods in Immunology and Immunocytochemistry." **5**. Academic Press, New York, 1976.

[21] T. H. Weller and A. H. Coons, *Proc. Soc. Exp. Biol. Med.* **86**, 789 (1954).

[22] L. M. Silver and S. C. R. Elgin, *in* "The Cell Nucleus" (H. Busch, ed.), **5**, p. 215. Academic Press, New York, 1978.

tight-sealing lid. Wet paper towels or cheesecloth should be placed at the bottom of the chamber and glass rods on clay holders are used to keep the slides perfectly horizontal and above the towels. Empty plastic pipette tip racks can also be used to support the slides in a box. Apply 100–200 μl of diluted antiserum with or without blocking agent to the squash area immediately after wiping the slide with the tissue and placing the slide in the humidity chamber. Be sure that there are no bubbles in the applied solution or some parts of the squash may not be exposed to the antibody. Be sure that the squash is kept wet at all times. Incubate the slides with the primary antibody for 15–120 min at room temperature.

When the primary incubation is complete, rinse the antiserum from the slide with cold TBS-Tween in a squirt bottle. Then wash the slide in cold TBS-Tween (with gentle agitation of the slide in a rack) three times for 5 min each. Wipe the slide with the tissue as before and place back in the humidity chamber. Add the correct dilution of secondary antibody to the slide in the same manner that the primary antibody was added and incubate for 15–120 min at room temperature as described earlier. Typical dilutions range from 1:50 to 1:5000. Keep the slides away from bright light when incubating with the fluorescently labeled secondary antibody, or the fluorescence will fade. Work quickly and use conditions of low light if possible. After the incubation is complete, wash the slides with cold TBS buffer as before and wipe the slide dry except for the region of the squash. Place one drop of mounting solution on the squash and then place a non-siliconized coverslip over it. Mounting solution may be made as shown in Table IV or purchased under the name of Vectashield Mounting Medium (cat# H-1000, Vector Labs). Blot excess solution from the slide and then, with a tissue, press the coverslip hard onto the slide without letting the coverslip slide. Blot excess solution from the sides once more and then use clear fingernail polish to seal the squash and to prevent the specimen from drying out. It is advantageous to use clear fingernail polish in case some of the chromosomes end up covered by the polish.

We generally use a Nikon E600 microscope with a 400× oil immersion lens, an epifluorescence attachment, and an Optronics digital camera for

TABLE IV

Final concentration	Volume/100 ml	Stock solution
Mounting solution		
90% glycerol	90 ml	100%
0.1 *M* Tris-HCl	10 ml	1 *M*, pH 7
0.2% *n*-propyl gallate	0.2 g	

viewing and photography. It is helpful to scan the slide with a 10× dry lens first to find a good squash and then move to a 400× oil immersion lens for photographs. The images are processed in Photoshop and printed on photo-quality glossy paper using a high-quality ink jet printer or laser printer. Different fluorophores fade at different rates. It is advantageous to minimize the amount of time spent viewing the chromosomes under the fluorescent microscope prior to photography to minimize fading. It is also advantageous to store the slides in the dark at 4°; this keeps the slides from drying out and the fluorophores from fading.

The methodology described thus far is for labeling one protein of interest. Multiple proteins may be viewed on polytene chromosomes by incubating with the primary antibodies at the same time (mixing them together in the diluent) or incubating sequentially with wash steps in between. The secondary antibodies may also be added together or sequentially.

Pulverize n-propyl gallate with a mortar and pestle before weighing, and allow it to dissolve by stirring overnight in the solution. n-Propyl gallate inhibits the loss of fluorescence during viewing.

Troubleshooting Guide

The following is a troubleshooting guide. For each problem stated, there is a possible explanation and an approach described to correct the problem. (Adapted from *Methods Cell Biol.* **35,** 214–216 (1991), with permission from Elsevier.)

a. Small Polytene Chromosomes

1. Larvae may be crowded or may have been harvested prior to the correct stage.
   Correction: Be sure that the larvae are not crowded; harvest them when they are their fattest and crawling up the side of the bottle.
2. Larvae may have been grown at too warm of a temperature.
   Correction: Growth at 18° is optimal for wild-type flies such as Oregon-R, but may vary for other fly lines.

b. Poor Chromosome Morphology (i.e., Indistinct Bands Under Phase Contrast, Smeared Staining Pattern)

1. Incubation too long in Cohen buffer, insufficient incubation in formaldehyde fixative, or incubation too long in 45% acetic acid.
   Correction: Check incubation periods. Different incubation periods may be optimal for some proteins of interest; this should be determined empirically.

c. Polytene Chromosomes Do Not Spread Well

1. Extended incubation in formaldehyde fixative.
   Correction: Incubate for 25 min or less.
2. Spreading incomplete.
   Correction: Using a phase-contrast microscope, constantly monitor the squash; tap the coverslip over the squash with the eraser end of a pencil until chromosome arms are well spread.
3. Squashed too gently.
   Correction: Apply more pressure in the final flattening.

d. Too Much Background Material

1. Glands were insufficiently washed in Cohen buffer.
   Correction: Incubate the glands for at least 8 min.
2. Fat body and other extraneous tissues were not removed.
   Correction: Remove extraneous tissues before squashing.
3. Dirty coverslips/and or slides.
   Correction: Clean slides and coverslips with 95% ethanol, dry with lens paper, handle glass at edges, and squash with a tissue between your thumb and the coverslip.

e. Broken Polytene Chromosomes

1. Squashing may have been done with too much force, or the coverslip may have moved during the procedure.
   Correction: Squash more gently and monitor under a phase-contrast microscope. Hold the coverslip while tapping with an eraser on the end of a pencil to prevent sliding.
2. Insufficient incubation in formaldehyde fixative, or too long in 45% acetic acid.
   Correction: See b1.

f. Refractile Polytene Chromosomes

1. Chromosomes squashed too gently. The chromosomes are not flat enough.
   Correction: See c3.
2. Allowed chromosomes to dry.
   Correction: Keep squash area wet at all times. If necessary, use a humidifier on the bench during squashing steps if the air is very dry.
3. After liquid nitrogen freezing, the chromosomes thawed.
   Correction: Immediately immerse slides in TBS after removing the coverslip.

g. *Staining Pattern Outlines the Chromosome*

1. Chromosomes dried (one observes fluid retracting at the corners of the coverslip).
   Correction: See f2.

h. *Stretched-Out Chromosomes*

1. Coverslip moved during flattening.
   Correction: See e1.
2. Insufficient formaldehyde fixation.
   Correction: See b1.

i. *No Intact Chromosomes in Squash Area*

1. Squashed gland floated out from under coverslip.
   Correction: Use less 45% acetic acid on coverslip.
2. Coverslip moved during squashing or squashed too hard.
   Correction: See e1.
3. Cannot find chromosomes in squash area at high magnification.
   Correction: With a waterproof marker, mark the position of the squashed gland on the edges of the coverslip.
4. Scraped off chromosomes with razor blade.
   Correction: Be sure to freeze the slide before removing the coverslip; insert the edge of a razor blade under a corner of the coverslip and flip off the coverslip with an upward motion.
5. Chromosomes fell off the slide after the slide was frozen.
   Correction: If the slides are not to be used within 2–4 h, keep the slides in storage medium.
6. Incomplete siliconization of the coverslip.
   Correction: Review methods of siliconization.[15]

j. *High Intensity of Staining of Debris*

1. Antiserum contains antibodies that cross-react with components other than the chromosomal protein of interest.
   Correction: Preabsorb or affinity purify the antiserum.
2. No blocking agents or insufficient amount of blocking agents used.
   Correction: Add 10% calf or goat serum, 5 mg/ml of γ-globulin, and/or 5% nonfat dried milk to the antibody. Increase the concentrations if needed.
3. Antibody dried onto chromosome squash.
   Correction: Use the humidity chamber during antibody incubations, be sure that the slides are level on application of antibody, and use a sufficient amount of antibody solution ($\sim$200 μl) to avoid drying.

4. Titer of antibody is too high.
Correction: Decrease titer of antibody.

k. Weak or Absent Signal on Chromosomes

1. Antibody does not react to the antigen on chromosomes.
Correction: Try variations in the fixation protocol.
2. Incubation in formaldehyde is too long.
Correction: See c1.
3. Weak or faded fluorescent label.
Correction: On incubation with fluorescent conjugated antibody, keep slides in the dark. Be sure *n*-propyl gallate was added to the mounting medium.
4. Titer of antibody is too low.
Correction: Increase antibody titer.
5. Incorrect secondary antibody is being used.
Correction: Check that the secondary antibody is against the animal that the primary antibody was derived from.

l. Chromosomes Are Present, but Not Coming into Focus or Are Blurry

1. Two coverslips may be stuck together on the slide.
Correction: Remove the upper coverslip with a razor blade.
2. Chromosomes were not squashed flat and are difficult to focus on due to multiple focal planes.
Correction: Be sure to sufficiently flatten the chromosomes when squashing.
3. The table that the microscope is on may be vibrating while you are viewing the chromosomes and taking pictures.
Correction: A vibration isolation table may be needed if a suitable stable surface cannot be found.

Applications and Results

Immunofluorescent staining can be used to analyze the global distribution of a specific protein on polytene chromosomes, as well as to compare that of two or more proteins simultaneously in wild-type or mutant larvae. It can also be used for comparative studies of *Drosophila*. Genetic stocks with rearrangements can provide important information regarding the autonomy of a protein's association within a region of the polytene chromosomes. The impact of mutations in chromosomal proteins on global organization of the chromosomes can also be examined. Immunological staining can indicate if proteins localize correctly in the mutant background or if they localize ectopically. Finally, we will discuss the value of transgenic fly lines that can be used to test the effect of a specific transcriptional regulator on gene expression.

Simultaneous Localization of Multiple Proteins on Polytene Chromosomes: HP1, HP2, and Modified Histones

An example of the use of immunofluorescence microscopy of polytene chromosomes to help to determine the function of an unknown protein comes from studies of Heterochromatin Protein 1 or HP1. Monoclonal antibodies were produced using a protein extract of *D. melanogaster* embryonic nuclei, and the antibodies were screened by staining polytene chromosomes as described earlier.[23] One of these antibodies, C1A9, recognizes a protein that is highly concentrated in the chromocenter of salivary gland polytene chromosome squashes of third instar larvae, HP1. HP1 is most prominently associated with the chromocenter, the telomeres, and the small fourth chromosome; some euchromatic sites are stained, albeit with less intensity.[9] The chromocenter, as mentioned previously, is formed from the fused centric heterochromatin of all four chromosomes. Thus, the staining pattern suggests that the antigen is primarily associated with heterochromatin.[24] The gene for HP1 was mapped to region 29A,[23,25] where a dominant suppressor of position effect variegation (PEV) had already been mapped.[26] PEV occurs when a gene that normally resides within euchromatin is placed near heterochromatin; the gene is expressed in some cells and repressed in others, giving rise to a variegated phenotype.[27] This phenotype is most commonly observed in the fly eye, when the *white* gene is translocated next to heterochromatin on the X chromosome. Mutations within HP1 have been found to be strong suppressors of PEV.[25] These results indicate that HP1 contributes to the heterochromatic structure responsible for silencing. In the last few years, it has been shown that HP1 binds to histone H3 methylated at lysine 9, a prominent marker of heterochromatic domains.[28,29]

[23] T. C. James and S. C. R. Elgin, *Mol. Cell. Biol.* **6,** 3862 (1986).

[24] T. C. James, J. C. Eissenberg, C. Craig, V. Dietrich, A. Hobson, and S. C. R. Elgin, *Eur. J. Cell Biol.* **50,** 170 (1989).

[25] J. C. Eissenberg, T. C. James, D. M. Foster-Hartnett, T. Hartnett, V. Ngan, and S. C. R. Elgin, *Proc. Natl. Acad. Sci. USA* **87,** 9923 (1990).

[26] D. A. R. Sinclair, R. C. Mottus, and T. A. Grigliatti, *Mol. Gen. Genet.* **191,** 326 (1983).

[27] J. B. Spofford, *in* "The Genetics and Biology of *Drosophila*" (M. Ashburner and E. Novitski, eds.), **2a,** p. 955. Academic Press, Orlando, FL, 1976.

[28] A. J. Bannister, P. Zegerman, J. F. Partridge, E. A. Miska, J. O. Thomas, R. C. Allshire, and T. Kouzarides, *Nature* **410,** 120 (2001).

[29] M. Lachner, D. O'Carroll, S. Rea, K. Mechtler, and T. Jenuwein, *Nature* **410,** 116 (2001).

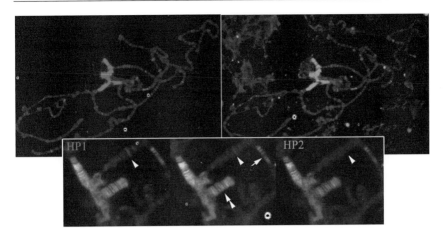

FIG. 3. Simultaneous immunolocalization of HP1 and HP2. *Upper:* polytene chromosomes stained with anti-HP1 (red, left) and anti-HP2 (Ab P-6; green, right). *Lower:* close-up view of the chromocenter. Single antibody signals are shown in black and white; the color merge of HP1 signal (red) and HP2 signal (green) is shown in the center. Note the staining of the chromocenter, in a banded pattern along the fourth chromosome (double arrowhead) and of 5–6 bands in 31B (arrow). The arrowhead identifies a euchromatic band positive for HP1 but not HP2. Adapted, with permission, from: C. D. Shaffer, G. E. Stephens, B. A. Thompson, L. Funches, J. A. Bernat, C. A. Craig, and S. C. R. Elgin, *Proc. Natl. Acad. Sci.* **99,** 14332 (2002). (See color insert.)

We are interested in other proteins that bind to HP1 and might contribute to a multiprotein complex required for heterochromatin-induced gene silencing. Through a yeast two-hybrid screen using HP1 as bait, an HP1-interacting protein, Heterochromatin Protein 2 (HP2), has been identified.[8] Upon staining of polytene chromosomes with a polyclonal antibody prepared against a peptide of HP2 from the C-terminal region, a polytene chromosome staining pattern nearly coincident with that of HP1 is seen (see Fig. 3). Mutations in HP2 result in suppression of PEV, suggesting that HP2 also has a role in heterochromatin structure. The nearly coincident staining pattern of these two proteins shows that they colocalize, thus suggesting that they may be part of a multiprotein complex.

Methylation of lysine 9 on histone H3 has been found to recruit HP1.[28,29] HP1 also interacts with the histone methyltransferase SU(VAR)3–9,[10,30,31] providing a mechanism for the spreading of heterochromatin. Various

[30] L. Aagaard, G. Laible, P. Selenko, M. Schmid, R. Dorn, G. Schotta, S. Kuhfittig, A. Wolf, A. Lebersorger, P. B. Singh, G. Reuter, and T. Jenuwein, *EMBO J.* **18,** 1923 (1999).

[31] S. Rea, F. Eisenhaber, D. O'Carroll, B. D. Strahl, Z. W. Sun, M. Schmid, S. Opravil, K. Mechtler, C. P. Ponting, C. D. Allis, and T. Jenuwein, *Nature* **406,** 593 (2000).

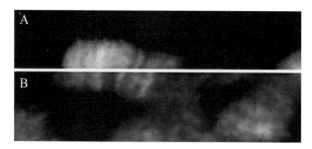

Fig. 4. Immunfluorescent staining for HP1 versus histone H3 acetylated at lysine 14 on the fourth chromosome of *D. melanogaster*. *Panel a:* HP1. *Panel b:* Histone H3 acetylated at lysine 14. Notice the nonoverlapping staining patterns of HP1 and this histone modification, suggesting different functions.

histone modifications have been found to recruit nonhistone chromosomal protein complexes. Antibodies specific for the modified histones have allowed us to analyze the global distribution of both the nonhistone chromosomal proteins and the modified histones. Figure 4 shows double labeling using antibodies for histone H3 acetylated on lysine 14, associated with active regions of the genome,[32] versus HP1, associated with inactive regions, on the fourth chromosome of *D. melanogaster*. Through double labeling, it is evident that HP1 and histone H3 acetylated on lysine 14 have different distribution patterns and therefore are likely to have different roles in gene regulation.

Analyzing the Distribution of Proteins on Rearranged Chromosomes and Chromosomes from Related Species

As mentioned previously, HP1 was found prominently associated with the fourth chromosome of *D. melanogaster*. This chromosome is immediately adjacent to the chromocenter. To show that the staining of HP1 on the fourth chromosome is not due to the chromosome's close proximity to the chromocenter, a fly line with a translocation of a portion of the fourth chromosome to the third chromosome was stained for HP1 association.[9] Salivary gland chromosome squashes were prepared from third instar larvae of the stock *T(3;4)f/In(3L)P*. A region of approximately seven bands of the fourth chromosome is translocated to position 65D1–2 on chromosome 3 in this stock. C1A9 antibody staining of this stock shows that the association of the fourth chromosome with HP1 is not dependent on its close association with the chromocenter.

[32] J. Nakayama, J. C. Rice, B. D. Strahl, C. D. Allis, and S. I. S. Grewal, *Science* **292,** 110 (2001).

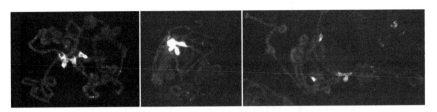

FIG. 5. Immunofluorescent staining of HP1 in three *Drosophila* species. *Left panel: D. melanogaster. Middle panel: D. sechellia. Right panel: D. pseudoobscura.* Notice the prevalent staining of the chromocenter and the dot chromosome.

In another line, $T(2;3)lt^{x13}$, chromosome arm 2L is broken within the β-heterochromatin at 2LH37, between the *light* gene and the chromocenter, and joined to chromosome arm 3R at map position 97D2. Thus, a portion of β-heterochromatin is now separated from the chromocenter by long segments of euchromatin. Upon immunofluorescent staining with the C1A9 antibody, this region is prominently stained, again indicating that the association of HP1 with this heterochromatic region is autonomous and not dependent on close association with the chromocenter.

The availability of fly stocks from other *Drosophila* species allows one to check for homologues and determine their localization. An HP1 homolog has been identified in *D. virilis* by preparing polytene chromosome squashes and staining with C1A9 antibody.[9] There are 40–60 million years of evolution between *D. melanogaster* and *D. virilis*. HP1 has been observed as a heterochromatic protein, predominantly associated with the pericentric heterochromatin found at the chromocenter in all *Drosophila* species examined to date (*D. pseudoobscura, D. sechellia*, and others) (see Fig. 5). Interestingly, the banded portion of the fourth chromosome or dot chromosome is not stained by HP1 antibodies in *D. virilis*.[9]

Analyzing the Impact of Mutations in Chromosomal Proteins

Su(var)3–9 encodes a histone methyltransferase that selectively methylates lysine 9 of histone H3 and binds to heterochromatin through direct interaction with HP1.[10] SU(VAR)3–9, like HP1, has been shown to be a dominant modifier of PEV, that is, heterochromatin-induced gene silencing.[33] Immunological assays of larval salivary gland polytene chromosomes done using a polyclonal antibody indicate that SU(VAR)3–9 associates with the chromocenter, the fourth chromosome, and more weakly at the telomeres and a few euchromatic sites.[10] GFP antibodies also recognize a

[33] G. Reuter and I. Wolff, *Mol. Gen. Genet.* **182**, 516 (1981).

SU(VAR)3-9-EGFP fusion protein. The role of HP1 in the association of SU(VAR)3-9 with heterochromatin was examined by staining HP1-deficient third instar larvae that express a SU(VAR)3-9-EGFP fusion protein. In these larvae, SU(VAR)3-9-EGFP remains associated with the heterochromatin and the fourth chromosome, but it is also associated generally with euchromatic regions due to methylation of lysine 9 in histone H3 at euchromatic sites. HP1 is thus essential for maintaining the correct localization of SU(VAR)3-9 to heterochromatin.

SU(VAR)3-9 null mutants were analyzed for the distribution of HP1 on polytene chromosomes. In these larvae, the association of HP1 with the chromocenter was dramatically reduced. Interestingly, the association of HP1 with the fourth chromosome was not affected, suggesting that HP1 binding to the fourth chromosome may be governed by a different methyltransferase activity localized to that portion of the genome.

Examining Protein Localization on Custom P-Elements

Drosophila polytene chromosomes offer the ability to monitor protein-DNA and protein-protein interactions that participate in chromatin packaging and gene expression *in vivo*. To test the effects of a specific transcriptional regulator on gene expression it is often desirable to tether the regulator upstream of a reporter gene. This can be accomplished by fusing the transcriptional regulator to a DNA binding domain that has a known target sequence. The target sequence is cloned upstream of a reporter gene whose expression can be easily monitored. This methodology has been applied to the study of HP1. Full-length HP1 was fused to the DNA binding domain of the LacI repressor (see Fig. 6, panel A). The LacI-HP1 fusion protein was expressed in *Drosophila* stocks carrying a mini-*white* reporter transgene with LacI binding sites positioned upstream of the transcription start site (see Fig. 6, panel A). Expression of the reporter gene provides pigment to the *Drosophila* eye. Association of the HP1 fusion protein with the LacI binding sites was observed in preparations of fixed and squashed polytene chromosomes (see Fig. 6, panel B) and correlated with the lack of eye pigmentation at 25 of 26 sites tested.[34] These results indicate that HP1 can establish silent chromatin at the majority of ectopic chromosome locations.

Cytological studies on fixed and squashed polytene chromosomes can also be used to monitor protein-protein interactions *in vivo*. In the case described earlier, it is possible to determine whether HP1 associates with known partner proteins when tethered at ectopic genomic locations. An

[34] Y. Li, J. R. Danzer, P. Alvarez, A. S. Belmont, and L. L. Wallrath, *Development* **130,** 1817 (2003).

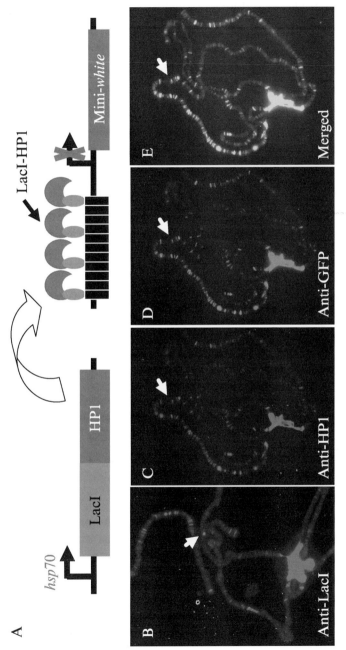

Fig. 6. *(continued)*

association between the histone H3 methyltransferase SU(VAR)3–9 and HP1[10] is thought to be responsible for the mechanism of spreading silent chromatin.[29] Polytene chromosomes from a stock containing tethered HP1 and expressing the GFP tagged SU(VAR)3–9 protein were fixed, squashed, and stained with antibodies to HP1 and GFP. The results indicate that SU(VAR)3–9 is recruited to the site of tethered HP1 as well as sites of endogenous HP1 (see Fig. 6, panels C–E).

Reporter transgenes have also been designed to help elucidate the issue of cellular memory. The Fab-7 element is used by *Drosophila* Polycomb and trithorax group proteins to maintain repressed or active gene expression, respectively, of the segmentation genes during embryogenesis.[35] Once the active or repressed state is established, the state is mitotically inherited throughout development and to some degree through meiosis. Histone H4 hyperacetylation of this element can be inherited epigenetically through mitosis and meiosis and can be followed by immunological staining of polytene chromosomes. Fab-7 has thus been deemed a cellular memory module.[36]

Conclusions

In conclusion, we have presented a method for determining the *in vivo* distribution of chromosomal proteins on *Drosophila* polytene chromosomes. Some applications of this technique have also been discussed. By combining genetic, biochemical, and molecular biology techniques with the cytological approach, a greater understanding of the molecular

[35] G. Cavalli and R. Paro, *Cell* **93**, 505 (1998).
[36] K. Ekwall, T. Olsson, B. M. Turner, and R. C. Allshire, *Cell* **91**, 1021 (1997).

FIG. 6. Determining protein-DNA and protein-protein interactions on chromosomes *in vivo*. (A) Diagram of the tethering system showing the LacI-HP1 fusion transgene expressed from an *hsp70* promoter (left) and the mini-*white* reporter transgene containing upstream *lacI* repeats (right). (B) Polytene chromosomes from a *Drosophila* stock expressing a LacI-HP1 fusion protein and carrying the mini-*white* reporter inserted at cytological position 93A/B. Localization of the HP1 fusion protein is detected using antibodies to the LacI DNA binding domain (green). LacI-HP1 associates with the *lacI* repeats (arrow) and with sites of endogenous HP1. (C)–(E) Polytene chromosomes from a stock expressing a LacI-HP1 fusion protein and a SU(VAR)3-9 protein tagged with GFP. HP1 localization is detected with antibodies to HP1 (red, panel C), SU(VAR)3-9EGFP localization is detected with antibodies to GFP (green, panel D), and colocalization is observed in the merged image (yellow, panel E). HP1 and SU(VAR)3-9 show colocalization over the mini-*white* reporter at cytological position at 1E (arrow) and at sites of endogenous HP1. (See color insert.)

mechanisms of gene regulation can be gained. Few other systems offer the well-developed genetic tools in combination with the ability to perform cytological studies as *Drosophila* does. For more information on the formation of polytene chromosomes and various other applications, see Schwartz *et al.*[4] this volume.

[26]   Indirect Immunofluorescent Labeling of *Drosophila* Polytene Chromosomes: Visualizing Protein Interactions with Chromatin *In Vivo*

By BRIAN E. SCHWARTZ, JANIS K. WERNER, and JOHN T. LIS

Background

Drosophila melanogaster was first recognized as a valuable experimental organism 100 years ago. One of the particularly attractive features of this model system for studies of genes and their regulation is the "giant" or polytene chromosomes that occur in the secretory glands of *Drosophila* (as well as other dipteran flies). Historically, polytene chromosomes provided an important link between the genetic map and the physical location of deletions, insertions, inversions, and translocations on the genetic map. More recently, these chromosomes have provided an efficient means of analyzing both the global distribution of particular proteins on chromosomes and the recruitment of particular proteins to specific chromosomal loci that are undergoing changes in activity.

Polytene chromosomes form in cells that grow in size without dividing during larval development. In these cells, homologous diploid pairs of chromosomes are tightly paired and undergo successive rounds of amplification. Each synapsed pair may replicate up to nine times forming around 1000 strands of DNA, which remain aligned and attached to each other. This precise alignment allows differences in chromatin compaction to be seen as a series of bands and interbands extending across the width of the chromosome arms. Electron microscopy studies show that about 95% of the DNA is concentrated in the bands, which are much more compacted regions of chromatin than the interbands, which separate the bands. Although the packing ratio of chromatin in polytene chromosomes varies regionally, the average packing ratio has been estimated to be between 30 and 57, about 100 times more extended than a metaphase chromosome. This is slightly less compact than a 30-nm chromatin fiber, which has a

METHODS IN ENZYMOLOGY, VOL. 376   0076-6879/04 $35.00

packing ratio of about 50. The amplification of aligned chromatids is responsible for the impressive width of the polytene chromosome, which is about 5 μm, or the length of a typical metaphase chromosome. A tight association is formed not only among the replicated chromatids but also among the four chromosomes of the *Drosophila* genome at the centromeres of each of the chromosomes. These chromosome centromeres are attached to each other and together comprise the chromocenter, a region of underreplicated heterochromatin that is diffuse and lacks the distinctive banding of the euchromatin arms.

Despite their unusual structure, polytenes function as interphase chromosomes and are transcriptionally active and responsive to hormonal and environmental signals. The most visible manifestations of this transcriptional activity are the puffs that can form at highly active loci. Puffs are localized regions of decondensed chromatin that, in *Drosophila*, occur in a developmental stage-specific pattern. They also appear at loci harboring stress-responsive genes following an environmental challenge, such as heat shock. Generally, puffing occurs at loci undergoing high levels of transcription, and the size of the puff roughly corresponds to the length of the transcription unit.[1] Nonetheless, the processes of transcription and puffing are not strictly coupled. For example, the drug sodium salicylate can elicit puffs at some heat shock gene loci, yet transcription of these genes is blocked by this drug.[2] This type of puffing is likely a consequence of salicylate-induced DNA binding of heat shock factor to the multiple regulatory elements associated with these loci.

Indirect immunofluorescence staining of polytenes provides a rapid way of determining the genomic distribution of chromosomal proteins. This was first recognized in early experiments in the laboratories of Elgin[3] and Bautz.[4] These early experiments revealed with spectacular clarity the location of RNA polymerase II and its strong association with chromosomal puffs and interbands, and histones and their concentration in DNA-rich chromosomal bands. More recently, antibodies have been used to investigate a variety of protein interactions with chromatin *in vivo*. For example, male-specific lethal (msl) proteins involved in X chromosome dosage compensation in males stain only the X chromosome.[5] Factors involved in gene silencing are found predominantly in heterochromatin.[6] Many

[1] J. A. Simon, C. A. Sutton, R. B. Lobell, R. L. Glaser, and J. T. Lis, *Cell* **4,** 805 (1985).
[2] N. A. Winegarden, K. S. Wong, M. Sopta, and J. T. Westwood, *J. Biol. Chem.* **271,** 2697 (1996).
[3] L. M. Silver and S. C. Elgin, *Chromosoma* **68,** 101 (1978).
[4] U. Plagens, A. L. Greenleaf, and E. K. Bautz, *Chromosoma* **59,** 157 (1976).
[5] J. R. Bone and M. I. Kuroda, *Genetics* **144,** 705 (1996).
[6] T. C. James, J. C. Eissenberg, C. Craig, V. Dietrich, A. Hobson, and S. C. Elgin, *Eur. J. Cell Biol.* **50,** 170 (1989).

transcription and RNA processing factors localize to sites of highly active transcription, the chromosomal puffs.[7] A TBP-related factor, TRF1, stains chromosomes in a manner that revealed it was used predominantly for Pol III gene promoters.[8]

The interpretation of a chromosomal staining pattern generated with a specific antibody depends on the antibody or antibodies used. Several steps can be taken to ensure that the pattern of antibody staining represents the true distribution of the protein used to generate the antibody. First, several animals can be used to generate antibody, and the specificity of the antibodies can be checked by probing western blots of nuclear proteins. Only those antibodies recognizing preferentially single bands corresponding to a peptide of the correct molecular weight should be used to probe chromosomes. Obviously, the more specific the western the better; however, a knowledge of the modification states, degradation products, and different size forms of the protein is useful in assessing the specificity of the antibody when more than one band is observed. Additionally, antibodies can be purified by affinity chromatography to the specific antigen to further enrich for the specific antibody.[9] Finally, the specificity of any staining pattern can be checked by double-staining with two antibodies raised to the same protein antigen but produced in different species (e.g., rat and rabbit[10]). The staining patterns of two specific antibodies should show significant overlay, or better, complete overlap.

The ability to immunostain for two or three factors at once allows one to compare factor distributions relative to one another. The colocalization of two factors may be indicative of a physical interaction *in vivo*. The staining pattern of the kinase cdk7, for example, largely colocalizes with its phosphorylated product, the carboxy-terminal domain of RNA polymerase II.[11] Similarly, the kinase P-TEFb can be seen to colocalize with its Pol II product and both can be compared to the promoter-bound transcription factor HSF in the triple staining shown in Fig. 1.

In addition to gaining a genome-wide perspective on protein distribution, immunostaining of polytene chromosomes can also provide higher resolution views of single genes. Now that the sequence of the genome is complete (http://www.fruitfly.org/), and the physical and polytene maps are more precisely correlated, the antibody staining of a particular band

[7] E. D. Andrulis, J. Werner, A. Nazarian, H. Erdjument-Bromage, P. Tempst, and J. T. Lis, *Nature* **420,** 837 (2002).

[8] S. Takada, J. T. Lis, S. Zhou, and R. Tjian, *Cell* **101,** 459 (2000).

[9] J. T. Lis, P. Mason, J. Peng, D. H. Price, and J. Werner, *Genes Dev.* **14,** 792 (2000).

[10] E. D. Andrulis, E. Guzman, P. Doring, J. Werner, and J. T. Lis, *Genes Dev.* **14,** 2635 (2000).

[11] B. E. Schwartz, unpublished results.

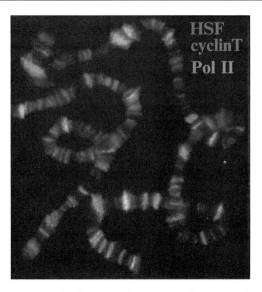

FIG. 1. Triple immuno-stain. Polytene chromosomes from heat shocked larvae were stained with antibodies against HSF (red), the cyclin T subunit of P-TEFb (blue), and the serine 2 phosphorylated form of RNA polymerase II (green). (See color insert.)

or interband can be interpreted with precision. The resolution depends on the chromatin environment of the gene; decondensed chromatin allows for better resolution than condensed chromatin. For this reason, genes residing within the interband regions are much better suited for this type of analysis than those found in bands. In puffs, the most extreme case of chromatin decondensation, resolution is maximized. Puffs are therefore ideally suited to the study of transcription factors and allow one to resolve factors associated with gene promoter regions from downstream transcribed regions. For instance, the activator of heat shock genes, HSF, binds to specific promoter DNA elements on heat stress and triggers the recruitment and elongation of RNA polymerase II through the body of the gene. The gene occupancy of these two factors can be resolved within the heat shock puff (see Fig. 2).

The resolution and sensitivity of polytene immunofluorescence can be further enhanced with the use of various transgenic fly lines. High-resolution studies of a single gene of interest are often aided by mapping a single-copy transgene that is removed from its native chromatin context. This is especially valuable if the native site contains several repeats of the gene, or nearby genes, which can complicate the assignment of a specific stained polytene band to a particular gene or DNA segment. Detecting a

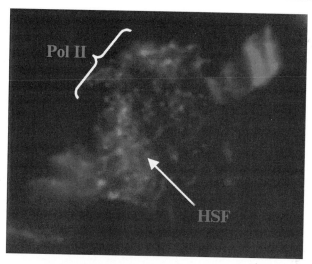

FIG. 2. Elongating form of Pol II (green) resolves from promoter-bound HSF (red) during heat shock. The heat shock puff from a single transgenic copy of an *Hsp70-lacZ* fusion gene at chromosomal locus 9D is shown. The HSF band at the left edge of the puff marks the promoter of the transgene, while elongating Pol II appears within the body of the puff. (See color insert.)

new labeled band created at a transgenic site (not present in the parental line) to which the gene or DNA segment has been moved provides unambiguous proof of the assignment. Creating transgenic sites with smaller and smaller segments can, in principle, pinpoint the DNA sequence with which the protein associates.

The sensitivity of the immunofluorescent signal can be further enhanced by examining polymeric sequences introduced on transgenes. We have used this approach to create a polymer of 40 copies of a 55-bp *hsp70* promoter fragment[12] that vigorously recruits HSF during heat shock.[13] By co-immunostaining with antibodies against other factors, we have identified proteins that interact with DNA-bound HSF (see Fig. 3). Therefore, this technique not only allows the amplification of otherwise weak immunofluorescent signals but also allows one to determine the protein recruitment ability of defined promoter elements.

[12] H. Xiao, Ph.D. Thesis, Cornell University, 1989.
[13] L. S. Shopland and J. T. Lis, *Chromosoma* **105**, 158 (1996).

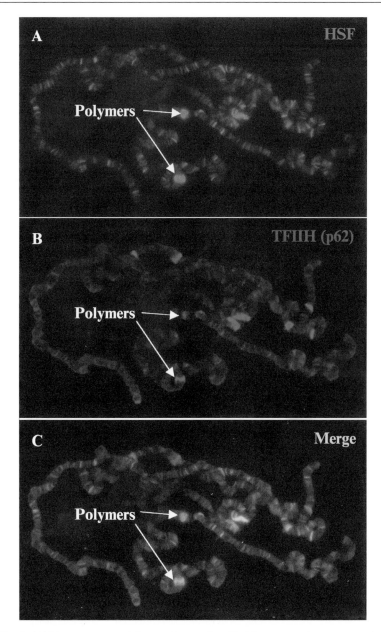

FIG. 3. (A) Transgenes containing a polymer of an *Hsp70* promoter fragment recruit HSF (green) during heat shock. (B) HSF can, in turn, recruit TFIIH (red). (C) Merge. (See color insert.)

Methods

Culturing of Drosophila *for Healthy Larvae*

The best source of polytene chromosomes is from the salivary glands. To obtain the best chromosome spreads, it is very important that healthy larvae are used. Grow animals in bottles of newly made yeast glucose media at about 23° and make sure the bottle is not overcrowded. The fly lines are grown at about 23°. Approximately 40 flies are put into the bottle for 2 days. New bottles are set up every 2 days. Larvae are collected from bottles that are 6–8 days old. At this point bottles should have light brown pupae as well as third instar larvae on the walls of the bottle. Wandering third instar larvae are used typically. If first or second instar larvae are climbing the walls, the culture is too dense or a media problem exists.

Heat Shocking of Larvae

Non-heat shock: Third instar larvae are taken directly from the culture bottle and placed in the dissection buffer and dissected immediately.

Heat shock larvae: Third instar larvae are taken from the culture bottle and placed on a piece of plastic wrap that has a small piece of moist filter paper on it. The plastic wrap is closed with a dialysis clamp and submerged in a 36.5° water bath for the prescribed amount of time. At the end of the heat shock time, very quickly remove the plastic wrap from the water bath and move the larvae quickly from the plastic wrap into the dissection buffer and dissect immediately.

Dissection of Salivary Glands from Larvae [Modifications from Protocol Received from Renato Paro]

The larvae are dissected in 100 μl of dissection buffer (0.7% NaCl), in a Boerner slide. To dissect the salivary glands from the third instar larvae, hold the larvae with one pair of forceps and then using another pair of forceps pull from the mouth parts. The salivary glands need to be identified and dissected free from all the other parts of the larvae. Fat bodies can be removed from the salivary gland, although not all need to be removed. Transfer the salivary glands to 40 μl of fix buffer[†] on Rain-X–treated coverslip.[‡] Incubate for 10–20 min before squashing.

[†] Fix Buffer: 50 μl 37% paraformaldehyde (1.85 g paraformaldehyde in 5 ml H_2O and 70 μl 1 N KOH). Heat to 80° for 30 min. Store at −20° in aliquots:

450 μl acetic acid
500 μl H_2O
Keeps for 2 h at room temperature

Squashing of Salivary Glands

Using a base-treated slide,[§] pick up the coverslip with the fix buffer and gland on it. Using a needle probe, tap up and down on the coverslip over the area where the salivary glands are located. It is acceptable to let the coverslip slide around as long as it stays in the same general area on the microscope slide. Holding an edge of the coverslip in place with one hand, using the needle probe, make circular motions on the coverslip, paying particular attention to the area where the salivary gland is located. Place slide, coverslip down, on a Kimwipe and apply pressure with your hand to remove excess moisture from the slide. Using a phase scope equipped with a 25× lens, scan the slide to find the chromosomes. Look for intact chromosomes that have strong banding and are well-spread, avoiding those slides that contain severely twisted or broken chromosomes. Put the slide and coverslip in liquid nitrogen until ready to process all slides.

Staining of Slides

1. Remove slides from liquid nitrogen. Blow breath over coverslip. Using a razor blade under one corner, flick the coverslip off quickly. Using a diamond-tip pen, make small lines on the microscope slide to indicate where coverslip was.
2. Wash:
   2× 10 min in PBS
   1× 10 min in PBS with 1% Triton.
3. Block 40–60 min in block solution (PBS with 5% non-fat milk powder).
4. Rinse well in PBS.
5. Add primary antibodies in PBS and 1% BSA, for a total of 20 μl per slide. Put antibody solution directly over the area where the squashes are. Cover with 22-mm^2 coverslip, using lines etched in the slide to position the coverslip. Leave overnight at 4° in humid chamber.

[‡] Rain-X–treated coverslips: using a Kimwipe saturated with Rain-X, coat 22 mm^2 coverslip with Rain-X. Lay coverslip on Kimwipe to dry. Rinse coverslip with Kimwipe saturated with water. Let it dry. Clean the coverslips with lens paper prior to using. These treated coverslips can be stored and used later.

[§] Base treated slides: dissolve 140 g NaOH in 560 ml dH$_2$O. Add 840 ml 95% EtOH. Put microscope slides in glass slide trays. Cover with NaOH solution. Leave shaking for 2 h. Do 4× 1 h rinses with dH$_2$O. Put in drying oven. The slides need to dry at least overnight before using. The slides should be stored in the oven until used. Slides that are removed from the oven and not used can be returned to the oven and used another time.

Concentration of primary antibody used varies for different antiserum preparations. We usually use them at 1:10 to 1:200.

6. Dip microscope slide in PBS to remove coverslip.
7. Rinse in PBS.
8. Wash 3× 5 min in block solution.
9. Rinse in PBS.
10. Add secondary antibody solution: antibodies in PBS, 1% BSA, and 2% donkey goat serum. Again have 20 μl per slide. Cover with 22-mm^2 coverslip, using lines etched in slide to position coverslip. Leave for 1 h in humid chamber at room temperature. Concentration of secondary antibodies is 1:200 to 1:500. Secondary antibodies are from Jackson ImmunoResearch, and we usually use RhodamineRedX and Cy2.
11. Dip microscope slide in PBS to remove coverslip.
12. Wash 10 min in "wash 300" (300 mM NaCl, 0.2% NP40, 0.2% Tween 20).
13. Wash 10 min in "wash 400" (400 mM NaCl, 0.2% NP40, 0.2% Tween 20).
14. Rinse in PBS.
15. Hoechst stain (4 μl Hoechst in 50 ml PBS for 20 min).
16. 2× 15 min washes in PBS.
17. Add about 20 μl of glycerol containing *n*-propylgallate (2.5%) to coverslip.
18. Tap-dry the slide and touch to the glycerol on the coverslip.
19. The coverslip can be held in place with a dab of nail polish.
20. Store slides at −20°.

Imaging of Chromosome Spread

Imaging is done on a Zeiss Axioplan Microscope. Images are taken with either 25× or 100× lens and are processed with Openlab software.

Analysis of Chromosome Staining Patterns

The successful experiment that produces 50 intense sites of chromosomal labeling is a joy to view, but this joy is soon replaced by the realization of the daunting task that remains—mapping these sites. The identification of the sites of antibody staining on polytene chromosome, even for the seasoned cytologist, usually requires comparison to reference photographs and drawings. A place to begin is with the photographs and maps of the arms and description of landmarks provided by Lefevre.[14] This combination of representative photos and drawings is valuable in making the initial assignments of labeled bands to a particular chromosomal arm and region. Comparing the stained chromosomes to photographs of chromosomal segments at different stages of development in Ashburner[15] helps one deal with the variations in banding patterns that occur in normal development. Defining the developmental stage of particular salivary gland chromosomal spread early in the process can direct one's attention to chromosomal maps representing the same stage. Finally, Sorsa's book of "Chromosome Maps of *Drosophila*" compiles detailed illustrations (drawings of Bridges and Bridges, photographic maps, and electron microscopic maps) of each of the 102 chromosomal numbered divisions, and is invaluable for high-resolution mapping.[16] We find it convenient to make Powerpoint slides of separate images and merged images (e.g., antibody and DNA stain). Lines are drawn to each labeled band in the antibody-stained image, and then this set of lines is copied and pasted on the DNA-stained pattern. The DNA-stained bands at and around each line are identified by comparing to reference maps and written onto the Powerpoint image as one proceeds, providing a "working map."

Advantages and Limitations

Immunofluorescent staining of polytene chromosomes provides a highly visual representation of transcription factor distribution *in vivo*. The primary advantage of this technique is the ease and quickness with which the distribution of a particular protein over the entire *Drosophila* genome can be assessed. This procedure can be completed in 1 or 2 days, although the detailed mapping and analysis of labeled sites can take much longer depending on the complexity of the labeling pattern and the type of information one needs to extract from the pattern.

The advantage of surveying the entire *Drosophila* genome ironically is accompanied by the biggest limitation: resolution. While the method is about 100 times higher in resolution than comparable analysis of metaphase chromosomes, the site of labeling can rarely be assigned to a specific gene or promoter without further genome manipulations (as in the use of transgenic lines discussed earlier). As mentioned, the best spatial resolution can be obtained within puffs, yet even in the largest transgenic puffs

---

[14] G. Lefevre, Jr., *in* "The Genetics and Biology of *Drosophila*" (M. Ashburner and E. Novitski, eds.), p. 31. Academic Press, New York, 1976.

[15] M. Ashburner, *in* "Developmental Studies on Giant Chromosomes" (W. Beerman, ed.), p. 101. Springer, Berlin, 1972.

[16] V. Sorsa, "Chromosome Maps of *Drosophila*." CRC Press, 1988.

it is only possible to resolve the 5' from the 3' regions of a single gene with certainty. This may be somewhat improved with a confocal microscope or by taking a photographic Z-series and using software that can subtract background fluorescence. If the goal is to map the position of a protein within a gene with very high resolution, a more appropriate assay is chromatin immunoprecipitation or ChIP, coupled with assays of the ChIP DNA on microarrays. In this approach, proteins are crosslinked to chromatin in living cells with formaldehyde, much as they are in polytene chromosome fixation. The protein of interest is immunoprecipitated. The DNA covalently attached to the immunoprecipitated protein is purified, PCR amplified, tagged with a fluorescent label, and used to probe a microarray (chip) that contains DNA representing various portions of the gene of interest. Microarrays containing spotted *Drosophila* cDNA libraries have been produced (BD Biosciences, Palo Alto, CA); however, the complete genome is not yet available.

Another limitation of polytene immunofluorescence analysis is the challenge of quantifying immunofluorescent signals. The amount of immunofluorescent signal at a particular locus can vary from squash to squash for several reasons. First, the extent of polytenization can differ in chromosomes derived from the same salivary gland. This could lead to exaggerated signals in over-replicated chromosomes and more modest signals in comparatively under-replicated chromosomes or chromosomal regions. Second, the preparation of the squash can lead to deformations such as over-stretching and twisting of the chromosome arms—aberrations that can in some cases significantly affect the measured amount of fluorescent signal. Due to the nature of the squashing procedure, these deformations almost always occur. It is therefore helpful to screen for damaged chromosomes with a general DNA stain such as DAPI or Hoechst (see also Stephens *et al.*[17] for more information on troubleshooting squashes). Although quantifying immunofluorescent bands presents the above-mentioned problems, it can be accomplished successfully if the chromosomes are carefully prepared and screened for those with crisp, undistorted banding patterns. We have used NIH Image software to quantify the intensity of signals at specified loci.[13] By measuring the signals of several squashes it is possible to make quantitative conclusions within acceptable limits of error. Typically, these measurements are borne out by other experimental approaches, such as ChIP analysis.

Not all chromosomal proteins show highly specific, strong-staining patterns. While antibodies to many chromosomal proteins that have been examined produced well-defined patterns of staining that can be readily

[17] G. E. Stephens, C. A. Craig, Y. Li, L. L. Wallrath, and S. C. R. Elgin, *Meth. Enzymol.* **376**, 372 (2004).

mapped, some do not (we estimate more than half of the proteins we tested work well). There are those that show no staining pattern, even when multiple antibodies are used, and those that produce detectable weak staining of specific sites and a diffuse staining of chromosomes. In some cases, examining multiple high-quality chromosome squashes can produce slides with sufficient signal to noise ratios to allow mapping.[18] The range of responses seen with different proteins and different antibodies can be simply explained in some cases by the quality of antibodies and abundance of a protein on a particular site, but in other cases, other factors such as the efficiency of fixing proteins to chromosomes, availability of the antigen to chromosome surface, or the ability of antigenic determinants to survive fixation will influence signal strength. Trying more than one slide preparation protocol can sometimes help solve these problems.

The study of transcription in *Drosophila* is often aided by the use of drugs that can inactivate or otherwise regulate the activity of specific factors. Because the reliable uptake of such drugs by whole flies can present dosage problems, it is worthwhile to note that explanted salivary glands have been successfully treated with drugs *in vitro*. In this way, the drug dosage can be precisely controlled and delivered directly to tissues containing polytene chromosomes. Glands cultured in buffered media remain physiologically active and are also capable of mounting a heat shock response. Therefore, it is possible to inactivate or regulate a specific transcription factor, apply a heat shock treatment or other inducement of particular gene expression, prepare a chromosome spread, and immunostain to determine the effect on transcription *in vivo*. This approach has been used to show that heat shock gene transcription is sensitive to alpha amanitin.[19] The localization of other factors may be indirectly influenced by inactivating the protein of interest, and this can also be assayed by immuno-staining.

In summary, immunofluorescent staining of polytene chromosomes is a reliable, fast, and convenient means of assessing the genome-wide distribution of chromatin binding proteins. The availability of transgenic flies and the ability to treat salivary glands with drugs makes it a powerful tool to investigate the function and dynamics of transcription and other factors *in vivo*.

[18] J. M. Park, J. Werner, J. M. Kim, J. T. Lis, and Y. J. Kim, *Mol. Cell* **8,** 9 (2001).
[19] J. L. Compton and B. J. McCarthy, *Cell* **14,** 191 (1978).

[27] X-Chromosome Inactivation in Mouse Embryonic Stem Cells: Analysis of Histone Modifications and Transcriptional Activity Using Immunofluorescence and FISH

By Julie Chaumeil, Ikuhiro Okamoto, and Edith Heard

X-chromosome inactivation provides a powerful model system with which to investigate the different steps in facultative heterochromatin formation. During early development, one of the two X chromosomes is transcriptionally silenced in every cell of a female embryo, thereby achieving dosage compensation between males and females for X-linked gene products.[1] The X inactivation process[2] is dependent on the action of a unique RNA, Xist, which coats the X chromosome in *cis* and induces its inactivation (see Brockdroff[3] for review). Once established, the inactive state of the X chromosome is highly stable in somatic cells and is normally only reversed in the female germ line. The inactive X is characterized by a number of features in addition to Xist RNA coating, such as asynchronous replication timing,[4] DNA methylation of promoter regions of house-keeping genes,[5] various modifications of core histone,[6–10] and association with a histone variant, macroH2A.[11,12] The facultative nature of the X inactivation process allows the kinetic and functional dissection of the role that these characteristics play in the establishment and maintenance of inactive chromatin. As X inactivation occurs during the earliest differentiation steps in embryogenesis (3.5–5.5 days postcoitum in the mouse), the number of

---

[1] M. F. Lyon, *Nature,* **190,** 372 (1961).
[2] P. Avner and E. Heard, *Nat. Rev. Genet.* **1,** 59 (2001).
[3] N. Brockdorff, *Trends Genet.* **18,** 352 (2002).
[4] N. Takagi, O. Sugawara, and M. Sasaki, *Chromosoma* **85,** 275 (1982).
[5] D. P. Norris, H. N. Brockdorff, and S. Rastan, *Mamm. Genome,* **1,** 78 (1991).
[6] P. Jeppesen and B. M. Turner, *Cell* **74,** 281 (1993).
[7] B. A. Boggs, P. Cheung, E. Heard, D. L. Spector, A. C. Chinault, and C. D. Allis, *Nat. Genet.* **30,** 73 (2002).
[8] A. H. Peters, J. E. Mermoud, D. O'Carroll, M. Pagani, D. Schweizer, N. Brockdorff, and T. Jenuwein, *Nat. Genet.* **30,** 77 (2002).
[9] J. Silva, W. Mak, I. Zvetkova, R. Appanah, T. B. Nestorova, Z. Webster, A. H. F. M. Peters, T. Jenuwein, A. P. Otte, and N. Brockdorff, *Dev. Cell* **4,** 481 (2003).
[10] K. Plath, J. Fang, S. Mlynarczyk-Evans, R. Cao, K. A. Worringer, H. Wang, C. C. de la Cruz, A. Otte, B. Panning, and Y. Zhang, *Science* **300,** 131 (2003).
[11] C. Costanzi and J. R. Pehrson, *Nature* **393,** 599 (1998).
[12] B. P. Chadwick and H. F. Willard, *Hum. Mol. Genet.* **10,** 1101 (2001).

embryonic cells available at these stages can be limiting. Embryonic stem (ES) cells, which are pluripotent cells derived from the inner cell mass of blastocysts, provide a useful tissue culture system for studying X inactivation. ES cells can be maintained in the undifferentiated state and differentiation can be easily induced *in vitro*. In the case of ES cells with more than one X chromosome, X inactivation occurs on differentiation. This system thus allows the successive steps of X inactivation to be followed[13–16] and has been shown to parallel closely the events that occur *in vivo*.[9,10]

In this section, we outline techniques for analyzing X-inactivation kinetics in differentiating ES cells. We focus particularly on changing patterns of histone modifications during X inactivation, using immunuofluorescence combined with RNA FISH on interphase nuclei, as well as metaphase chromosome staining combined with DNA FISH. We also describe a protocol for assaying late replication timing of the inactive X chromosome. The techniques and tools presented should provide a basis for defining potential causal relationships between different events, not only during X inactivation but also during the establishment of other patterns of gene activity.

Background to the Early Events in the X Inactivation Process

In undifferentiated ES cells, all X chromosomes present are active and Xist is expressed at very low levels, detectable by RNA FISH as a punctate signal (or "dot") at its site of transcription (see Fig. 1A). The onset of X inactivation requires an increase in steady-state levels of *Xist* from the chromosome that will be inactivated. Thus, when differentiation of female ES cells is induced, Xist RNA accumulates in *cis* over the territory of the X chromosome in interphase nuclei and this "coating" can be detected by RNA FISH as a domain equivalent to the X-chromosome territory (see Fig. 1B). This is followed (1–2 days later) by transcriptional silencing of X-linked genes, based on the observed disappearance of primary transcript signals detected by RNA FISH.[15,17] Using immunofluorescence combined with RNA FISH, it has been found that X-chromosome–wide changes in

---

[13] A. M. Keohane, L. P. O'Neill, N. D. Belyaev, J. S. Lavender, and B. M. Turner, *Dev. Biol.* **180,** 618 (1996).

[14] A. Wutz and R. A. Jaenisch, *Mol. Cell* **5,** 695 (2000).

[15] E. Heard, C. Rougeulle, D. Arnaud, P. Avner, C. D. Allis, and D. L. Spector, *Cell* **107,** 727 (2001).

[16] J. E. Mermoud, B. Popova, A. H. F. M. Peters, T. Jenuwein, and N. Brockdorff, *Curr. Biol.* **12,** 247 (2002).

[17] J. Chaumeil, I. Okamoto, M. Guggiari, and E. Heard, *Cytogenet. Res.* **75,** (2002).

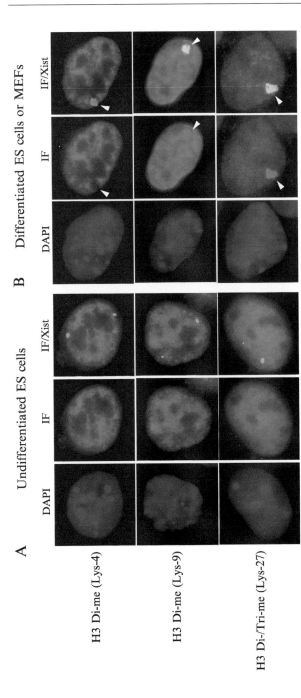

Fig. 1. Examples of immunofluorescence combined with RNA FISH on interphase nuclei. Modifications of the N-terminal histone tails of histone H3 on the inactive X chromosome in differentiating female ES cells. In each case, nuclei are shown with typical patterns observed in undifferentiated female ES cells (panel A) or during differentiation (panel B). Immunodetection with Alexa GAR 568 conjugated secondary antibody (red, column 2 of each panel) was combined with Xist RNA FISH (Spectrum Green labeled probe, green, column 3 of each panel). Three antibodies were used here, which detect di-methylation of H3 lys-4 (from Upstate Biotechnology), di-methylation of H3 lys-9 (gift from D. Allis, also available from Upstate Biotechnology), and di-/tri-methylation of H3 lys-27 (gift from D. Reinberg, see Sarma and Reinberg, this volume). Prior to the onset of X inactivation, in undifferentiated ES cells, Xist is transcribed at a low level from both X chromosomes and the primary transcripts can be detected as a "dot" at each Xist locus (green, A). At the beginning of inactivation, Xist RNA starts to accumulate, over the future inactive X chromosome, and can be detected as a green domain in the nuclei (green, B). In this

the modifications status of histones, such as di- and tri-methylation and hypoacetylation of H3 Lys-9,[15,16] tri-methylation of H3 Lys-27,[9,10] as well as hypomethylation of H3 Lys-4, Lys-36, and Arg-17,[15,17] are early events in X inactivation, occurring immediately after Xist RNA coating begins. The early appearance of these global histone changes on the X chromosome undergoing inactivation has been correlated with the Xist RNA-dependent recruitment of the Polycomb group proteins, Eed, and Enxl (Ezh2) to the X chromosome,[9,10] the latter being a histone methyltransferase with mainly H3 Lys-27 and some H3 Lys-9 activity in mammalian cells.[18] Other changes, such as late replication timing,[4] appear to occur within a similar window of time,[13,17] while DNA methylation of promoter regions of housekeeping genes[13] and association with the histone variant macroH2A[19] appear to be later events. Histone modifications thus represent strong candidates for the early changes responsible for transcriptional silencing of the X chromosome. Furthermore, as histone modifications are found on the X chromosome during mitosis, they may carry part of the mitotically heritable epigenetic signature that renders the inactive state clonally stable.

Cell Culture

Culture of Mouse Embryonic Fibroblasts

In order to assess the status of the fully inactive X chromosome in somatic cells, female mouse embryonic fibroblasts (MEFs), prepared from 13.5-day embryos (sexed by gonad inspection) can be used.[20] These can be cultured in DMEM with GlutMAX (GIBCO/Invitrogen, cat. no. 31966–021) supplemented with 10% fetal bovine serum (FBS; Invitrogen,

[18] A. Kuzmichev, K. Nishioka, H. Erdjument-Bromage, P. Tempst, and D. Reinberg, *Genes Dev.* **16**, 2893 (2002).

[19] T. P. Rasmussen, M. A. Mastrangelo, A. Eden, J. R. Pehrson, and R. Jaenisch, *J. Cell Biol.* **150**, 1189 (2000).

[20] B. Hogan, R. Beddington, F. Constantini, and E. Lacy, "Manipulating the Mouse Embryo: A Laboratory Manual." Cold Spring Harbor Laboratory Press, Cold Spring Harbor, NY, 1994.

same window of time, some histone modifications begin to appear on the inactive X chromosome (red, B). In merge images of immunofluorescence and RNA FISH (third column, B), green coloration shows exclusion of the modification or yellow coloration shows the enrichment on the Xist domain. The inactive X chromosome is depleted in H3 Lys-4 di-methylation (first row) and enriched in H3 Lys-9 di-methylation (second row) and Lys-27 di-/tri-methylation (third row) (B, arrowheads). DNA is stained with DAPI (blue, first column of each panel). (See color insert.)

cat. no. 10270106). It should be noted that only low passage number (no higher than 4–5) cells should be used if characteristics of primary, somatic cells wish to be examined.

ES Cell Culture

For detailed protocols on ES cell culture, readers are urged to consult dedicated source of information, Hogan *et al.*[20] and Robertson.[21] For proper maintenance of ES cell lines in their undifferentiated, pluripotent state, a well-equipped tissue culture facility and rigorous culture conditions are required. Some general recommendations include the use of sterile, disposable plasticware or glassware that has never been exposed to detergent and is kept separate from general laboratory supplies, as ES cells are highly sensitive to trace levels of detergent. Only high-quality water, classified as "Type I Reagent Grade Water" (ASTM standard), should be used. Fetal bovine serum should be purchased as "ES cell grade" or various batches must be tested in parallel for quality control. Tissue culture grade plastic flasks or plates, or glass coverslips (ESCO, cat. no. 9611301) are used for ES cell culture and in all cases they have to be first coated in filter-sterilized 0.1% gelatin (Type A, Sigma cat. no. G-2500) in PBS 1X (PBS 10X, GIB-CO, cat. no. 70013–016; sterile water, Invitrogen, 15230–089). The gelatin solution is left in contact with the plastic for a minimum of 30 min (it should not be allowed to dry out) and should be aspirated just prior to addition of cell medium. We grow ES cells at 37° in 8% CO_2.

ES cells can be maintained in the undifferentiated state either through culture on mitotically inactivated feeder cells (such as mouse embryonic fibroblasts) and/or in the presence of leukemia inhibitory factor (LIF), depending on the cell line used. It should be noted that, for X inactivation studies, ES cells should prefererably be cultured on male rather than female feeder cells in order to avoid interference during analysis due to the presence of the inactive X chromosome in female fibroblast cells. For X inactivation kinetics studies, we have mainly used two female ES cell lines, LF2 (a gift from Dr. Austin Smith) and PGK12.1 (a gift from Dr. Neil Brockdorff), both of which grow on gelatin-coated flasks or plates,[22] without the need for feeder cells. It should be noted that similar X inactivation kinetics have been found using feeder-dependent ES cell lines. ES cells are maintained as an undifferentiated culture in DMEM with GlutMAX, 15% fetal calf serum (GIBCO, cat. no. 16141–079), 10^{-4} mM 2-mercaptoethanol

[21] E. J. Robertson, *in* "Teratocarcinomas and Embryonic Stem Cells: A Practical Approach" (E. J. Robertson ed.), p. 71. IRL Press, Oxford, England, 1987.
[22] A. G. Smith, *J. Tiss. Cult. Meth.* **13**, 89 (1991).

(Sigma, cat. no. M7522), and 1000 U/ml LIF (Chemicon, cat. no. ESG1107). ES cells should be plated and maintained at relatively high density and passaged at 70–80% confluence. This usually means passaging every couple of days (cell doubling time can be between 8 and 22 h depending on the ES cell line and serum used). ES cell colonies should be monitored daily for density and also any signs of differentiation (see refs. 20 and 21 for feeder cell–dependent ES cells and Smith[22] for non–feeder-dependent cell lines). The medium should be changed 3 h prior to trypsinization to increase viability of the cells upon passaging or cryopreservation. To passage, the flask or plate is rinsed with PBS, then trypsin-EDTA is added (GIBCO, cat. no. 25200–072). After 8 min in the incubator, an equal volume of ES cell medium is added and the cells are dispersed into a single cell suspension by vigorous pipetting (20–40 times) with a plugged Pasteur pipette, then centrifuged. This resuspension step is important, as undissociated clumps of ES cells will rapidly form large colonies and begin to differentiate before the next passage. Cells should then be plated at 2–4 × 10^4 cells/cm^2. For freezing, 1 × 10^6 cells should be resuspended in 1 ml of FCS/10% DMSO, then put at $-80°$ O/ N then into liquid nitrogen. The number of passages should always be kept to a minimum for ES cells in order to avoid genetic and karyotypic abnormalities. If possible, multiple vials should be frozen at every passage.

ES Cell Differentiation

Differentiation of ES cells can be induced using a variety of strategies. The most classical method involves differentiation into embryoid bodies (EBs) by removal of feeder cells,[21] LIF withdrawal, and culture of aggregated (lightly trypsinized) cells in non-adherent Petri dishes (plastic dishes used for bacterial agar plates are ideal) in DMEM supplemented with 10% FBS. EBs have to be cultured in suspension, at least for the first 4 days, which is the period during which X inactivation is established. In order to perform immunofluorescence and/or RNA FISH on EBs, cytocentrifugation is required, which can disrupt interphase chromatin structure and nuclear architecture. An alternative strategy (based on ref. 22), which we prefer as it allows cells to be grown directly on coverslips during differentiation, involves the use of retinoic acid. ES cells are plated at a low density of 10^4 cells/cm^2 (if necessary after removal of feeder cells by adsorption). Once the majority of cells have attached (this usually takes 8 h–overnight), the ES cell medium is removed, the cells are washed three times in PBS to eliminate LIF, and differentiation medium is added. Differentiation medium is DMEM supplemented with 10% FBS, 100 nM

all-*trans*-retinoic acid (RA) (Sigma, cat. no. R2625; a 10^{-3} M stock solution is prepared in ethanol and stored at $-20°$), and 10^{-4} mM 2-mercaptoethanol. Addition of the latter has been found to minimize cell death. Differentiation medium is changed daily and can be preceded by one or two washes in PBS to remove debris if necessary. Changes in cell morphology can be detected by day 2, with clearly fibroblast-like and neuronal-like cells by day 4.

Combined Immunofluorescence and RNA FISH Analysis

Numerous methods involving a variety of fixation and permeabilization techniques are available for performing immunofluorescence (IF), the choice depends on the cell type, the epitope, and the antibody being used.[23] However, the combination of IF with RNA FISH is not always successful and depends on the nature of the proteins and/or transcripts being detected. Furthermore, the conditions that preserve nuclear architecture and chromatin structure often render RNA FISH ineffecient. Conversely, conditions that are ideal for RNA FISH (involving extraction in cytoskeletal buffer[24,25]) often result in poor-quality immuno-detection, presumably through loss of the protein during extraction or damage to the epitope. Thus, the aim of a combined immunofluorescence and RNA FISH analysis is to preserve nuclear structure and the antibody's epitope as far as possible but at the same time to enable penetration of the FISH probe for adequate detection of nuclear transcripts. We have tested a number of different methods and conditions and describe a method we have found to be optimal for Xist RNA FISH combined with immuno-detection of histone modifications. We prefer to perform immunofluorescence (under RNAse-free conditions) prior to RNA FISH, as the formamide treatment during the RNA FISH procedure is sometimes incompatible with preservation of the epitopes detected by some antibodies.

Immunofluorescence

Sub-confluent fibroblasts or ES cells cultured on gelatin-coated coverslips are briefly washed once in PBS and then fixed in freshly prepared 3% formaldehyde (paraformaldehyde, RE/PURO cat. no. 387507, in PBS, pH 7.2) for 10 min at RT. Following two washes in PBS at RT,

---

[23] D. L. Spector, R. D. Goldman, and L. A. Leinwand, "Cells: A Laboratory Manual." Cold Spring Harbor Laboratory Press, Cold Spring Harbor, NY, 1998.

[24] R. H. Singer, J. B. Lawrence, and C. Villnave, *Biotechniques* **4**, 230 (1986).

[25] A. M. Femino, F. S. Fay, K. Fogarty, and R. H. Singer, *Science* **280**, 585 (1998).

permeabilization of the cells is then performed on ice, in PBS containing 0.5% Triton X-100 (ICN, cat. no. 807423), 2 mM Vanadyl Ribonucleoside Complex (an RNase inhibitor, NEB, cat. no. S1402S) for 3–5 min. The exact time of permeabilization depends on the antibody and cell type. It should be noted that shorter permeabilization times (less than 3 min) results in less efficient RNA FISH. After rinsing in PBS, the cells are incubated in 1% BSA (GIBCO, cat. no. 15260–037) in PBS for 15 min, to block non-specific sites of antibody binding. The primary antibody, diluted in blocking buffer and 0.4 U/μl RNAguard (RNase inhibitor, Amersham/ Pharmacia, cat. no. 270815–01), is then applied. The coverslips are placed cell-side down, avoiding the formation of air bubbles, onto a drop of antibody solution on a sterile glass slide. The volume depends on the size of coverslip used (we routinely use 18 × 18 mm coverlips and 35 μl of antibody solution). Incubation is performed in a humid chamber at room temperature for 45 min (the temperature and length of incubation can vary between antibodies). The coverslips are then carefully removed with forceps. If any resistance is encountered, the coverslip should be flooded with PBS so that it floats in order to avoid damaging the cells. The coverslip is quickly put back into a well containing PBS and four or more washes in PBS, of 5 min each, are performed with rotational shaking. Blocking and incubation in secondary antibody is then performed following the same procedure as for the primary antibody earlier. After washing in PBS, preparations can be postfixed in 3% paraformaldehyde for 10 min at RT and rinsed twice in 2× SSC prior to RNA FISH. For immunofluorescence alone, postfixation and SSC washes are unnecessary and the sample can be DAPI stained, washed, and mounted for observation as described later. Note that the sample should never be allowed to dry at any stage, as this can lead to background problems and artifacts.

The specificity of the primary antibodies used should be verified using western blotting. For detailed protocols on the handling of antibodies and controls of specificity the reader is referred to Harlow and Lane.[26] A list of the antibodies directed against histone modifications that we have used (most of which are available from Upstate Biotechnology) and their dilutions are listed in Table I.

Secondary antibodies must be targeted against the appropriate class of immunoglobulin of the species in which the primary was made and generated from a separate species. A control of the specificity of the secondary antibody should always be included (i.e., an IF without prior primary antibody incubation). For combined IF and RNA or DNA FISH, the choice of

[26] E. Harlow and D. Lane, "Antibodies: A Laboratory Manual." Cold Spring Harbour Laboratory Press, Cold Spring Harbor, NY, 1989.

TABLE I

Antibody	Enriched (+) depleted (−) on the Xi	Dilution	Source
Di-me(Lys4) H3	−	1/300	Upstate Biotechnology
Ac(Lys9) H3	−	1/300	Upstate Biotechnology
Di-me(Lys9) H3	+	1/300	Gift from D. Allis/ Upstate Biotechnology
Di-me(Arg17) H3	−	1/300	Upstate Biotechnology
Di-/tri-me(Lys27) H3[a]	+	1/50	Gift from D. Reinberg
Di-me(Lys36) H3	−	1/300	Upstate Biotechnology
Ac(penta) H4	−	1/300	Upstate Biotechnology
Ac(Lys5) H4	−	1/100	Upstate Biotechnology
Ac(Lys8) H4	−	1/300	Upstate Biotechnology
Ac(Lys12) H4	−	1/300	Upstate Biotechnology
Ac(Lys16) H4	−	1/100	Upstate Biotechnology

[a] Note: The reader is referred to the chapter by Samra and Reinberg for characterization of the anti-H3 di-/tri-methylated Lys-27 antibody.

fluorochrome to which the secondary antibody is conjugated will depend on the fluorochrome with which the FISH probe is labeled and on the filter sets available on the microscope. For information on the choice of fluorophores the reader is referred to catalogs of manufacturers. We commonly use Molecular Probes (e.g., Alexa Fluor 568 or 488, goat anti-rabbit or goat anti-mouse IgGs) for combination with Spectrum Red or Spectum Green labeled probes, respectively (see later). In the case of triple labelings, in which two different primary antibodies are used in conjunction with RNA FISH, or else one antibody but two different transcripts are detected by RNA FISH, fluorophores in the infrared range of the spectrum can be used, such as Alexa 680, if the appropriate filter sets are available (note that these fluorophores are not visible to the human eye). In the case of a double IF experiment, high-affinity purified secondary antibodies should be used which minimize cross-species reactivity, and the appropriate controls (i.e., each primary antibody alone with both secondaries) should always be performed.

RNA Fluorescence In Situ Hybridization

For a general description and discussion of RNA FISH protocols, the reader is referred to Spector et al.[23] To detect the primary transcripts of genes, genomic probes several kilobase pairs long or fluorescently labeled oligos must be used. Probes spanning introns and exons will detect both the processed mRNA and the primary transcript. Oligonucleotides within intronic sequences will clearly be specific for the primary transcript. For Xist RNA detection we have used several genomic DNA probes, spanning a minimum of 3 kb with success. Labeling of DNA probes to be used for *in situ* detection of transcripts can be performed either by nick translation or random priming or else fluorescently tagged oligonucleotides can be used. The latter avoids the labeling step and also enables discrimination between sense or antisense transcripts (double-stranded DNA probes will of course detect both), but is costly. We routinely use nick translation (Vysis, cat. no. 32-801300) with fluorescent dUTPs (Spectrum Green, Spectrum Red, Vysis, cat no. 6J9410-6J9420 or Cy5 Amersham, cat. no. PA55026) to label genomic probes (plasmids, lambda clones, or BACs). 1–2 μg of DNA is labeled in a 50-μl reaction. Following the reaction, the size range of the labeled DNA is always checked by electrophoresis on an agarose gel. The optimal range size of the FISH probe is between 50 and 200 bp, short enough to enter the nucleus and long enough to be specific. Fluorescently labeled probes of this kind can be directly stored at $-20°$ for a few weeks.

Prior to hybridization, 0.1 μg of probe (usually 5 μl of a standard nick translation reaction of 50 μl) is precipitated with 10 μg of salmon sperm (per 18 $\times$ 18 mm coverlip) and washed twice in 70% ethanol (to remove unincorporated nucleotides). The dried pellet is thoroughly resuspended in 5 μl formamide (Sigma, cat. no. F5786) by vortexing and denaturation is then performed at 75° for 7 min. Following this 5 μl of a 2$\times$ hybridization buffer is immediately added, to give a final solution of 2$\times$ SSC, 20% dextran sulfate, 1 mg/ml BSA (NEB, cat. no. B9001S), 200 mM VRC. The 10 μl of probe solution is deposited onto a sterile glass slide and the coverslip is removed from the 2$\times$ SSC solution with forceps and carefully lowered cell-side down avoiding the formation of air bubbles. Once the coverlip has made contact with the probe solution, it should not be moved to avoid damaging the cells. Hybridization is overnight at 37° in a dark, humid chamber (paper tissues soaked in 50% formamide/2$\times$ SSC). After three washes in 50% formamide/2$\times$ SSC (adjusted to pH 7.2) and three washes in 2$\times$ SSC at 42°, DNA is counterstained for 2 min in 2$\times$ SSC containing 0.2 mg/ml DAPI, followed by a final wash in 2$\times$ SSC. Coverslips are then mounted in glycerol-based medium (90% glycerol (Sigma, cat. no. G5516), 0.1$\times$ PBS, 0.1% *p*-phenylenediamine (Sigma, cat. no. 27–515–8), pH 9). We use a Leica DMR fluorescence microscope with a Cool SNAP fx camera (Photometrics) and Metamorph software (Roper) for image acquisition. An example of Xist RNA FISH combined with immunofluorescence is shown in Fig. 1.

Immunofluorescence and DNA FISH on Metaphase
Chromosome Spreads

Immunofluorescence

The technique we describe involves IF performed on unfixed metaphase chromosomes and is based on a previously published protocol of ref 13. Metaphase chromosomes are prepared from cells at 80% confluence, following a 1-h incubation with 0.1 μg/ml colcemid (Roche, 295892). To maximize the collection of mitotic cells the supernatant is collected and adherent cells are trypsinized, pelleted, resuspended in PBS, and pelleted (this wash performed twice). The cells are then resuspended (10^5 cells/ml) carefully in 75 mM KCl, incubated for 10 min at RT and then placed on ice for between 5 min and 1 h. Cytocentrifugation is then performed (Shandon), with the time and speed of centrifugation depending on the cells used. For ES cells and early stages of differentiation, 8 min at 1200 rpm is usually appropriate. After air-drying, the slides are incubated in freshly prepared KCM buffer (120 mM KCl, 20 mM NaCl, 10 mM Tris-HCl, pH 8, 0.5 M EDTA, 0.1% Triton X-100) for 15 min at RT, and incubated successively with primary and secondary antibody in KCM, 1% BSA for 40 min in a humidified dark chamber at RT (immunofluorescence protocol as earlier). Several washes in KCM buffer are performed. A postfixation step in 3% formaldehyde can be performed at this stage to prevent signal loss if this is a problem. The slides should again be washed in KCM containing DAPI (0.2 mg/ml DAPI) for 2 min, washed once more in KCM and mounted in medium (see earlier) for visualization on a fluorescence microscope. An example is shown in Fig. 2.

DNA Fluorescence In Situ *Hybridization*

Xist RNA does not associate with the mitotic X chromosome efficiently during the early stages of ES cell differentiation (J. Chaumeil and E. Heard, unpublished observations). Therefore, in order to identify the X chromosome on mitotic chromosome spreads, DNA FISH using an X-chromosome paint can be performed following immunofluorescence. This does not of course allow the inactive X chromosome to be identified. A double immunofluorescence of a mark known to characterize the inactive X chromosome (such as H3 Lys-4 hypomethylation) could therefore be used. As the DNA denaturation step can destroy the immunofluorescence signal in some cases, images should be recorded prior to the DNA FISH experiment. To remove the coverslip and the mounting medium from the sample following imaging, slides are washed 3–4 times in KCM (if the coverslip does not fall off by itself during the first wash, it should carefully

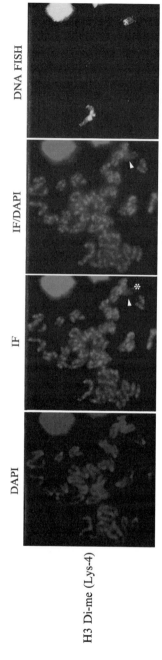

FIG. 2. Example of immunofluorescence on metaphase spreads. Histone H3 modifications on the inactive X chromosome in differentiating female ES cells. Metaphase spreads are shown with typical patterns observed in female ES cells from day 3 of differentiation onward. A single X chromosome shows specific hypomethylation on H3 Lys-4 (red, arrowhead). X chromosomes were detected by X chromosome DNA FISH following immunofluorescence (yellow). The distal part of the X chromosome is methylated on H3 Lys-4 and probably corresponds to the pseudoautosamal region (asterisk). DNA is stained with DAPI (blue). (See color insert.)

be removed using forceps before the second wash). The samples should then be dehydrated through a series of 3-min washes in 70%, 80%, and 95% ethanol. Denaturation of the DNA is performed by plunging the slide in a 50-ml falcon tube or coplin jar containing 70% formamide/2× SSC (pH 7.2) for 3 min at 75°. The slides should immediately be plunged into ice-cold 70% ethanol and then dehydrated (as earlier). Denaturation of chromosome paint probes and pre-annealing of competitor DNA as well as *in situ* hybridization are performed exactly as recommended by the supplier (mouse X-chromosome paint, Cambio, cat. no. 1187-XMB-02). Hybridization is performed overnight at 42°. Posthybridization washes are at 45°, three times in 50% formamide/2× SSC and three times in 2× SSC. Slides are counter-stained with DAPI, mounted, and viewed under the fluorescence microscope (as earlier). An example is shown in Fig. 2.

Replication Timing Studies Using BrdU Incorporation and Acridine Orange Staining

In order to determine the onset of late replication timing of the inactive X chromosome, we have tested a number of techniques, all of which depend on the incorporation of the thymidine analog, 5-bromo-2-deoxyuridine (BrdU), to detect replicated DNA. The technique that proved most robust in our hands is in fact one of the most classic techniques to have been used for studies of replication timing and involves acridine orange staining.[4,27,28] In this method, DNA in which BrdU has been incorporated into DNA will stain weakly red, whereas DNA with no BrdU incorporation will stain green. This technique has the advantage that both BrdU-negative and BrdU-positive regions can be simultaneously visualized, allowing identification of mitoses with an overall "late" replicating banding pattern (i.e., red G-bands and green R-bands). Undifferentiated or differentiating ES cells are incubated for 8–9 h in medium supplemented with 100 μg/ml BrdU (Sigma). Colcemid (Roche) at a concentration of 0.025 μg/ml is added to the medium for the final 0.5–1 h of incubation. The cell suspension is centrifuged for 2 min at 400g. The pellet is gently resuspended in 2 ml of freshly prepared 0.075 M KCl. The cell suspension is then incubated at RT for 10 min and 3 ml of freshly prepared fixative (3:1 methanol: acetic acid) is then added directly to the hypotonic medium and incubated for a further 10 min. The cell suspension is then centrifuged for 5 min at 150g and the pellet is dispersed in 10 ml of fixative. After a second centrifugation of 2 min at 400g, the pellet is dispersed in 1–2 ml of fixative and

[27] B. Dutrillaux, C. Laurent, J. Couturier, and J. Lejeune, *Acad. Sc. Paris* **276**, 3179 (1973).
[28] N. Takagi and M. Oshimura, *Exp. Cell. Res.* **78**, 127 (1973).

the tubes are kept on ice. Clean and dry slides are placed on wet tissues (Kimwipes, Kimberly-Clark). The fixed cell suspension is spread by dropping one drop onto an ethanol-cleaned slide from a height of about 10 cm. The slides are allowed to dry very quickly. They are then stained with Sorensen's phosphate buffer (pH 6.8) supplemented with 0.05 mg/ml acridine orange solution. The slides are incubated for 10–15 min in this solution, rinsed immediately in running tap water for 1–2 s, then rinsed in phosphate buffer (pH 6.8). They are mounted with a coverslip in phosphate buffer (pH 6.8) and excess buffer is removed with blotting paper. The metaphase spreads are then immediately examined under a fluorescence microscope[4,27,28] using a filter block for acridine orange (blue filter, exc. 490 nm, emiss. 530 nm/640 nm for single- and double-stranded DNA detection). Whole chromosomes or chromosome segments replicated in the presence of BrdU show a weak red fluorescence whereas, those that have not incorporated BrdU show bright green fluorescence. The autosomes and the active, synchronously replicating X chromosome have certain regions that replicate late in the S phase of the cell cycle as well as others that replicate early. When BrdU is present only during the latter half of S phase, an unequivocal red and green banding pattern is observed in these chromosomes in contrast with the late-replicating inactive X chromosome, which is stained homogeneously red. An example is shown in Fig. 3.

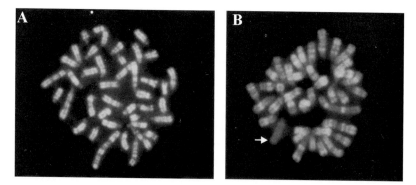

Fig. 3. Detection of asynchronous replication of the X chromosomes using Acridine Orange staining. Metaphase spreads showing the typical green/orange banding pattern representative of BrdU incorporation in the latter half of S phase are shown. In (A) no single orange-staining X chromosome can be detected, indicating that both X chromosomes replicate synchronously. The first sign of a single uniformly stained orange chromosome (the late-replicating X chromosome, see arrow) was observed at day 3 of ES cell differentiation, where 6.6% ($n > 200$) of metaphases displayed this pattern. The proportion rose to almost 30% ($n > 200$) by day 7. (See color insert.)

Conclusion

The techniques we describe here should provide a basis for investigating the onset of X inactivation or other developmentally regulated processes in differentiating ES cells. Using these techniques we have been able to determine the relative order of the following events: Xist RNA coating occurs within the first 24–48 hours of differentiation; histone H3 modifications such as hypomethylation of Lys-4, hypoacetylation of Lys-9, and hypermethylation of Lys-9 and Lys-27 (see Fig. 1) are detectable on the X chromosome in a proportion of interphase cells as soon as Xist RNA accumulates. Histone H4 hypoacetylation of lysines 5, 8, 12, and 16 is also observed from this time, although with slightly slower kinetics.[15,17] The appearance of these marks on the X chromosome during mitosis (see Fig. 2) is found from days 2–3, suggesting that they represent true epigenetic marks. On the other hand, a late-replicating X chromosome is only detected from day 3.[17] Thus, histone modifications appear to be direct consequences of Xist RNA's action, while late replication timing seems to be a slightly later event. Using this as a basis, the relationship between these different events can now be defined and tested, using for example, genetically modified ES cells or siRNA-induced silencing of candidate modifying enzymes.

[28] X Inactivation in Mouse ES Cells: Histone Modifications and FISH

By BARBARA PANNING

In mammals, X inactivation is used to equalize X-linked gene dosage between females, with two X chromosomes, and males, with only one, and it provides a particularly dramatic example of gene regulation by alterations in chromatin structure. X inactivation can be divided into several phases: choice, initiation, spread, and maintenance.[1] At around the time of gastrulation, when totipotent embryonic cells differentiate into more restricted somatic lineages, one X chromosome per diploid genome is randomly chosen to remain active and all additional X chromosomes are altered from euchromatin to transcriptionally inactive heterochromatin. Once an X chromosome has been inactivated, it remains silenced in all

[1] K. Plath, S. K. Mlynarczyk-Evans, D. A. Nusinow, and B. Panning, *Annu. Rev. Genet.* **36,** 233 (2002).

subsequent generations. As a result, every female is a mosaic of clonal groups of cells in which either the maternally inherited X chromosome or the paternally inherited X chromosome is silenced. This indicates that the repressed chromatin structure of the inactive X chromosome (Xi) is extraordinarily stable, since it can be propagated through all the cell divisions in the lifetime of an individual. Thus, the study of X inactivation addresses two equally interesting issues in the study of gene expression. First, what activities are used to alter chromatin structure and how are these activities restricted to particular portions of the genome? Second, how are different chromatin states stably maintained through cell division?

The *Xist* gene is necessary for X inactivation as X chromosomes bearing *Xist* deletions are not silenced.[2,3] *Xist* encodes a 17.5-kb, spliced, polyadenylated, non-coding RNA that is expressed exclusively from the Xi in female somatic cells.[4–6] *Xist* RNA is unusually stable and remains in the nucleus where it coats the Xi.[7–9] In embryonic cells that are poised to undergo X inactivation *Xist* RNA can be detected only at its site of transcription on the single active X chromosome in male cells and from both Xa's in female cells.[10] When embryonic cells are differentiated and X inactivation is initiated, *Xist* transcripts spread in *cis* from the *Xist* gene to coat the entire X chromosome. The spread of *Xist* RNA correlates temporally with the spread of silencing along the X chromosome and occurs within a single cell cycle.[11–13] The Xi in female somatic cells is characterized by increased methylation of CpG islands,[14] a decrease in

[2] G. D. Penny, G. F. Kay, S. A. Sheardown, S. Rastan, and N. Brockdorff, *Nature* **379**, 131 (1996).

[3] Y. Marahrens, B. Panning, J. Dausman, W. Strauss, and R. Jaenisch, *Genes Dev.* **11**, 156 (1997).

[4] G. Borsani, R. Tonlorenzi, M. C. Simmler, L. Dandolo, D. Arnaud, V. Capra, M. Grompe, A. Pizzuti, D. Muzny, C. Lawrence *et al.*, *Nature* **351**, 325 (1991).

[5] N. Brockdorff, A. Ashworth, G. F. Kay, P. Cooper, S. Smith, V. M. McCabe, D. P. Norris, G. D. Penny, D. Patel, and S. Rastan, *Nature* **351**, 329 (1991).

[6] C. J. Brown, A. Ballabio, J. L. Rupert, R. G. Lafreniere, M. Grompe, R. Tonlorenzi, and H. F. Willard, *Nature* **349**, 38 (1991).

[7] C. J. Brown and S. E. Baldry, *Somat. Cell Mol. Genet.* **22**, 403 (1996).

[8] C. M. Clemson, J. C. Chow, C. J. Brown, and J. B. Lawrence, *J. Cell Biol.* **142**, 13 (1996).

[9] C. M. Clemson, J. A. McNeil, H. F. Willard, and J. B. Lawrence, *J. Cell Biol.* **132**, 259 (1996).

[10] B. Panning and R. Jaenisch, *Genes Dev.* **10**, 1991 (1996).

[11] B. Panning, J. Dausman, and R. Jaenisch, *Cell* **90**, 907 (1997).

[12] S. A. Sheardown, S. M. Duthie, C. M. Johnston, A. E. Newall, E. J. Formstone, R. M. Arkell, T. B. Nesterova, G. C. Alghisi, S. Rastan, and N. Brockdorff, *Cell* **91**, 99 (1997).

[13] A. Wutz and R. Jaenisch, *Mol. Cell* **5**, 695 (2000).

[14] S. F. Wolf, D. J. Jolly, K. D. Lunnen, T. Friedmann, and B. R. Migeon, *Proc. Natl. Acad. Sci. USA* **81**, 2806 (1984).

abundance of acetylated isoforms of core histones,[15] changes in histone H3 methylation patterns,[16–19] an increase in amounts of the histone H2A variant, macroH2A1.2,[20] replication late in S phase,[21] and localization at the nuclear periphery.[22] X inactivation occurs in mouse embryonic stem (ES) cells when they are induced to differentiate in culture, providing a model system in which to assay the timing of alterations in chromatin structure relative to the *cis*-spread of *Xist* RNA that occurs at the onset of X inactivation. Fluorescence in situ hybridization (FISH) for *Xist* RNA when combined with immunostaining provides a powerful tool to analyze the alterations in chromatin structure that occur during X chromosome silencing. Figure 1 provides an example results that can be achieved using immunostaining and FISH in differentiating female ES cells.

Differentiation of ES Cells

For X inactivation studies, differentiation of ES cells into embryoid bodies (EBs) using hanging drop cultures is very effective. When the resulting EBs are placed on a tissue culture substrate coated slides the EBs attach to the slides. The cells that delaminate and migrate away from the EBs can be analyzed at various intervals after plating to visualize the alterations in chromatin structure that occur during X chromosome silencing.

1. Trysinize and pellet female ES cells.
2. Resuspend the cell pellet in 1 ml of differentiation media.
   Differentiation media:
   DMEM
   15% fetal bovine serum
   Non-essential amino acids
   0.1 mM β-mercaptoethanol

[15] P. Jeppesen and B. M. Turner, *Cell* **74,** 281 (1993).
[16] E. Heard, C. Rougeulle, D. Arnaud, P. Avner, C. D. Allis, and D. L. Spector, *Cell* **107,** 727 (2001).
[17] J. E. Mermoud, B. Popova, A. H. Peters, T. Jenuwein, and N. Brockdorff, *Curr. Biol.* **12,** 247 (2002).
[18] J. Silva, W. Mak, I. Zvetkova, R. Appanah, T. B. Nesterova, Z. Webster, A. H. Peters, T. Jenuwein, A. P. Otte, and N. Brockdorff, *Dev. Cell* **4,** 481 (2003).
[19] K. Plath, J. Fang, S. K. Mlynarczyk-Evans, R. Cao, K. A. Worringer, H. Wang, C. C. de la Cruz, A. P. Otte, B. Panning, and Y. Zhang, *Science* **300,** 131 (2003).
[20] C. Costanzi and J. R. Pehrson, *Nature* **393,** 599 (1998).
[21] N. Takagi, *Exp. Cell Res.* **86,** 127 (1974).
[22] M. L. Barr, C. D., *Acta Cytol.* **6,** 34 (1961).

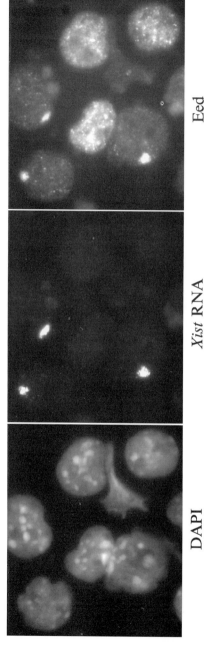

DAPI   *Xist* RNA   Eed

Fig. 1. Immunostaining for Eed and FISH for *Xist* RNA in differentiating female ES cells. In a differentiating population of female ES cells, a fraction of cells exhibit *Xist* RNA coating of the Xi (middle panel) in nuclei, visualized DAPI staining (left panel). In these cells Eed, a component of a Polycomb Group complex that includes the histone methyltransferase Ezh2, is also enriched on the Xi (right panel). Recruitment of the Eed-Ezh2 complex correlates with an increase in histone H3 lysine 27 methylation on the Xi.[18,19] In the cells that do not yet show *Xist* RNA coating, the levels of Eed are much higher, suggesting that abundance of Eed decreases when ES cells are differentiated.

3. Determine cell number, and dilute cells to a final concentration of 10 cells per microliter.
4. Place 100–200 10–20 μl drops of diluted ES cells on the inside of a Petri dish lid.
5. Return the lid of the Petri dish to its base; the base should be filled with PBS to hydrate the hanging drops.
6. Incubate hanging drops for 2 days.
7. After 2 days flood the lid of the Petri dish with a small volume of differentiation media to suspend the EBs forming in the hanging drops.
8. Transfer the EBs to a Petri dish filled with differentiation media.
9. Incubate the EBs in suspension culture for 2 more days.
10. The EBs should now be large enough to be seen by eye. Use a P20 pipetman to remove individual EBs from the suspension culture and plate them on commercially available slides that have tissue culture coatings. Multichamber slides can be very useful if several antibodies or FISH probes will be employed. The EBs will adhere to the slide in approximately 24 h.
11. At 24-h intervals after plating fix samples for immunostaining and FISH as described later.

Immunostaining

To simultaneously visualize protein and RNA, immunostaining is followed by FISH. Many antisera contain RNases, resulting in loss of RNA during the immunostaining procedure. Addition of tRNA and RNase inhibitors to the immunostaining buffers to suppress the RNases in the sera preserves RNA during immunostaining.

1. Rinse slides with 1× PBS.
2. Incubate slides for 30 s in ice-cold cytoskeletal buffer.
   Cytoskeletal buffer:
       100 mM NaCl
       300 mM sucrose
       3 mM MgCl$_2$
       10 mM PIPES, pH 6.8
3. Incubate slides for 30 s in ice-cold cytoskeletal buffer + 0.5% Triton X-100.
4. Incubate slides for 30 s in ice-cold cytoskeletal buffer.
5. Fix in 4% paraformaldehyde/1× PBS, 10 min, room temperature.
6. Wash slides two times, 5 min each in room temperature PBS-Tween. Slides can be stored for several days in PBS-Tween at 4° prior to immunostaining.

PBS/Tween
  1× PBS
  0.2% Tween 20

7. Block slides for 30 min at 37° in blocking buffer in a humid chamber. The humid chamber should contain PBS/Tween. Blocking buffer:
  1× PBS
  5% goat serum (heat inactivated)
  0.2% Tween 20
  0.2% fish skin gelatin
  4–5 units/ml RNAsin
  1 mg/ml molecular biology-grade yeast tRNA

8. Incubate with primary antibody diluted in blocking buffer, 1 h, 37° in a humid chamber. Temperature may be varied depending on the primary antibody.

9. Wash slides three times, 5 min each with PBS/Tween.

10. Block slides for 5 min at 37° in blocking buffer in humid chamber.

11. Incubate with secondary antibody diluted in blocking buffer for 30 min, 37° in a humid chamber.

12. Wash slides three times, 5 min each with PBS/Tween. Minimize exposure to light to limit photobleaching.

13. Fix in 2% paraformaldehyde/1× PB for 10 min at room temperature.

14. Transfer slide to 70% ethanol. Slides can be stored in 70% ethanol at 4° for several weeks prior to carrying out FISH.

FISH

The FISH procedure involves generating a labeled probe, hybridizing the probe to the fixed and immunostained sample, and then detection of the labeled probe. The specific activity of RNA probes is highest, and since the immunostaining may cause a slight reduction in the amount of RNA in the sample, using high specific activity probe generates the best results with FISH. All stock solutions are autoclaved to minimize likelihood of RNase contamination, and all stock solutions are diluted in autoclaved water.

Generate Template

Templates can be generated by cloning fragments into plasmids with T3, T7, or SP6 promoters or by appending T3, T7, or SP6 promoters to PCR products. The maximum size of an *in vitro* transcribed RNA that will penetrate into a fixed, permeabilized cell efficiently is approximately 500 nucleotides. Therefore, multiple 500 nucleotide RNAs spanning the length of the cDNA will allow for the most robust detection of any RNA.

Prepare Template

If using plasmid template, digest with appropriate restriction enzyme to linearize. Phenol-chloroform extract, ethanol precipitate twice, and dry. Resuspend in water at a concentration of approximately 1 mg/ml.

If using PCR product, set up PCR reaction and run fraction out on gel. If the band is clearly visible, then about 1 μl of PCR product is enough for the *in vitro* transcription reaction. It does not need any further manipulation.

In Vitro *Transcription*

1. Mix the following at room temperature:
5× buffer	20 μl
100 mM DTT	10 μl
4000 units/ml RNA sin	2.5 μl
2.5 mM A, G, UTP mix	20 μl
10 mM CTP	1 μl
10 mM bioCTP	4 μl
DNA, 1 mg/ml	2 μl
Polymerase	2 μl
Water	38.5 μl
2. Incubate at 37° for 1 h.
3. Add 1 μl of 1 U/μl RNase free DNase 1.
4. Remove unincorporated nucleotides using G-50 Sephadex spun columns, prepared with TE.
5. After RNA has spun through column, run 5 μl on agarose minigel to ensure that *in vitro* transcription has worked. The remaining RNA should be ethanol precipitated with 10 μg of molecular biology-grade yeast tRNA and stored at −80° or −20° under ethanol.

Prepare Probe Mix

1. Aliquot 1/20 of the biotinylated RNA produced from a 2-μg labeling reaction.
2. Add:
   10 μl of molecular biology-grade yeast tRNA, 10 mg/ml
   15 μl sheared boiled salmon sperm DNA, 10 mg/ml
   10 μl 3 M Na acetate
   45 μl water
   250 μl 100% ethanol

3. Mix well and pellet.
4. Wash precipitated RNA with 70%, then 100% ethanol.
5. Dry pellet in speed vacuum to remove all traces of ethanol.
6. Resuspend pellet in 25 μl of ultrapure formamide (Sigma). Ensure that the pellet is completely dissolved by using pipette tip to break up any large pieces and vortexing vigorously.
7. Mix with 25 μl of 2× hybridization buffer.
   2× hybridization buffer:
      1 part 20× SSC
      2 parts 10 mg/ml BSA, molecular biology-grade (New England Biolabs)
      2 parts 50% dextran sulphate
      Mix very well, store at −20°
8. Denature by incubation at 95–100° for 5–10 min.
9. Quench on ice immediately.
10. Store at −20°.

FISH

1. Dehydrate cells on slides through an ethanol series, 2 min each in 85%, 95%, and 100% ethanol, air-dry.
2. Add probe to air-dried cells.
3. Place coverslip over probe. For a 12-mm circle coverslip use 3 μl of probe; for a 22-mm square coverslip use 10 μl.
4. Incubate in a humid chamber at 37° overnight. The humid chamber should contain 2× SSC/50% formamide.

Washing and Detection

1. Immerse slide with coverslip in 2× SSC/50% formamide. Once the slide is wet carefully slide off the coverslip and continue washing for 5 min.
2. Wash in 2× SSC/50% formamide, once more for 5 min each at 39°.
3. Wash in wash buffer II, three times for 5 min each at 39°.
   Wash buffer II:
      0.5 M NaCl
      10 mM Tris, pH 7.5
      0.1% Tween-20
4. Dilute RNase A to 25 μg/ml in wash buffer II.
5. Put 150 μl of RNase A in wash buffer II on a 24 × 60 mm coverslip.
6. Place slide, cell-side down over coverslip and use capillary action to draw the coverslip to the slide.

7. Flip slide and place coverslip-side up in humid chamber filled with wash buffer II for 30 min.
8. Wash in wash buffer II, two times for 5 min each at 39°.
9. Wash in 2× SSC/50% formamide, two times for 5 min each at 39°.
10. Wash in 2× SSC, three times for 5 min each at 39°.
11. Wash in 1× SSC, two times for 5 min at 39°.
12. Wash in 4× SSC, once for 5 min at 39°.
13. Dilute fluorochrome-conjugated avidin in detection buffer, 1:200. Detection buffer:
    1 part 20× SSC
    1 part BSA, 10 mg/ml, molecular biology grade (New England Biolabs)
    3 parts water
14. Put 150 μl of detection buffer with avidin fluorochrome on 24 × 60 mm coverslip.
15. Place slide, cell-side down, over coverslip and use capillary action to draw coverslip up to slide.
16. Flip slide and place coverslip-side up in humid chamber filled with 4× SSC.
17. Incubate for 30 min at 37°.
18. Wash in 4× SSC, for 5 min each at 39°.
19. Wash in 4× SSC+ 0.2% Tween with DAPI, for 5 min at 39°. DAPI:
    Stock of 1 mg/ml, dissolved in dimethyl formamide. Use 2 μl per 50 ml of 4× SSC.
20. Wash in 4× SSC for 5 min at 39°.
21. Add antifade mounting media. Any commercially available antifade media is appropriate.
22. Place a 24 × 60 mm coverslip on the slide.
23. Blot off excess mounting media.
24. Slides may be visualized immediately or stored at 4°. If slides will be stored for longer than a week, use nail polish to seal the coverslip to the slide.
25. Gather images on a fluorescent microscope equipped with appropriate filter sets and a camera.

Troubleshooting

Generally it is best to test the FISH probe on fixed material prior to doing immunostaining in combination with FISH. Follow the first five fixation steps of the immunostaining protocol. After the paraformaldehyde fixation, transfer slides to 70% ethanol. Follow the FISH protocol. If the

FISH does not work, then troubleshooting of the probe and solutions is necessary. Introduction of RNases is one likely cause of failure, and using freshly prepared, autoclaved solutions can solve this problem. The amount of labeled probe that is best for detection of different RNAs varies and testing a range of concentrations of labeled probe can be used to identify an optimal probe concentration.

RNA has been implicated in regulating chromatin structure at centromeres[23–25] and in recruiting chromatin modifying enzymes to the dosage compensated X chromosome in *Drosophila*,[26,27] indicating that RNA may be more generally used to regulate chromatin structure. Thus, the techniques described in this chapter may be relevant to the study of other examples of regulated changes in chromatin structure.

Acknowledgments

Many thanks to Hannah Cohen, Susanna Mlynarczyk-Evans, and Kathrin Plath for critical reading of this manuscript.

[23] C. Maison, D. Bailly, A. H. Peters, J. P. Quivy, D. Roche, A. Taddei, M. Lachner, T. Jenuwein, and G. Almouzni, *Nature Genetics* **30,** 329 (2002).
[24] T. A. Volpe, C. Kidner, I. M. Hall, G. Teng, S. I. Grewal, and R. A. Martienssen, *Science* **297,** 1833 (2002).
[25] I. M. Hall, G. D. Shankaranarayana, K. Noma, N. Ayoub, A. Cohen, and S. I. Grewal, *Science* **297,** 2232 (2002).
[26] V. H. Meller, K. H. Wu, G. Roman, M. I. Kuroda, and R. L. Davis, *Cell* **88,** 445 (1997).
[27] H. Amrein and R. Axel, *Cell* **88,** 459 (1997).

Author Index

Numbers in parentheses are footnote reference numbers and indicate that an author's work is referred to although the name is not cited in the text.

Subject Index

A

Acetylation microarray
applications, 302–304
chromatin immunoprecipitation
cell growth and harvesting, 292
cross-linking
double cross-linking, 294–295
formaldehyde, 292
immunoprecipitation reaction, 293
polymerase chain reaction, 295–297
sonication, 292–293
troubleshooting, 293–294
DNA microarray analysis
data quantification, normalization, and
analysis, 299–302
Klenow labeling of probe and
hybridization, 297–299
yeast intergenic microarray
preparation, 302
principles, 290–292
Acid-urea-triton gel electrophoresis, histone
deacetylase assay for inhibitor
characterization, 201–202
AFM, *see* Atomic force microscopy
Agarose multigel electrophoresis
chromatin structure analysis
composition analysis of genomic
fragments, 27–28
data interpretation, 28–29
flexibility assay, 27
overview, 17–18, 20–21
secondary chromatin structure
formation, 26
data analysis, 25–26
electrophoresis, 24
equipment, 21
gel preparation, 22, 24
principles, 18–19
Southern blot, 25
AME, *see* Agarose multigel electrophoresis
Atomic force microscopy
chromatin fiber imaging
glutaraldehyde fixation, 78–79

image acquisition, 79
materials, 78
surface preparation and sample
deposition, 79
instrumentation, 76–78
principles, 73–74
AUT gel electrophoresis, *see*
Acid-urea-triton gel electrophoresis

B

Brahma, bromodomain structure, *see*
Nuclear magnetic resonance
Bromodomain structure, *see* Nuclear
magnetic resonance

C

ChIP, *see* Chromatin immunoprecipitation
Chromatin, *see also* Histone; Nucleosome
agarose multigel electrophoresis analysis
of structure, *see* Agarose multigel
electrophoresis
antibodies, *see also* Histone antibodies
applications, overview, 209–210
histone antigen interactions, 209–212
immunogen preparation
dehistonized chromatin, 214–215
high mobility group proteins, 215–216
histone fractions, 213–214
nucleic acids, 216
nucleosomes, 214
synthetic histone peptides, 212–213
preparation, 216–217
specificity and affinity, 217–220
electron microscopy three-dimensional
structure determination
adhesion to support films, 31
buffer components, 30
concentration of sample, 30
cryomicroscopy, 36
electron tomography
alignment and reconstruction, 46–47

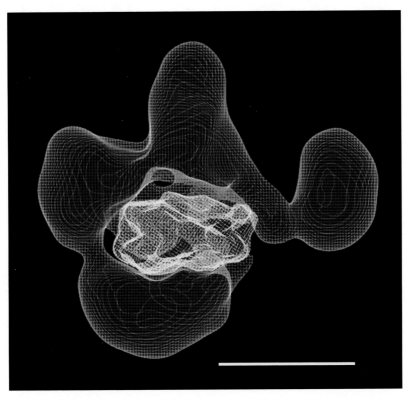

ASTURIAS *ET AL.*, CHAPTER 4, FIG. 5. Possible mode of RSC-nucleosome interaction. An X-ray structure of the nucleosome was filtered to 25 Å and manually fitted in the central cavity of the RSC structure [using the program O (T. A. Jones, J. Y. Zou, S. W. Cowan, and M. Kjeldgaard, *Acta Crystallogr. A* **47** (Pt 2), 110 (1991).)]. The close fit between the nucleosome and the RSC cavity is apparent. The scale bar corresponds to approximately 100 Å.

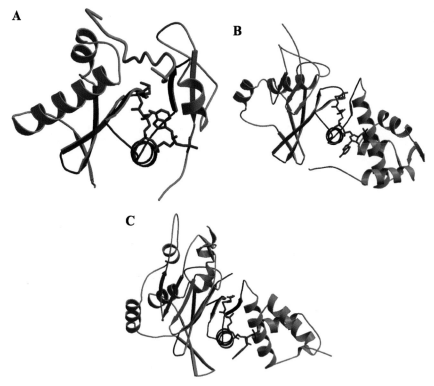

A

B

C

MARMORSTEIN, CHAPTER 7, FIG. 2. Overall structure of HAT proteins. (A) Schematic structure of the Gcn5/PCAF HAT domain. tGcn5 on complex with CoA (red) and a 19-residue histone H3 peptide (green). The structurally conserved catalytic core domain is colored in blue, and the structurally variable N- and C-terminal domains are colored in aqua. (B) Schematic structure of the yEsa1 member of the MYST HAT domain in complex with CoA (red). The color coding is as indicated in (A). (C) Schematic structure of the yHAT1 HAT domain in complex with Acetyl-CoA (red). The color coding is as indicated in (A).

MARMORSTEIN, CHAPTER 7, FIG. 4. Substrate binding by HAT proteins. (A) tGcn5/CoA/ histone H3 complex. CoA is colored in purple and the ordered regions of the 19-residue histone H3 peptide (residues 7–21) are colored in green. An overlay of the histone H3 peptide from the ternary complex with the 11-residue histone H3 peptide (residues 9–19) is shown in grey. (B) Schematic tGcn5-histone H3 interactions in the ternary tGcn5/CoA/histone H3 complex shown in (A). Regions of the histone H3 peptide that are ordered in the ternary complex with the 19-residue peptide, but not ordered in the ternary complex with the 11-residue peptide, are colored in grey. (C) Superposition of putative substrate binding sites of HAT proteins. The superposition is generated by superimposing the core domains from the Gcn5/PCAF member, PCAF (pink), and the MYST member, yEsa1 (blue), with yHAT1 (green). Only the core domain and CoA of yEsa1 (blue) is shown for clarity.

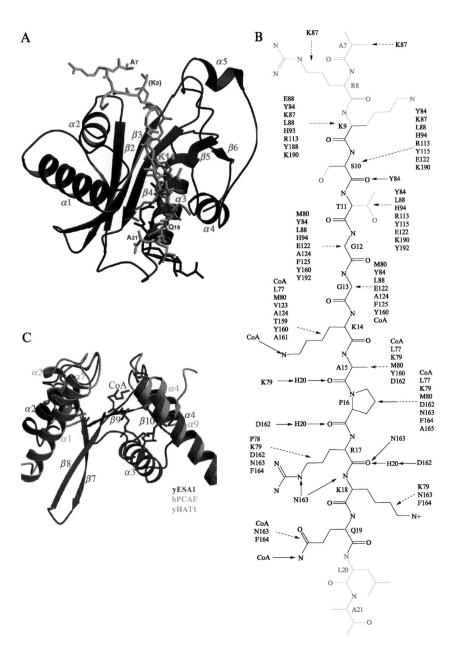

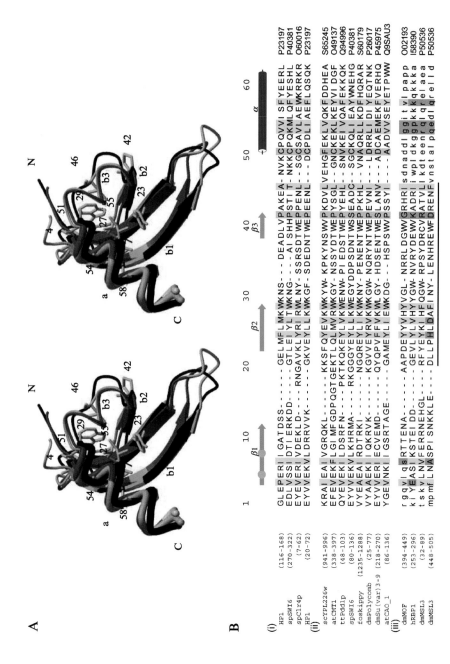

A

B

(i)

HP1   (116-168) GLEPERIIGATDSS------GELMFLMKWKNS---DEADLVPAKEA-NVKCPQVVISFYEERL P23197
spSWI6   (270-322) EDLVSSIDTIERKDD-----GTLEIYITWKNG---AISHHPSTIT-NKKCPQKMLQFYESHL P40381
spClr4p   (7-62)   EYEVERIVDEKLD------RNGAVKLYRIRWLNY-SSRSDTWEPPENL--SGCSAVLAEWKRKR O60016
HP1   (20-72)   EYVVEKVLDRRVVK-----GKVEYLLKWKGF-SDEDNTWEPEENL--DCPDLIAEFLQSQK P23197

(ii)

scYPL226w (941-996) KRAFEAIVGRQKL------KKSFQYEVKWKYW-KPKYNSWPKDVLVEHGFEKLVQKFDDHEA S65245
atCMT1   (338-397) EFEVEKFLGIMFGDPQGTGEKTLQLMVRWKGY-NSSYDTWEPYSGL--GNCKEKLKEYVIDGF O49137
ttPdd1p   (48-103)   QYEVEKILDSRFN------PKTKQYELVKWENW-PIEDSTWEPYEHL--SNVKEIVQAFEKKQK Q94996
spSWI6   (80-136)   EYYVEKVKHRMA------RKGGGYEYLLKWEGYDDPSDNTWSSEADC-SGCKQLIEAYWNEHG P40381
foskippy   (1235-1288) VYAAEAIRDTRKI------NGQREYLIKWKNY-PENENTWEPPKHL--VNAQRLLKDFHQRAR S60179
dmPolycomb (25-77)   EYYVERIQKRVK------KGVVEYVRVKVKGW-NQRYNTWEPEVNI--LDRRLIDIYEQTNK P26017
dmSu(var)3-9 (218-270) EYYVENKIGSRTAGE-----QYQPVFFVKWLGY-HDSENTWESLANV--ADCAEMEKFVERHQ P45975
atCAO_1   (86-136)   YGEVNKIGSRTAGE-----GAMEYLIEWKDG--HSPSWWPSSYI---AADVVSEYETPWW Q9SAU3

(iii)

dmMOF   (394-449) rgqvqsRTTENA------AAPDEYYVHYVGL-NRRLDGWGRHRIsdnaddIggitvIpapp O02193
hRBP1   (253-296) kiYEASiKSTEIDD-----GEVLYLVHYGW-NVRYDEWVKADRIiwplldkggpkkkka I58390
dmMSL3   (32-89)   tskvLNVFERRNEHGL---RFYEYKLHFQGW-RPSYDRCVRATVIikdteenqIqrelaea P50536
dmMSL3   (448-505) mpmfLNASPISNKKLE---DLLPHLDAFINY-LENHREWFDRENFvnstalpqedIqrelld P50536

NIELSEN *ET AL*., CHAPTER 10, FIG. 1. Sequence alignment and structures of chromodomains. (A) Superposition of the structures of the HP1β chromodomain (black, PDB code 1apo), the HP1β chromo shadow domain (grey, PDB code 1dz1), and the Swi6 chromo shadow domain (white, PDB code 1e0b). (B) Comparison of sequences containing the chromo box consensus motif (i) chromodomains with known structures, (ii) selected chromodomain sequences that clearly fit the requirements of the structure, and (iii) the putative chromodomains from MSL3, MOF1, and RBP1. In (B) the positions of secondary structure elements are indicated by green arrows (*extended* strands) and dark blue cylinders (α-helices). The blue dots indicate the positions of the conserved bulges in the first strand. Conserved hydrophobic core residues that define the fold are shown in yellow, while those determining the borders of the turn between the β-strand and the helix are shown in green. In MSL3, MOF1, and RBP1, the grey residues indicate absence of conserved residues important for the formation of the 3D chromodomain fold. The chromo box used in previous sequence analysis is indicated by a black line. For each sequence the residue and SWISS-PROT/TrEMBL access numbers are given. Residues are color coded similarly in (A) and (B)—the residue numbers in (A) correspond to the positions of the amino acids in the sequence alignment. Panels (A) and (B) were produced using MOLSCRIPT [P. J. Kraulis, *J. Appl. Crystallogr.* **24,** 946 (1991).] and ALSCRIPT [G. J. Barton, *Protein Eng.* **6,** 37 (1993).], respectively.

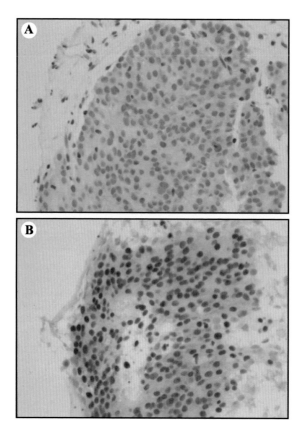

RICHON *ET AL.*, CHAPTER 13, FIG. 3. SAHA induces histone H3 acetylation in prostate tumor biopsies. 40× magnification of prostate tumor biopsies from a patient before (A) and after (B) treatment with 900 mg/m^2/day for 3 days with the HDAC inhibitor, SAHA. Paraffin-embedded biopsy tissue was sectioned and stained using a 1:2500 dilution of an anti-acetylated histone H3 antibody (Upstate). Positive staining was visualized with diaminobenzidine (brown color). Cell nuclei are counterstained with hematoxylin (blue/gray color).

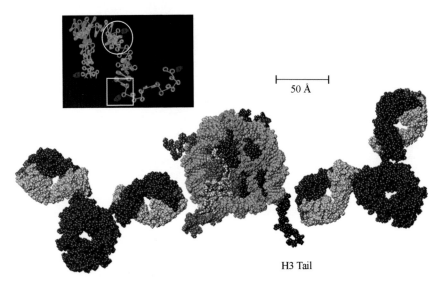

50 Å

H3 Tail

BUSTIN *ET AL.*, CHAPTER 14, FIG. 1. Molecular model of the interaction of antibodies with a nucleosome. Structures shown are based on coordinates from the Protein Data Bank for nucleosome core particle (1KX5), an antibody Fc fragment (1DN2), and an antibody–peptide complex represented by the Fab of antibody PC283 bound to a peptide derived from a hepatitis B virus surface antigen (1KC5). Antibody heavy and light chains are blue and yellow, respectively. A peptide bound in the antibody-binding site is shown in red, and illustrates the location and relative size of an active site. Nucleosomal DNA is cyan, the H3 tail is black, and the histones are represented by different colors. The model illustrates the size relationship between an antibody and a nucleosome core. The *inset* illustrates the size relationship between antibody probes (in red) and chromatin (model provided by C. Woodcock). A representative compacted and decompacted region is denoted by a circle and square, respectively; note that antigens within these regions exhibit differential accessibility to antibody probes.

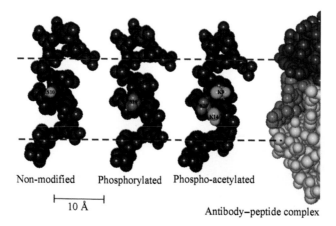

Non-modified   Phosphorylated   Phospho-acetylated

├─── 10 Å ───┤

Antibody–peptide complex

Bustin *ET AL.*, Chapter 14, Fig. 2. Models of an antibody–peptide interaction, and a fragment of the amino-terminal tail of histone H3 with posttranslational modifications. For illustrative purposes, residues 1 to 20 are arbitrarily shown in an α-helical conformation. *From left to right:* the unmodified fragment with the oxygen of Ser-10 denoted by cyan; the fragment phosphorylated at Ser-10, with the phospho moiety denoted by red; the fragment phosphorylated at Ser-10 and acetylated at Lys-9 and Lys-14 (the acetyl moiety is denoted green). The representative antibody active site is rightmost, and is composed of regions from heavy chain (blue) and light chain (yellow). The antigen (violet) is a peptide derived from the hepatitis B virus surface. Note the complementarity between the residues in the binding surfaces of the antigen and the antibody, which results in an exact fit. The distance between the dashed lines denotes the approximate size of an epitope.

A   B

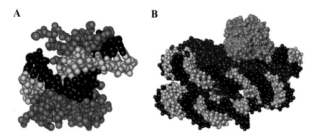

Bustin *ET AL.*, Chapter 14, Fig. 3. Effect of macromolecular interactions on accessibility of antigenic binding sites in chromatin. (A) Model of an HMGBox–DNA complex, based on coordinates of Sry (1J46). Regions in the interface between protein (violet) and DNA (black and yellow) are inaccessible to antibody. (B) Globular region of histone H1 (1GHC) positioned on the nucleosome core. Note that complex formation blocks antibody access to binding surfaces on histone H1 (green) and nucleosomal DNA (black and yellow) and proteins (blue).

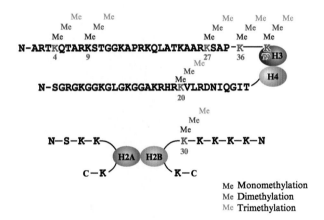

SARMA *ET AL.*, CHAPTER 17, FIG. 1. A schematic representation of the various histone lysine methylation sites and their potential degrees of methylation identified to date *in vivo*.

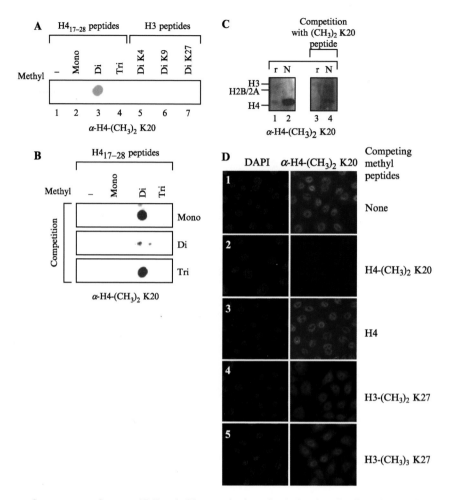

SARMA *ET AL.*, CHAPTER 17, FIG. 4. Characterization of polyclonal antibodies raised against di-methyl H4-K20 antibodies (Upstate 07-367) (A) Dot blot analysis to check antibody specificity using methyl peptides as indicated above the panel. (B) Competition dot blot performed in the presence of 1 μg/ml of various peptides as indicated to the right. The di-methyl signal is lost only in the middle panel where di-methyl H4-K20 is the competing peptide. (C) Western blot analysis of the antibody using recombinant nucleosomes labeled r (left panel, lane 1) and native oligonucleosomes isolated from HeLa cells labeled N (left panel, lane 2) shows reactivity with native H4 and not with recombinant H4. Competition with di-methyl H4-K20 peptide resulted in loss of signal on native oligonucleosomes (compare lane 2 on left and right panels). The antibody dilution used was the same as in the dot blot analyses and competition dot blots. (D) Immunofluorescent staining of HeLa cells with these antibodies at 1:100 dilution show a prominent nuclear signal (panel 1) as is expected for antibodies raised against histones. This signal is completely obliterated when the antibody is used in the presence of di-methyl K20 peptide (panel 2). There is no change in signal when unmodified H4 peptide (panel 3) or non-specific peptides like di- and tri-methyl K27 peptides (panels 4 and 5) are used.

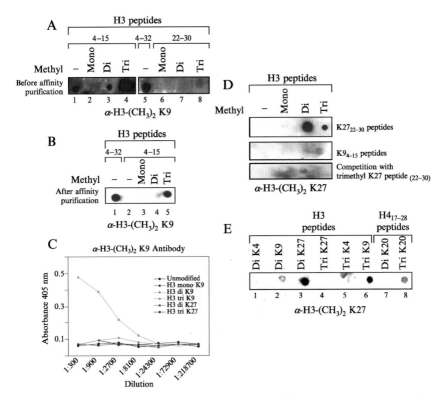

SARMA *ET AL.*, CHAPTER 17, FIG. 5. (A) Polyclonal antibodies raised against di-methyl H3-K9 before affinity purification showed strong reactivity with the tri-methyl K9 peptide within the same context (lane 4) and weak reactivity with either unmodified H3 within the same context (lane 1), di-methyl K9 (lane 3), or the longer unmodified H3 peptide (lane 5). There was no cross-reactivity seen with mono-methyl K9, mono-, di-, or tri-methyl K27 (lanes 2, 6, 7, and 8, respectively). (B) After affinity purification on the di-methyl H3K9 column, the antibody is able to recognize predominantly tri-methyl K9 (lane 5) and almost no reactivity is seen with di-methyl K9 (lane 4). A strong signal is still seen with the longer unmodified H3 (lane 2) while the shorter peptide showed no cross-reactivity. This antibody was used at a dilution of 1:1000. (C) ELISA analysis with the same antibody after affinity purification also confirmed that it recognizes the tri-methyl K9 modification but not di-methyl K9, K27 peptides with different degrees of methylation or the shorter unmodified H3 peptide. (D) Polyclonal antibodies raised against the di-methyl H3-K27 showed specific reactivity with di- and tri-methyl K27 (top panel), but not against methyl K9 peptides (middle panel). Competition with the tri-methyl K27 peptide completely eliminates the signal with di-methyl K27 peptide also (bottom panel). (E) Polyclonal antibodies raised against di-/tri-methyl H3-K27 are unable to recognize or react to an insignificant level with di- and tri-methyl K4 and K20 peptides, as indicated above the panel.

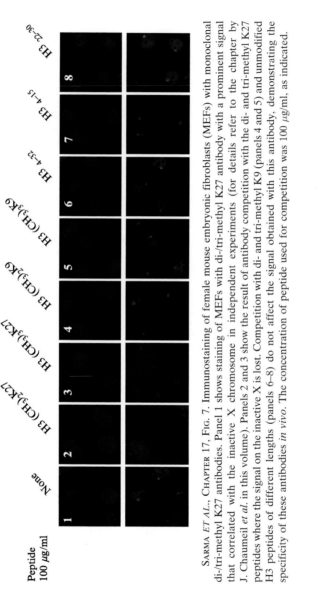

SARMA *ET AL.*, CHAPTER 17, FIG. 7. Immunostaining of female mouse embryonic fibroblasts (MEFs) with monoclonal di-/tri-methyl K27 antibodies. Panel 1 shows staining of MEFs with di-/tri-methyl K27 antibody with a prominent signal that correlated with the inactive X chromosome in independent experiments (for details refer to the chapter by J. Chaumeil *et al.* in this volume). Panels 2 and 3 show the result of antibody competition with the di- and tri-methyl K27 peptides where the signal on the inactive X is lost. Competition with di- and tri-methyl K9 (panels 4 and 5) and unmodified H3 peptides of different lengths (panels 6–8) do not affect the signal obtained with this antibody, demonstrating the specificity of these antibodies *in vivo*. The concentration of peptide used for competition was 100 μg/ml, as indicated.

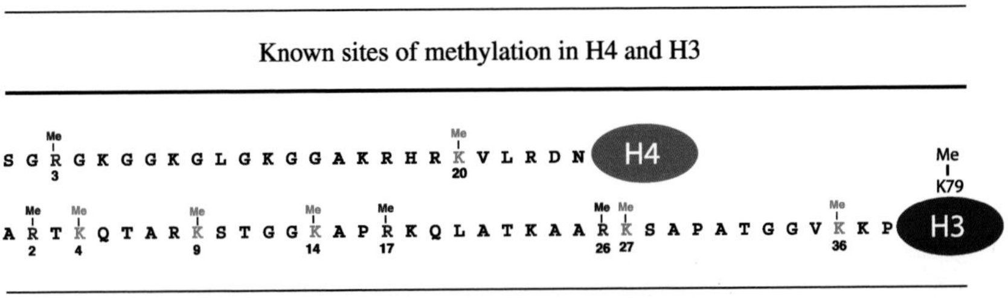

Known sites of methylation in H4 and H3

BANNISTER AND KOUZARIDES, CHAPTER 18, FIG. 1. Known sites of methylation within mammalian histone H3 and histone H4 N-terminal tails. Note that lysines may be mono-, di-, or tri-methylated, whereas arginines may be mono-, symmetric di-, or asymmetric di-methylated (see Fig. 2).

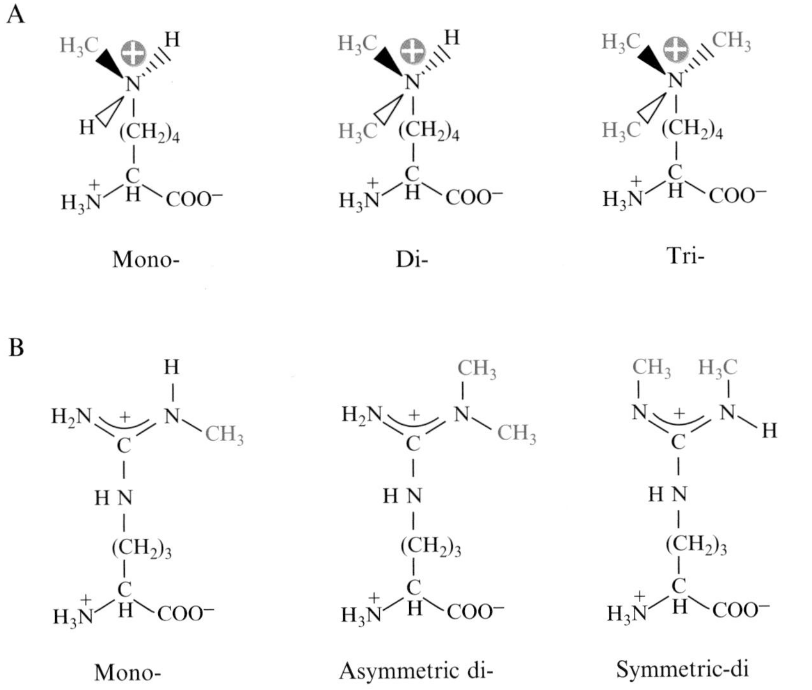

A

B

BANNISTER AND KOUZARIDES, CHAPTER 18, FIG. 2. Chemical structures of methylated states of lysine (A) and arginine (B).

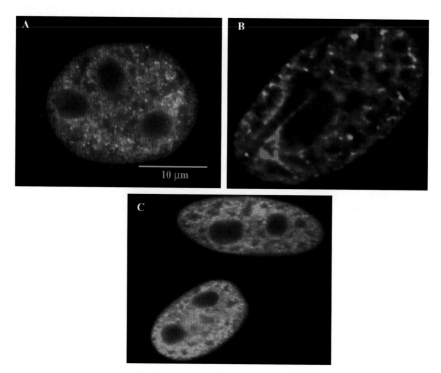

Bannister and Kouzarides, Chapter 18, Fig. 4. Immunoflourescence of mammalian cells using anti-methylated K4 of histone H3 antibodies. (A) Indian muntjac fibroblast cell stained with anti-monomethylated K4 of H3 antibody (green) and DAPI (red). (B) Indian muntjac fibroblast cell stained with anti-dimethylated K4 of H3 antibody (green) and DAPI (red). (C) Indian muntjac fibroblast cells stained with anti-trimethylated K4 of H3 antibody (green).

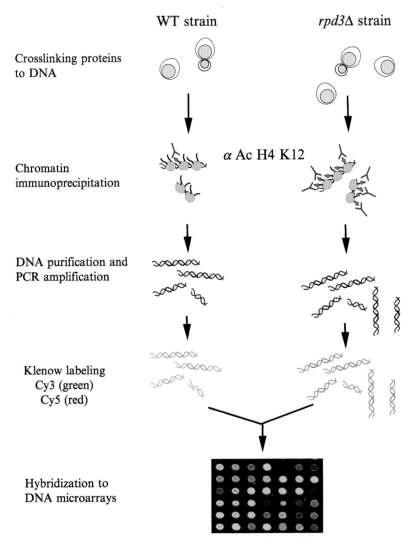

ROBYR *ET AL.*, CHAPTER 19, FIG. 1. Acetylation microarrays. Chromatin fragments from crosslinked mutant cells (*rpd3Δ*) and their isogenic WT counterparts were immunoprecipitated using highly specific antibodies raised against acetylated histone sites. DNA from enriched chromatin fragments was purified, amplified by PCR, and labeled with a fluorophore (Cy3 or Cy5). Probes from both sets of labeled DNA were then combined and hybridized to a DNA microarray containing either intergenic regions, ORFs, or both. For a given region on the microarray, the ratio of the normalized fluorescent intensities between the two probes indicates whether the analyzed lysine residue is hypo- or hyper-acetylated in the experiment strain. Reprinted with permission from D. Robyr and M. Grunstein, *Methods* **31** 83–89 (2003).

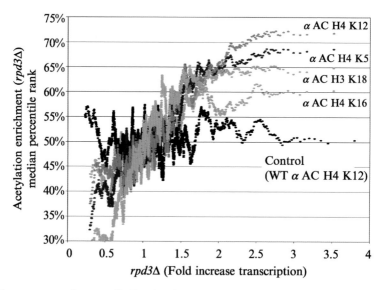

Acetylation enrichment (*rpd3Δ*)
median percentile rank

α AC H4 K12

α AC H4 K5

α AC H3 K18

α AC H4 K16

Control
(WT α AC H4 K12)

rpd3Δ (Fold increase transcription)

ROBYR *ET AL.*, CHAPTER 19, FIG. 2. Histone acetylation sites correlate differently with transcription resulting from *rpd3Δ*. This figure illustrates the scaling of different acetylation data sets using percentile ranking in order to compare them with a transcription data set. Due to inherent noise in microarray data, absolute correlation analysis between acetylation and transcription is not very informative. A moving average analysis that greatly reduces noise can be used to extract general trends. The moving average (window size, 100 data point; step, 1 data point) percentile rank of acetylation enrichment is plotted as a function of transcription increase resulting from *rpd3Δ* (B. E. Bernstein, J. K. Tong, and S. L. Schreiber, *Proc. Natl. Acad. Sci. USA* **97**, 13708 (2000)). Acetylation data are plotted for H4 K5 (dark blue), H4 K12 (red), H4 K16 (orange), and H3 K18 (light blue). Control corresponds to a comparison of two sets of probes amplified from the immunoprecipitation of acetylated H4 K12 in the WT strain and labeled separately with Cy3 and Cy5 prior to hybridization. Data show that increased acetylation at histone H4 K5 and K12 is associated most directly with increased gene transcription in rpd3Δ. H4 K16 show the poorest correlation with gene activity (*RPD3* disruption has no significant effects on the status of H4 K16 acetylation) (D. Robyr, Y. Suka, I. Xenarios, S. K. Kurdistani, A. Wang, N. Suka, and M. Grunstein, *Cell* **109**, 437 (2002)). Reprinted with permission from D. Robyr *et al.*, *Cell* **109**, 437–446 (2002).

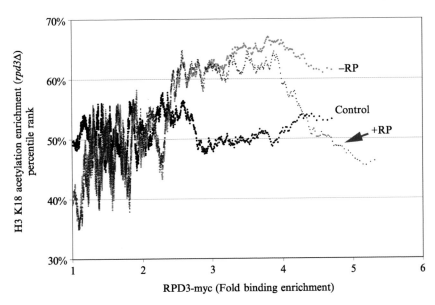

ROBYR *ET AL.*, CHAPTER 19, FIG. 3. Binding arrays are complementary to acetylation arrays. The moving average (window size, 100 data point; step, 1 data point) of Rpd3 enrichment (binding) over intergenic regions is plotted as a function of percentile rank of H3 K18 acetylation in *rpd3Δ*. Data sets with (+RP) and without (−RP) ribosomal protein genes are plotted as indicated. Rpd3 binds strongly at the promoter of ribosomal protein genes in logarithmically growing cells where these genes are highly active but under the same conditions, has little or no effect on acetylation of these promoters (S. K. Kurdistani, D. Robyr, S. Tavazoie, and M. Grunstein, *Nat. Genet.* **31**, 248 (2002); D. Robyr, Y. Suka, I. Xenarios, S. K. Kurdistani, A. Wang, N. Suka, and M. Grunstein, *Cell* **109**, 437 (2002)) or the expression of the RP genes (B. E. Bernstein, J. K. Tong, and S. L. Schreiber, *Proc. Natl. Acad. Sci. USA* **97**, 13708 (2000)). Thus, the ribosomal protein gene promoters as targets of Rpd3 are only detectable by binding arrays. This clearly illustrates the importance of combining different types of arrays (binding, acetylation, and expression) to fully comprehend HDAC function. Reprinted with permission from S. K. Kurdistani *et al.*, *Nat. Gen.* **31**, 248–254 (2002).

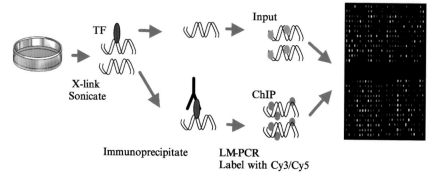

Ren and Dynlacht, Chapter 20, Fig. 1. A schematic diagram of genome-wide location analysis with mammalian transcription factors. The method can in principle be generalized for any DNA-binding regulatory protein.

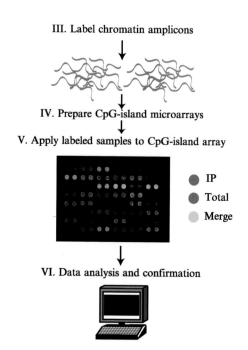

III. Label chromatin amplicons

IV. Prepare CpG-island microarrays

V. Apply labeled samples to CpG-island array

IP
Total
Merge

VI. Data analysis and confirmation

Oberley *et al.*, Chapter 21, Fig. 2. Preparation and hybridization of CpG arrays (Steps IV–VI). Step IV: CpG islands are spotted onto glassslides and post-processed according to the manufacturer's protocols. Step V: The Cy5- and Cy3-labeled samples are hybridized to the CpG-island microarray in the presence of CoT-1 DNA overnight at 60°. The slides are then washed to remove unbound labeled sample and then dried. Step VI: The fluorescent intensity of each feature is read with an appropriate scanner and data are imported into analysis program for further data analysis. Features with intensities significantly higher in the ChIP-labeled samples relative to the input sample are regarded as putative binding targets. However, if a feature also is positive on the IgG versus input control array, it is removed from the list of targets.

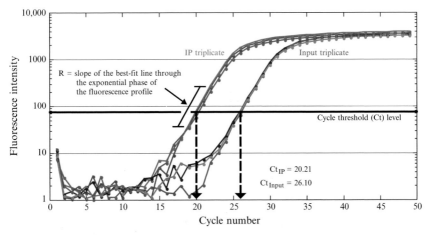

$$\text{Fold enrichment} = R^{\left(Ct_{Input} - Ct_{IP}\right)} = 1.91^{\left(26.10 - 20.21\right)} = 45.68$$

CICCONE *ET AL.*, CHAPTER 22, FIG. 2. A graphic demonstration of the data generated by real-time PCR and the calculations used in determining the enrichment of specific DNA sequences in an IP sample relative to an Input sample. Fluorescence data generated in real time were plotted as a function of the cycle number during PCR amplification of a particular target sequence from representative IP and Input triplicate reactions. The cycle number is listed along the *x*-axis and the fluorescence intensity is plotted on the *y*-axis. For this example, the Ct level was set to a fluorescence of 80, which resides within the exponential phase of amplification. Extrapolation of the fluorescence data from the set Ct level down to the *x*-axis provides Ct values for both the IP and Input samples. Taking into account the rate of amplification (*R*), the fold enrichment of a particular DNA sequence in the IP sample relative to the Input sample is 45.68.

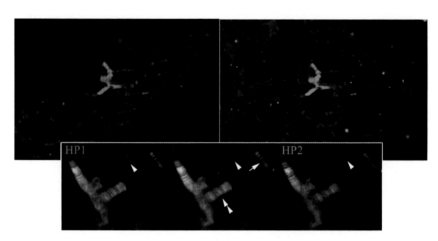

STEPHENS *ET AL.*, CHAPTER 25, FIG. 3. Simultaneous immunolocalization of HP1 and HP2. *Upper:* polytene chromosomes stained with anti-HP1 (red, left) and anti-HP2 (Ab P-6; green, right). *Lower:* close-up view of the chromocenter. Single antibody signals are shown in black and white; the color merge of HP1 signal (red) and HP2 signal (green) is shown in the center. Note the staining of the chromocenter, in a banded pattern along the fourth chromosome (double arrowhead) and of 5–6 bands in 31B (arrow). The arrowhead identifies a euchromatic band positive for HP1 but not HP2. Adapted, with permission, from: C. D. Shaffer, G. E. Stephens, B. A. Thompson, L. Funches, J. A. Bernat, C. A. Craig, and S. C. R. Elgin, *Proc. Natl. Acad. Sci.* **99,** 14332 (2002).

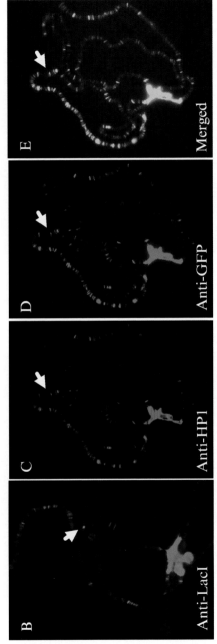

STEPHENS *ET AL.*, CHAPTER 25, FIG. 6. Determining protein-DNA and protein-protein interactions on chromosomes *in vivo*. (A) Diagram of the tethering system showing the LacI-HP1 fusion transgene expressed from an *hsp70* promoter (left) and the mini-*white* reporter transgene containing upstream *lacI* repeats (right). (B) Polytene chromosomes from a *Drosophila* stock expressing a LacI-HP1 fusion protein and carrying the mini-*white* reporter inserted at cytological position 93A/B. Localization of the HP1 fusion protein is detected using antibodies to the LacI DNA binding domain (green). LacI-HP1 associates with the *lacI* repeats (arrow) and with sites of endogenous HP1. (C)–(E) Polytene chromosomes from a stock expressing a LacI-HP1 fusion protein and a SU(VAR)3-9 protein tagged with GFP. HP1 localization is detected with antibodies to HP1 (red, panel C), SU(VAR)3-9EGFP localization is detected with antibodies to GFP (green, panel D), and colocalization is observed in the merged image (yellow, panel E). HP1 and SU(VAR)3-9 show colocalization over the mini-*white* reporter at cytological position at 1E (arrow) and at sites of endogenous HP1.

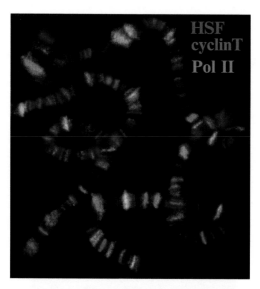

SCHWARTZ *ET AL.*, CHAPTER 26, FIG. 1. Triple immuno-stain. Polytene chromosomes from heat shocked larvae were stained with antibodies against HSF (red), the cyclin T subunit of P-TEFb (blue), and the serine 2 phosphorylated form of RNA polymerase II (green).

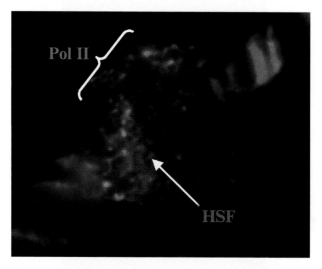

SCHWARTZ *ET AL.*, CHAPTER 26, FIG. 2. Elongating form of Pol II (green) resolves from promoter-bound HSF (red) during heat shock. The heat shock puff from a single transgenic copy of an *Hsp70-lacZ* fusion gene at chromosomal locus 9D is shown. The HSF band at the left edge of the puff marks the promoter of the transgene, while elongating Pol II appears within the body of the puff.

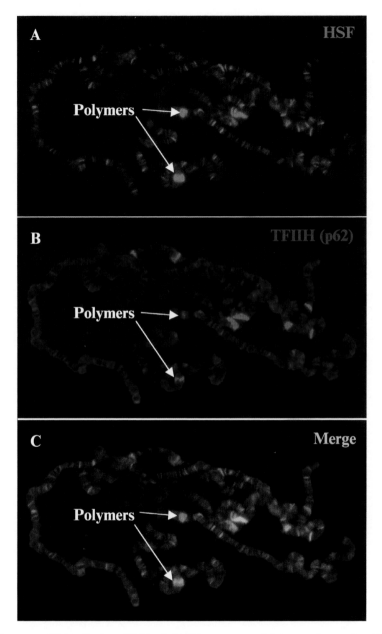

SCHWARTZ *ET AL.*, CHAPTER 26, FIG. 3. (A) Transgenes containing a polymer of an *Hsp70* promoter fragment recruit HSF (green) during heat shock. (B) HSF can, in turn, recruit TFIIH (red). (C) Merge.

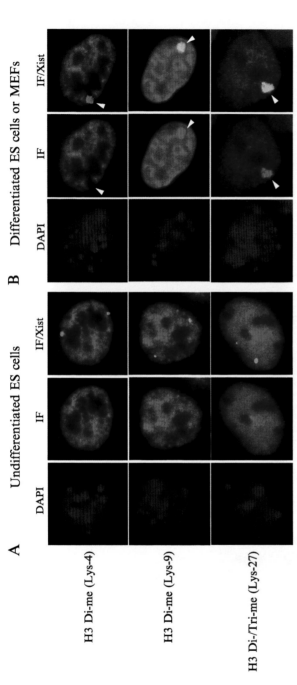

CHAUMEIL *ET AL.*, CHAPTER 27. FIG. 1. Examples of immunofluorescence combined with RNA FISH on interphase nuclei. Modifications of the N-terminal histone tails of histone H3 on the inactive X chromosome in differentiating female ES cells. In each case, nuclei are shown with typical patterns observed in undifferentiated female ES cells (panel A) or during differentiation (panel B). Immunodetection with Alexa GAR 568 conjugated secondary antibody (red, column 2 of each panel) was combined with Xist RNA FISH (Spectrum Green labeled probe, green, column 3 of each panel). Three antibodies were used here, which detect di-methylation of H3 lys-4 (from Upstate Biotechnology), di-methylation of H3 lys-9 (gift from D. Allis, also available from Upstate Biotechnology), and di-/tri-methylation of H3 lys-27 (gift from D. Reinberg, see Sarma and Reinberg, this volume). Prior to the onset of X inactivation, in undifferentiated ES cells, *Xist* is transcribed at a low level from both X chromosomes and the primary transcripts can be detected as a "dot" at each *Xist* locus (green, A). At the beginning of inactivation, Xist RNA starts to accumulate, over the future inactive X chromosome, and can be detected as a green domain in the nuclei (green, B). In this same window of time, some histone modifications begin to appear on the inactive X chromosome (red, B). In merge images of immunofluorescence and RNA FISH (third column, B), green coloration shows exclusion of the modification or yellow coloration shows the enrichment on the Xist domain. The inactive X chromosome is depleted in H3 Lys-4 di-methylation (first row) and enriched in H3 Lys-9 di-methylation (second row) and Lys-27 di-/tri-methylation (third row) (B, arrowheads). DNA is stained with DAPI (blue, first column of each panel).

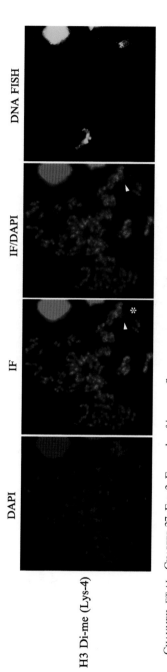

H3 Di-me (Lys-4)

DAPI   IF   IF/DAPI   DNA FISH

C<small>HAUMEIL</small> *ET AL.*, C<small>HAPTER</small> 27, F<small>IG.</small> 2. Example of immunofluorescence on metaphase spreads. Histone H3 modifications on the inactive X chromosome in differentiating female ES cells. Metaphase spreads are shown with typical patterns observed in female ES cells from day 3 of differentiation onward. A single X chromosome shows specific hypomethylation on H3 Lys-4 (red, arrowhead). X chromosomes were detected by X chromosome DNA FISH following immunofluorescence (yellow). The distal part of the X chromosome is methylated on H3 Lys-4 and probably corresponds to the pseudoautosomal region (asterisk). DNA is stained with DAPI (blue).

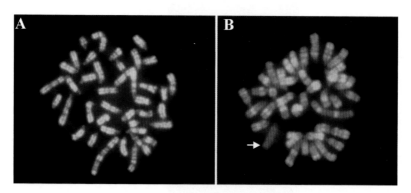

CHAUMEIL *ET AL.*, CHAPTER 27, FIG. 3. Detection of asynchronous replication of the X chromosomes using Acridine Orange staining. Metaphase spreads showing the typical green/orange banding pattern representative of BrdU incorporation in the latter half of S phase are shown. In (A) no single orange-staining X chromosome can be detected, indicating that both X chromosomes replicate synchronously. The first sign of a single uniformly stained orange chromosome (the late-replicating X chromosome, see arrow) was observed at day 3 of ES cell differentiation, where 6.6% ($n > 200$) of metaphases displayed this pattern. The proportion rose to almost 30% ($n > 200$) by day 7.